GASES IN METALS

PHILIPS TECHNICAL LIBRARY

GASES IN METALS

J. D. FAST

Professor at the Technical University of Eindhoven

M

An updated and reorganised edition of *Interaction of Metals and Gases: Vol. 2, Kinetics and Mechanisms*, by J. D. Fast, published by The Macmillan Press Ltd, 1971

First published 1976 by
THE MACMILLAN PRESS LTD
London and Basingstoke
Associated companies in New York Dublin Melbourne Johannesburg and Madras

ISBN 0 333 21214 2

PHILIPS

Printed in Great Britain by
Unwin Brothers Limited
The Gresham Press, Old Woking, Surrey

CONTENTS

Preface ix

Chapter 1. Solutions of Gases in Metals

1.1. Foreign interstitial solute atoms in metals 1
1.2. Size of the interstices 3
1.3. Effect of the size factor 7
1.4. Redistribution of interstitial atoms following lattice distortion 15
1.5. Electronic interaction 21
1.6. Chemical factors 24
1.7. Brillouin zones and energy bands 27
1.8. Fermi surface and electronic specific heat 36
1.9. Hydrogen in palladium 39
1.10. Hydrogen in palladium alloys 48
1.11. Hydrogen in nickel 64
1.12. The nature of bonding in transition-metal hydrides 73
1.13. Confrontation of proton theory and anion theory 79
1.14. Oxygen in transition metals and in transition-metal alloys 84

Chapter 2. Diffusion in Interstitial Alloys

2.1. Introduction 88
2.2. Steady-state diffusion 88
2.3. Non-steady-state diffusion 92
2.4. Measurement of diffusivities of gases in metals 97
2.5. Dependence of diffusivity on concentration 100
2.6. Concentration and chemical potential 108
2.7. The statistical nature of diffusion 112
2.8. Atomic and macroscopic treatment of diffusion problems 116

2.9. Interstitial diffusion in fcc and bcc metals 120
2.10. The influence of temperature 123
2.11. The value of D_0 for the diffusion of hydrogen 130
2.12. The entropy of activation for interstitial diffusion 132
2.13. Diffusion of hydrogen in iron 133
2.14. Diffusion of carbon in iron 142
2.15. Diffusion along grain boundaries and dislocations 148

Chapter 3. INTERNAL FRICTION

3.1. Introduction 159
3.2. Relaxation 159
3.3. Hysteresis 160
3.4. Damping due to interstitials in bcc metals 163
3.5. Damping due to interstitials in fcc metals 165
3.6. The cold-work peak 169
3.7. Exchange of solutes at dislocations 176

Chapter 4. THE SNOEK EFFECT AND ITS SIGNIFICANCE FOR METAL–GAS PROBLEMS

4.1. Relaxation time 180
4.2. Compliance 181
4.3. Time-dependent fields or stresses: derivation of the Debye equations 183
4.4. Derivation of the Debye equations without recourse to complex variables 186
4.5. The significance of the Debye equations for the Snoek effect 190
4.6. Checking the elementary theory with experimental data 193
4.7. Thermodynamic considerations 197
4.8. Relations between measures of internal friction 203
4.9. Determination of diffusivities of interstitial atoms 205
4.10. Determination of solubilities 210
4.11. Discovery of new phases 217
4.12. Study of surface reactions 221
4.13. Analytical determination of small quantities of carbon and nitrogen in iron 227
4.14. Interaction of interstitial atoms 229

4.15. Study of the kinetics of precipitation 238
4.16. Interaction between interstitial and substitutional atoms 240
4.17. Interaction between interstitial and substitutional atoms in the presence of a precipitate 250
4.18. Influence of quantum effects on diffusion 254

Index 257

PREFACE

My earlier book *Interaction of Metals and Gases*, Volume 2 (Macmillan, 1971) was generally reviewed very favourably. From one of the reviews I quote the following comments: 'The book is aimed at "metallurgical students, industrial metallurgists and chemists"; however, while a few specialists might prefer a greater in-depth treatment, this volume probably has a much wider value and may be recommended to anyone interested in solid-state theory, ultra-high-vacuum attainment and gas purification through metal membranes, and to those interested in heterogeneous metal catalysis, as almost essential background material. Typically, the volume would make an excellent reference text for related graduate or even undergraduate courses.'

Moreover, several reviewers expressed the wish for a less expensive book and I have therefore amended and reorganised the contents of the earlier book so as to provide a completely self-contained text at a lower price. I hope that the resulting new book, now appearing in paperback and bearing the title *Gases in Metals*, will be received as favourably as the more comprehensive *Interaction of Metals and Gases*.

Eindhoven, 1976 J. D. Fast

1

SOLUTIONS OF GASES IN METALS

1.1 Foreign interstitial solute atoms in metals

The noble gases are for practical purposes insoluble in metals[(1)]. On the other hand, hydrogen, nitrogen and oxygen are soluble in many metals. As long as the concentration of the dissolved gas is low, the solubility at constant temperature is proportional to the square root of the gas pressure. From this it follows that the gas is present in the metal, not in the form of molecules but as atoms.

The question of the nature of these solutions immediately presents itself. Are they substitutional or interstitial solutions; or, in other words, do the gas atoms replace some of the metal atoms in the lattice or do they lodge in the spaces between them? In all instances where an answer to this question could be given experimentally, the latter was found to be the case when the material was in a state of thermodynamic equilibrium. These experiments, in many cases, consist of a combination of X-ray investigation of the lattice constants and direct density measurements. As an illustration, it can be stated that the density of zirconium increases through absorption of oxygen, although oxygen atoms are much lighter in weight than zirconium atoms. This indicates inclusion of gas atoms in the interstices and not substitution of metal atoms, unless one assumes a contraction of the lattice due to the presence of oxygen. The X-ray investigation, however, shows the distances between the zirconium atoms to be greater in the oxygen-containing metal than in the pure metal. There is even a quantitative agreement between the density as determined directly by means of a hydrostatic balance or a pyknometer, and the density as determined by calculation from the lattice constants

1. A. van Wieringen and N. Warmoltz, *Physica, 's Grav.* **22**, 849 (1956), proved that at high temperature helium diffuses through single-crystal walls of the non-metals germanium and silicon. Their experiments show that helium has a noticeable solubility in these elements which crystallize with the loosely-packed diamond structure.

and the known composition of the material, when in this calculation it is assumed that all the oxygen is accommodated in the interstices. If part of the dissolved oxygen substitutes metal atoms, then this part is too small to be shown experimentally.

Other experiments supporting the interstitial character of the gas–metal solutions are measurements of diffusivities of gases in metals (see Chapter 2). They are much greater than the diffusivities in substitutional solutions. Only relatively small atoms, especially those of hydrogen, carbon, nitrogen and oxygen form interstitial solutions with metals. The atoms of most other non-metals as for instance sulphur and phosphorus are too big for this. From the available data it is found that the diffusion coefficients of nitrogen and carbon in iron at 700°C are approximately ten thousand times larger than that of sulphur and even a hundred thousand times larger than that for phosphorus under the same conditions.

Hydrogen is a special case of the non-metals (H, C, N, O) that dissolve interstitially in metals. Its diffusivity is much greater than that of the other elements mentioned; for instance in iron at 20°C it is 10^{12} times that of carbon or nitrogen. This exceptionally fast diffusion can be explained by assuming that the hydrogen moves from one interstice to another as protons since the diameter of a proton is only 10^{-5} of an atom or ion of carbon or nitrogen. This does not exclude the possibility that the hydrogen is present as atoms in the interstices but jumps as protons. It is possible that inside the metal an equilibrium exists of the simple form

$$H \rightleftharpoons p + e$$

where p and e represent a proton and an electron, respectively. In the past it was believed that this view is supported by the fact that electrolytic transport of hydrogen takes place in the direction of the negative electrode when a direct current is passed through palladium(1,2), tantalum(3), iron or nickel(4). Similarly, carbon migrates in the direction of the negative electrode in iron(5,6). However, oxygen in solution in zirconium moves towards the positive electrode(7). Today it is known that these experimental results give no decisive information on the state of ionization of the interstitial atoms (see the article of Oriani and Gonzalez(4), and also Sections 1.12 to 1.14).

1. A. Coehn *et al.*, *Z. Phys.* **62,** 1 (1930); **71,** 179 (1931); **83,** 291 (1933).
2. J. Knaak and W. Eichenauer, *Z. Naturf.* **23A,** 1783 (1968).
3. J. Wesolowski, J. Jarmula and B. Rozenfeld, *Bull. Acad. pol. Sci. Sér. Sci. chim.* **9,** 651 (1961).
4. R. A. Oriani and O. D. Gonzalez, *Trans. AIME* **239,** 1041 (1967).
5. W. Seith and O. Kubaschewski, *Z. Elektrochem.* **41,** 551 (1935).
6. P. Dayal and L. S. Darken, *Trans. AIME* **188,** 1156 (1950).
7. J. H. de Boer and J. D. Fast, *Recl Trav. chim. Pays-Bas Belg.* **59,** 161 (1940).

1.2. Size of the interstices

It has been mentioned above that only relatively small atoms dissolve interstitially in metals. This therefore suggests that the size of the interstices present in a metal is one of the factors determining its capacity for taking atoms of other elements into interstitial solution. Before this supposition can be checked (see following sections), it will be necessary first to consider the sizes of the interstices in the most frequently encountered crystal structures of metals.

The reader will be familiar with the fact that most metals have a simple crystal structure. Their lattice is generally either hexagonal close-packed (hcp), face-centred cubic (fcc) or body-centred cubic (bcc). One can obtain a rough impression of the size of spaces between the metal atoms by regarding these atoms as rigid spheres and the crystals as stacks of these spheres. When spheres of equal size are stacked in such a way that they fill the available space as completely as possible then 74% of the space is occupied by the spheres. The reader is reminded that both the hexagonal close-packed (hcp) and the face-centred cubic (fcc) structures are configurations where the spheres are as tightly packed as possible. One can imagine them being built up by stacking layers of tightly packed flat arrangements of spheres. Fig. 1 shows that each sphere in such a layer is surrounded by six cavities but every cavity by three spheres. Therefore when a second layer has been placed in close contact with the first one only half the number of cavities will be

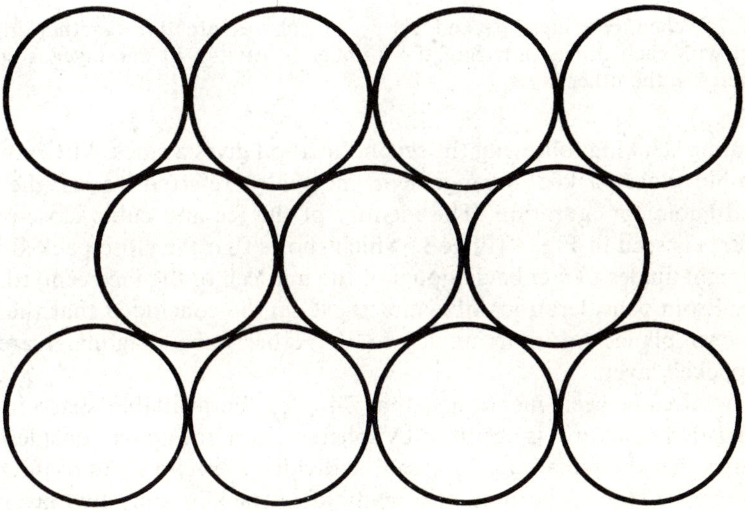

FIG. 1. Close-packed layer of spheres.

covered by spheres (Fig. 2). When adding a third layer there is a choice of two possibilities: either it can be positioned in such a way that each sphere in the third layer is exactly over a sphere in the first layer, as shown in Fig. 3 (Plate I), or in such a way that each sphere is above a cavity in the first layer. The first arrangement can be indicated as a stacking ABA, the second as a stacking ABC. Continuing stacking following the first method results in a stack ABABAB . . ., the hexagonal close-packed configuration;

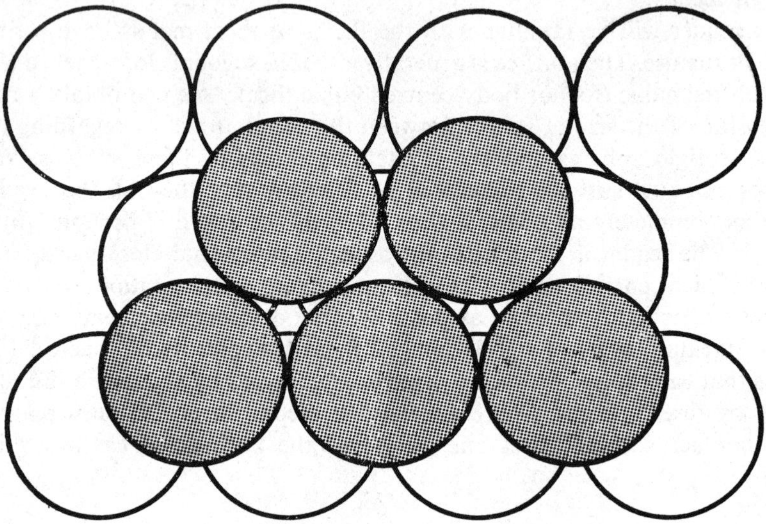

FIG. 2. When two close-packed layers of spheres are put together in close contact with each other, only half the number of cavities in one layer is covered by spheres in the other layer.

continuing stacking following the second method gives a stack ABCABC . . ., the cubic close-packed arrangement, normally referred to as the face-centred cubic configuration. The identity of the fcc and cubic close-packed structures is seen in Fig. 4 (Plate I) which shows that the close-packed layers are at right angles to a cube diagonal of the unit cell of the face-centred cubic lattice. From considerations of symmetry it can be concluded that the other octahedral planes (at right angles to the other cube diagonals) are also close-packed layers.

It has already been mentioned that 74% of the available space in both close-packed structures is occupied by spheres. From the above considerations it follows that the remaining 26% can be divided into two types of interstices which can be seen to be present already after stacking only two layers and which, consequently, are identical for the hcp and fcc structures. Every

FIG. 3. Close-packed hexagonal stacking of spheres.

FIG. 4. Close-packed cubic stacking of spheres.

[facing p. 4

a

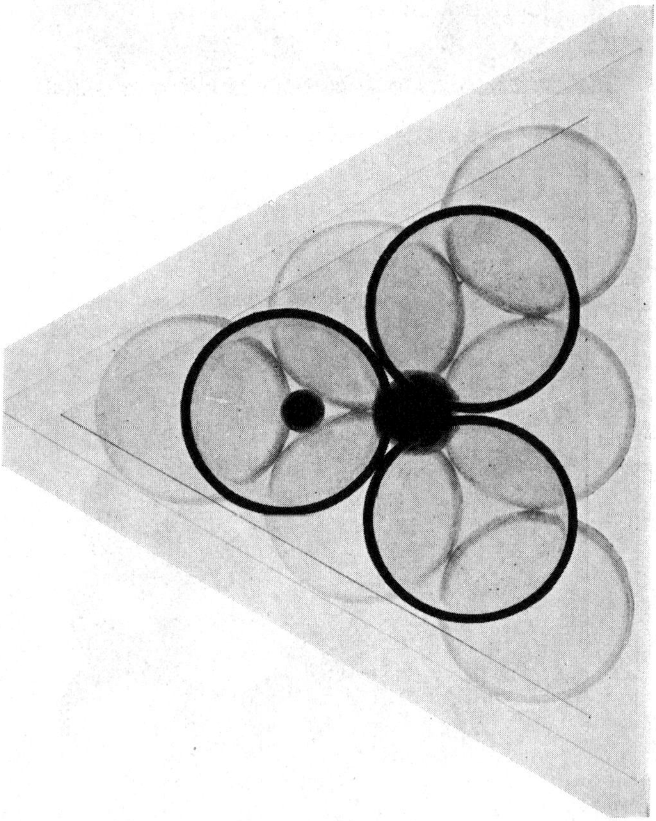

b

FIG. 5. An octahedral cavity in a close-packed stack of spheres can accommodate a larger sphere than a tetrahedral cavity as shown (a) from the side and (b) from above. In (b) the light source is below the model.

facing p. 5]

sphere of the second layer touches three spheres of the first layer; these four surround a so-called *tetrahedral interstice*, the centres of the four spheres being the corners of a regular tetrahedron. A cavity of the first layer not being covered by a sphere forms, together with a cavity in the second layer, a so-called *octahedral interstice*. Each one of these is surrounded by six spheres whose centres are the corners of a regular octahedron. Fig. 5 (Plate II) shows two tightly packed half layers of transparent spheres containing two smaller opaque spheres. The larger of the opaque spheres just fits in an octahedral interstice whilst the smaller just fits in a tetrahedral one.

In the unit cell of the face-centred cubic structure the centres of the octahedral interstices are located at the centre of the cell and at the midpoints of its edges. The centres of the tetrahedral interstices coincide with the centres of the eight cubes into which the unit cell can be divided by halving its edges. The four metal atoms surrounding such an interstice are situated at a corner of the unit cell and at the centres of the three faces meeting in that corner.

If R is the radius of the large spheres, representing the metal atoms, then the radii of the inscribed spheres in the octahedral and tetrahedral interstices are:

$$\text{fcc and hcp:} \quad r_6 = 0{\cdot}414R \quad \text{and} \quad r_4 = 0{\cdot}225R \qquad (1.2.1)$$

where the subscript of r refers to the coordination number, i.e. the number of spheres surrounding an interstice. The introduction of larger spheres into the interstices will lead to distortion of the stack of spheres.

The *body-centred cubic* (*bcc*) *structure* shows a less closely packed stacking of spheres than the two structures discussed above. The filling factor is 68 %, i.e. 6 % smaller than for the closest possible packing. The remaining 32 % can again be divided in octahedral and tetrahedral interstices. The centres of the octahedral interstices are located at the midpoints of the edges and at the centres of the faces of the cube representing a unit cell (see Fig. 6). In contrast to the close-packed structures, the centres of the six atoms surrounding an octahedral interstice form the corners of an *irregular octahedron*. Fig. 6 shows the distance AB to be equal to the lattice parameter a whilst the distances CE and DF are equal to $a\sqrt{2}$. The short axis AB is in the same direction as one of the cube axes. It is now clear that in a bcc crystal three groups of octahedral sites can be distinguished. Their short axes are at right angles to each other and they exist in equal numbers. They are indicated as x, y and z sites conforming to the x, y and z axes of the cubic unit cell.

The centres of the tetrahedral interstices in the bcc structure are located in the faces of the unit cell. Each face contains four of these centres, located at the midpoints of the perpendiculars from the centre of the face to the four edges. They are centres of *irregular tetrahedrons* which have as their corners

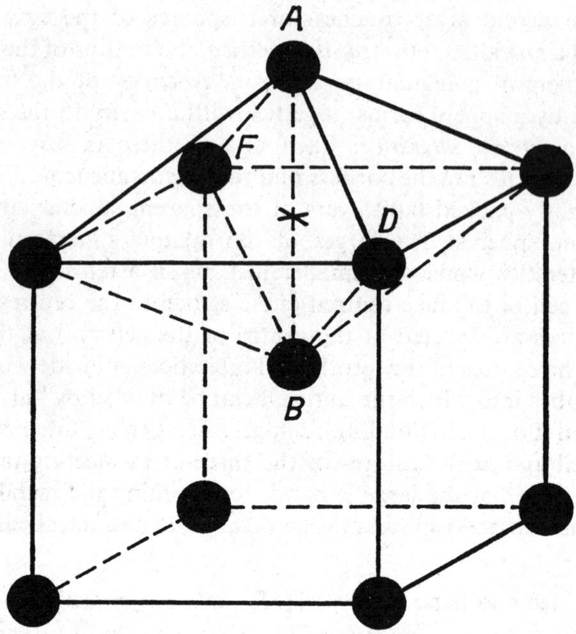

FIG. 6. The cross marks the centre of an octahedral interstice in the unit cell of a body-centred cubic metal. Such a cavity has a distorted octahedral symmetry: the distances CE and DF are $\sqrt{2}$ times greater than the distance AB. For clarity the spheres have been drawn too small. In the solid-sphere model they touch on the space diagonals. The true relative dimensions are shown in Fig. 9.

two corners and two centres of unit cells. The radii of the largest spheres that can be accommodated in the octahedral and tetrahedral interstices without distorting the bcc structure are:

$$\text{bcc:} \quad r_6 = 0{\cdot}155R \quad \text{and} \quad r_4 - 0{\cdot}291R \qquad (1.2.2)$$

A comparison of equations (1.2.1) and (1.2.2) shows the important fact that the closest packed sphere-stacks have interstices that are appreciably larger than those in the less closely packed bcc stack. The total cavity volume in the latter is greater than in the fcc and hcp stacks because the number of interstices is three times as large, i.e. three octahedral and six tetrahedral interstices for each metal atom against only one octahedral and two tetrahedral interstices per metal atom in the close-packed configurations.

Table 1 gives the sizes of the interstices in some metals, expressed in terms of the radii of the inscribed spheres, r_6 and r_4. The table also gives the radii R of the corresponding metal atoms; these are obtained from a paper by

Laves[1]. The values of r_6 and r_4 have been calculated from the R values, using the expressions (1.2.1) and (1.2.2).

Considering the atoms as if they were rigid spheres, as was done above, can be very useful in some cases. It should, however, be borne in mind that this simplification is not always permissible. Table 1 already indicates that the radii of the atoms are not invariable quantities; they are larger in the close-packed fcc and hcp structures than in the less closely packed bcc structure. For instance the radius of an iron atom in bcc iron is 0·124[2] nm whereas in fcc iron it is 0·127 nm (extrapolated to room temperature). The volume reduction which occurs in the transition from bcc iron (α iron) to fcc iron (γ iron) at 910°C is therefore smaller than would be expected from the model of rigid spheres. The limited value of the rigid sphere models will become still clearer in the following sections.

TABLE 1

Data of atomic radii and sizes of interstices in some metals (*nm*)

Metal	R	r_6	r_4
Cu (fcc)	0·128	0·053	0·029
Ag	0·144	0·060	0·032
Au	0·144	0·060	0·032
Fe	0·127	0·053	0·029
Ti (hcp)	0·145	0·060	0·033
Zr	0·160	0·066	0·036
Hf	0·158	0·065	0·036
Ti (bcc)	0·142	0·022	0·041
Zr	0·157	0·024	0·046
Hf	0·155	0·024	0·045
V	0·133	0·021	0·039
Nb	0·144	0·022	0·042
Ta	0·143	0·022	0·042
Fe	0·124	0·019	0·036

1.3. Effect of the size factor

Initially we shall only consider interstitial solutions of carbon, nitrogen and oxygen in metals. According to the published literature the atomic radii of these elements are 0·077 nm, 0·072 nm and 0·070 nm.To us these numbers have only relative value since they do not refer to C, N and O when in solid

1. F. Laves, *Trans. Am. Soc. Metals* **48A,** 124 (1956).
2. In the SI system of units 1 nanometre (nm) is equivalent to 10Å.

solution in metals. What is certain, is that these atoms are too big to be accommodated interstitially in a metal crystal without distorting it (Table 1). This is already evident from the fact that the volume of a metal increases when it absorbs carbon, nitrogen or oxygen. The smaller the interstice in which the C, N or O atom is located the greater is the lattice distortion and the greater is the distortion energy which must be supplied. A high distortion energy affects the solubility adversely. The importance of this size factor is illustrated by the following facts.

In the first place we mention that C, N and O, insofar as they dissolve in cubic or hexagonal close-packed metals, always occupy the large octahedral interstices (see Table 1). The second illustration of the importance of the size factor is the much greater solubility of carbon, nitrogen and oxygen in the hcp forms of titanium, zirconium and hafnium than in their bcc forms

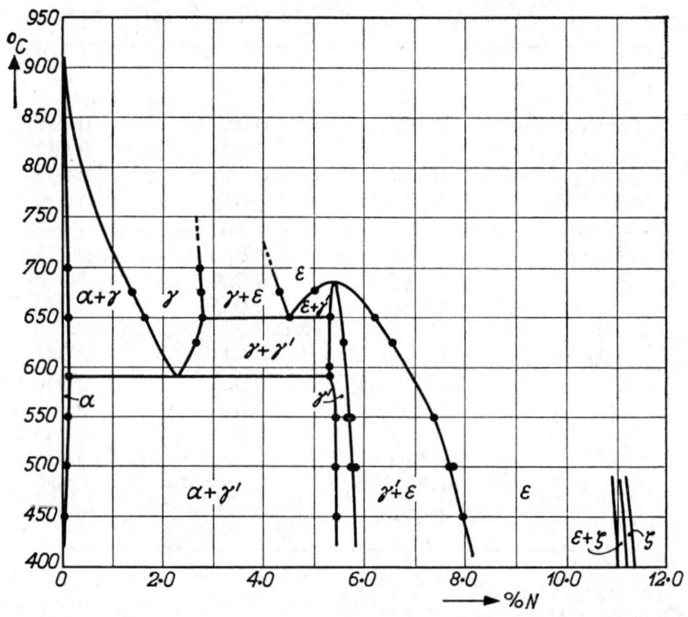

FIG. 7. Equilibrium diagram for the iron–nitrogen system according to PARANJPE, COHEN, BEVER and FLOE, *Trans. AIME* **188**, 261 (1950). The nitrogen content expressed in percentages by weight is plotted on the horizontal axis. The modification of the metal that is stable above 900°C is not called the β phase but the γ phase; this is for historical reasons since in the past it was incorrectly thought that the transition of iron from ferromagnetic to paramagnetic at about 760°C was caused by transition to a phase with a different crystal structure which was then called the β phase.

with much smaller interstices (Table 1). Similarly the solubility of carbon and nitrogen is greater in fcc iron than in bcc iron. This is confirmed by the phase diagrams of the systems Ti–C; Ti–N; Ti–O; Zr–C; . . . Fe–C; Fe–N. For instance the Fe–N and Ti–N diagrams given in Figs. 7 and 8 show that iron

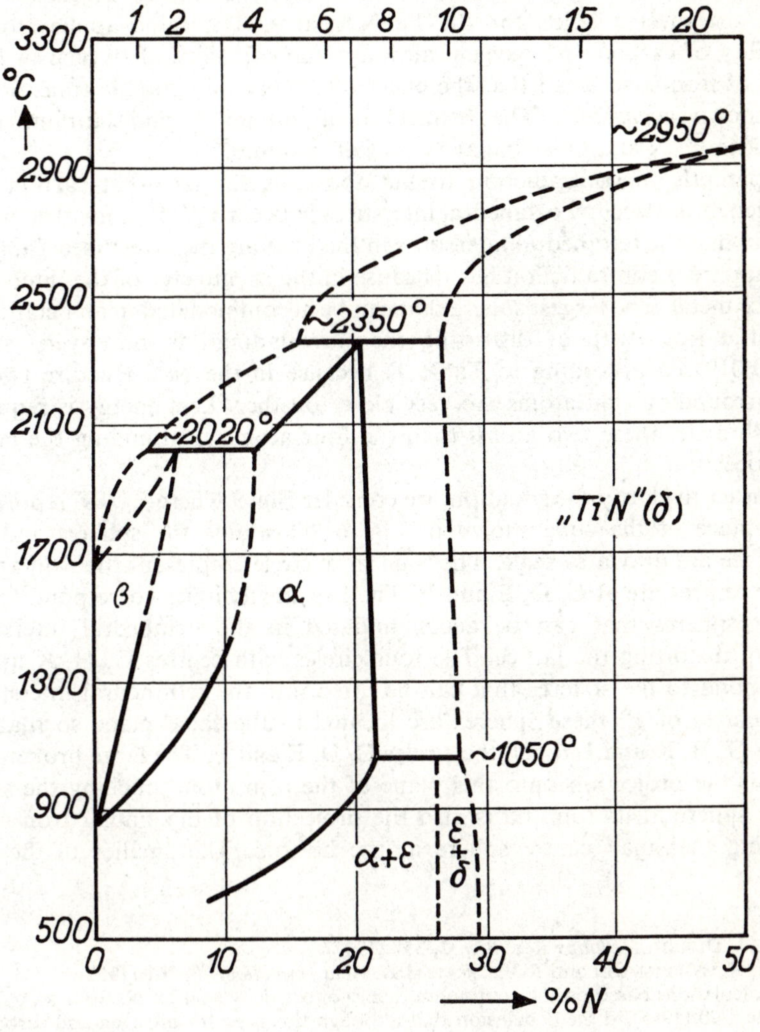

FIG. 8. Equilibrium diagram of the titanium–nitrogen system up to 50 atom per cent nitrogen according to PALTY, MARGOLIN and NIELSEN, *Trans. Am. Soc. Metals* **46,** 312 (1954). The nitrogen content is plotted on the lower horizontal axis in atom per cent and on the upper one in percentage by weight.

and titanium have a crystallographic transformation point at about the same temperature (circa 900°C). It is of interest to note that the closest packed phase with its large octahedral interstices is stable below the transformation temperature in the case of titanium but above that temperature in the case of iron. In agreement with this, the solubility of nitrogen in titanium is appreciably greater below the transformation point than above it whilst the reverse holds for the Fe–N system. The same applies to the solubility of carbon and oxygen, although the solubility of oxygen in both forms of iron is so small that the effect of the crystallographic transition is not known accurately. The behaviour of zirconium and hafnium with respect to C, N and O is similar to that of titanium.

Apparently in contradiction to the above, is the fact that carbon and nitrogen atoms occupy octahedral interstices in bcc iron[1,2], notwithstanding the fact that the tetrahedral interstices in this case are the larger (see Table 1). This apparent contradiction has its cause in the asymmetry of the interstices in a bcc metal. The largest sphere that can be accommodated in an octahedral site of α iron, without distorting the surroundings, is only very small ($r = 0{\cdot}019$ nm according to Table 1) because in the bcc structure two of the surrounding iron atoms are very close together. Less energy is required to push apart these two atoms than the four atoms surrounding the larger interstice[3].

In order to clarify the situation, we consider Fig. 9 where CDEF represents the topface of the cube shown in Fig. 6. This time the spheres and the cube-face are drawn to scale. The four large circles represent the iron atoms whose centres are at C, D, E and F. The five small circles correspond to the largest spheres that can be accommodated in the octahedral interstices without distorting the lattice. The four circles with centres G, H, K and L correspond to the spheres that can be fitted into the tetrahedral interstices. The centres of all these spheres are located in the same plane so that the circles G, H, K and L touch the circles C, D, E and F. The large broken-line circle is the projection onto this plane of the iron atom touching the small centre sphere at its top and is also the projection of the similar iron atom touching the small centre sphere at the bottom. The smaller of the two

1. L. J. Dijkstra, *Philips Res. Rep.* **2,** 357 (1947).
2. G. K. Williamson and R. E. Smallman, *Acta crystallogr.* **6,** 361 (1953).
3. Calculations based on the continuum theory of elasticity lead D. N. Beshers, *J. appl. Phys.* **36,** 290 (1965) to the conclusion that although this is so for nitrogen and carbon in iron, it does not apply to vanadium where these elements would occupy tetrahedral interstices which, according to Table 1, are larger than those in iron. According to his calculations oxygen and nitrogen in niobium and tantalum would also prefer tetrahedral sites to octahedral sites. It is doubtful whether these conclusions are justified (see Sections 1.5 and 1.6).

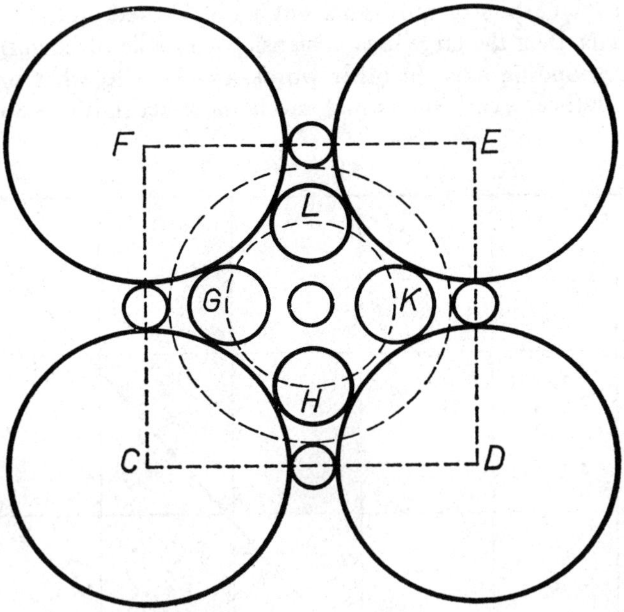

FIG. 9. The square CDEF corresponds to the top face of the unit cell of bcc iron in Fig. 6. The circles with centres C, D, E and F represent the four iron atoms, those with centres G, H, K and L represent the largest spheres that can be accommodated in the four corresponding tetrahedral interstices without distorting the lattice. The five smallest circles correspond to the largest spheres that can be fitted in the octahedral interstices. The large broken-line circle represents the iron atom in the cell centre below the plane of the drawing as well as the one above that plane. The small broken-line circle represents a nitrogen atom with its centre in the plane of the drawing coinciding with the centre of the central octahedral interstice. The figure is to scale. The size of the interstitial nitrogen atom has been derived from the lattice parameters of nitrogen martensite. The figure shows that this atom has to push the iron atoms, above and below it, so far apart that it will occupy not only an octahedral interstice but also the four surrounding tetrahedral interstices.

broken-line circles corresponds to a nitrogen atom whose centre is located in the plane of the drawing.

Fig. 9 shows that the presence of a carbon or nitrogen atom in an octahedral interstice of bcc iron amounts to using, not only this octahedral site but also the four surrounding tetrahedral sites. An appreciable local distension of the lattice in the direction perpendicular to the plane CDEF is therefore to be expected. We designate this direction the z direction (B–A direction in Fig. 6). Along the x and y directions, on the other hand, a small contraction may be expected. If the carbon or nitrogen atom is not located on the z axis but on the x or y axis, then the large local distension will take place in the direction of the corresponding axis. In other words, contrary to what happens in a fcc or hcp lattice, every interstitial atom in a bcc lattice causes a local

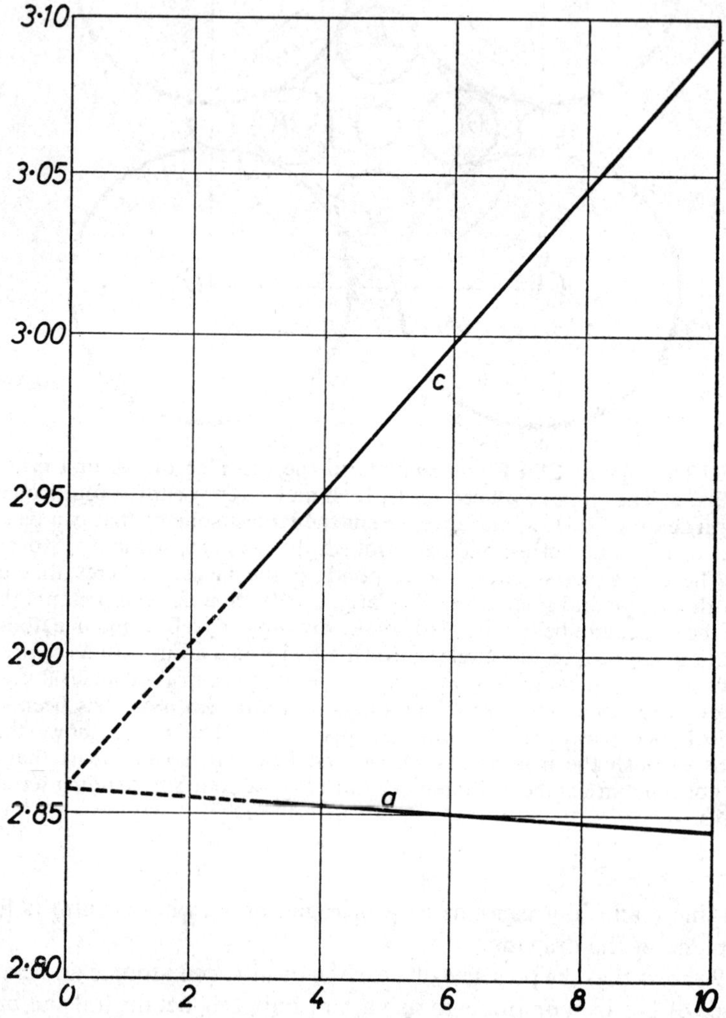

FIG. 10. Unit-cell dimensions ***a*** and ***c*** of nitrogen martensites and carbon martensites according to JACK, *Proc. R. Soc.* **A 208,** 200 (1951). The number of interstitial atoms (N or C) per hundred atoms of iron is plotted horizontally and the unit-cell dimensions in kX units vertically. (1 kX unit = 0·1002 nm).

tetragonal distortion. The large expansion has the direction of the axis on which the atom is located. This is in agreement with the above designation of the octahedral sites as *x*, *y* and *z* interstices.

The local asymmetric distortion of the iron lattice occurring on the introduction of a carbon or nitrogen atom means that the introduction of a second atom into one of the neighbouring interstices requires a different distortion energy. Occupation of a *z* site will clearly inhibit occupation of the *z* sites just above and below it. The introduction of a second atom into one of the neighbouring *z* sites in the same horizontal plane, however, will require a smaller distortion energy because the first atom has already caused some local distension of the lattice. From the above one could expect the existence of strongly supersaturated solutions of carbon and nitrogen in α iron where all foreign atoms occupy similar interstices. It should then be possible to measure the above mentioned tetragonal distortion. Such solutions do indeed exist and are known as martensites (carbon and nitrogen martensites). The tetragonality, found in these by means of X-ray measurements, completely agrees with what would be expected: the lattice parameter is greatly increased in the direction of one of the axes and somewhat decreased in the two other directions (Fig. 10). According to these measurements[(1)] the atomic radii of nitrogen and carbon in martensite have the same value, 0·071 nm, which is only a fraction smaller than the values mentioned in the beginnning of this section and which were determined in another way[(2)].

As is generally known, the formation of martensite plays an important part in steel hardening. Starting with iron that, in addition to other elements, contains a relatively large amount of carbon (for instance 1 wt %) it is heated to a temperature sufficiently high to bring it into the γ state (fcc). Because of the high solubility in the γ phase all the carbon then goes into homogeneous solution (see Fig. 11). Subsequently the metal is very rapidly cooled, so quickly indeed that the separation into α crystals and iron carbide (Fe_3C), which should occur according to the equilibrium diagram, cannot take place. Below 200°C, through mutually supporting shear-like movements of the atoms, the lattice transforms with extreme speed into the α form without diffusion or intermixing of atoms taking place. In the crystals, produced in this manner, the carbon is present in the above mentioned ordered distribution in strongly supersaturated solution. The collective stabilizing effect of this ordered distribution prevents its being lost in a short time. Although these martensitic (body-centred tetragonal) solutions of carbon in iron are metastable with respect to ferrite plus carbide, they are not necessarily so as regards random solutions with the same carbon content. In connection with this it is of interest to note the existence of a stable

1. K. H. Jack, *Proc. R. Soc.* **A 208,** 200 (1951).
2 See also: M. Cohen, *Trans. AIME* **224,** 638 (1962).

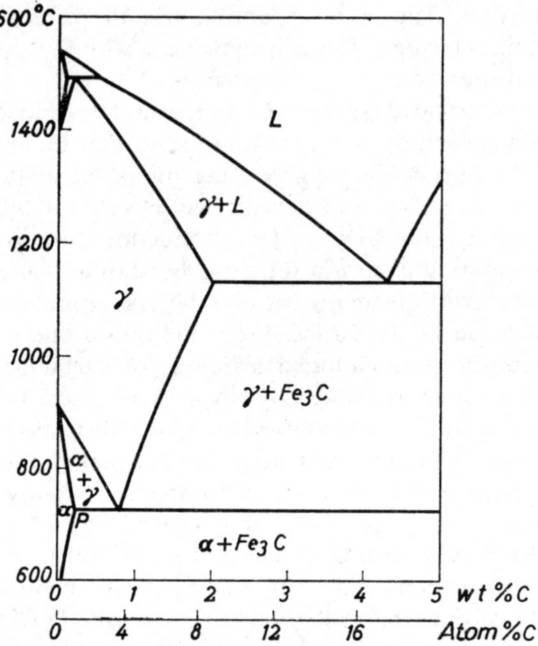

FIG. 11. Part of the phase diagram iron–Fe_3C. For clarity the point P is drawn too far to the right; in fact the maximum solubility of carbon in α iron corresponding to this point amounts to only 0·02% by weight.

body-centred tetragonal (bct) phase in the vanadium–oxygen system. Its composition is in the neighbourhood of an atom ratio O to V = 0·25.

The above implies that the cooperative energy effect under discussion may in some instances lead to a martensitic structure which is stable with respect to the bcc structure if the interstitial concentration is not too small and the temperature not too high. Zener[1] has shown that the transformation from bct to bcc at rising temperature is a first-order transformation. In addition to body-centred tetragonal phases with $c/a > 1$, one of which occurs in the vanadium–oxygen system, there also exist interstitial tetragonal phases with $c/a < 1$ and even interstitial body-centred orthorhombic (bco) structures as for instance in the systems Ta–O and Ta–H, respectively. To account for this, Meijering[2] has extended Zener's calculations to larger concentrations. If the temperature is not too high, then with increasing interstitial concentration the body-centred structure should pass through the following transformations:

$$\text{bcc} \rightarrow \text{bct}\ (c/a > 1) \rightarrow \text{bco} \rightarrow \text{bct}\ (c/a < 1) \rightarrow \text{bcc}$$

1. C. ZENER, *Trans. AIME* **167**, 550 (1946).
2. J. L. MEIJERING, *Phase Stability in Metals and Alloys* (editored by P. S. Rudman, J. Stringer and R. I. Jaffee), McGraw-Hill, New York (1967), p. 359.

In the latter bct phase two types of sites are preferentially occupied, while in the bco phase all three types (x, y and z) are unequally occupied. The transformations bct $\rightarrow$ bco are second order. The calculations leading to these results are based on the supposition that only the strain interactions under discussion are important. In many cases one or more of the structures mentioned will not be stable with respect to breakdown into two phases. It is also possible that face-centred or hexagonal phases will appear at the higher concentrations. Another complication may be the occurrence of superstructures. For instance, in Fe_8N there is not only strain order (z sites occupied preferentially), but the occupied z sites themselves show long-range order.

In the literature, difference of opinion exists on whether boron, whose atoms are larger than those of carbon, nitrogen and oxygen, belongs to the group of elements that can form primary interstitial solutions with metals. According to the literature[1,2,3] the solubility of boron in α iron (bcc) is greater than in γ iron. This is not in agreement with what would be expected for interstitial solutions but could be explained by assuming that boron dissolves interstitially in fcc iron and, because of its large size, substitutionally in bcc iron.

1.4. Redistribution of interstitial atoms following lattice distortion

The effect of the size factor on interstitial solubility was discussed above. The effect of the deformation energy is also evident at every lattice distortion that enlarges part of the interstices and reduces others. Such a distortion results in a change in the distribution of the atoms that are interstitially in solution, in such a way that the number occupying the reduced interstices declines whilst the number of those occupying the enlarged interstices increases. The reduction in entropy which accompanies this deviation from the statistical distribution will in this case be overcompensated by the simultaneous reduction in enthalpy, such that the Gibbs' free energy drops. Distortions of the lattice, leading to redistribution, can be caused by both internal and external stresses.

Distortion caused by external stresses

Let us consider solutions of nitrogen or carbon in α iron, not formed martensitically, i.e. very dilute solutions. In the case of nitrogen these contain less than 0·1% N and in the case of carbon less than 0·02% C (the

1. P. E. Busby, M. E. Warga and C. Wells, *Trans. AIME* **197,** 1463 (1953).
2. M. E. Nicholson, *Trans. AIME* **200,** 185 (1954).
3. C. C. McBride, J. W. Spretnak and R. Speiser, *Trans. Am. Soc. Metals* **46,** 499 (1954).

maximum solubilities at the eutectoidic temperatures of 590° and 720°C respectively; see Figs. 7 and 11). In these solutions the interstitial atoms are, on average, relatively far apart. They are distributed evenly over the

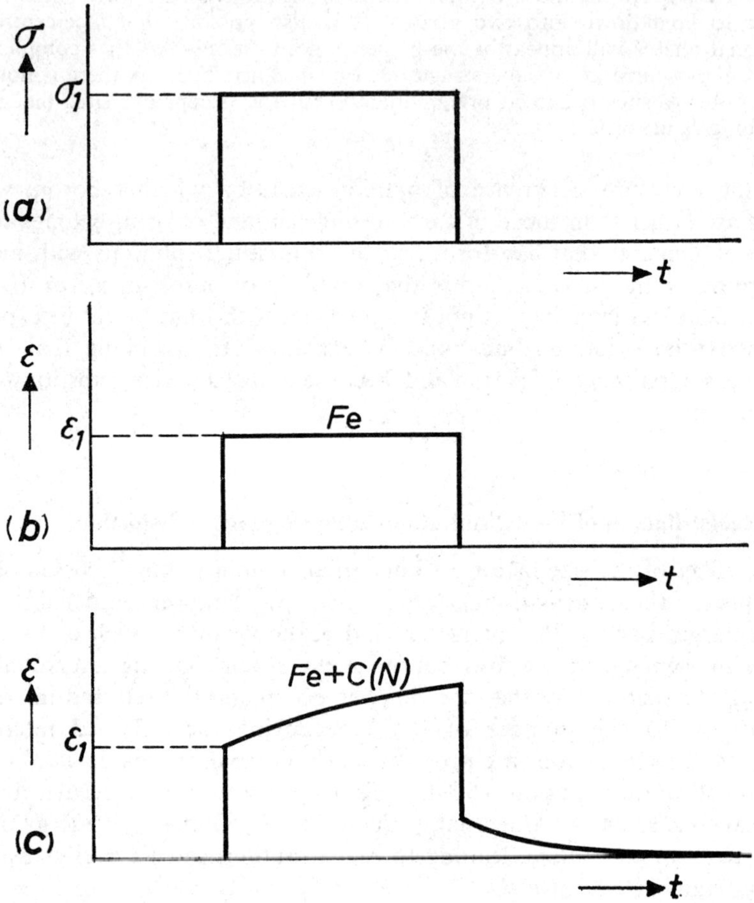

FIG. 12. (a) At a given moment a constant tensile stress σ_1 (much smaller than the yield stress) is applied to an iron crystal in the direction of one of the cube axes.

(b) The immediate reaction of the metal is an elastic strain ε_1, which remains constant in the case of pure iron.

(c) If the iron contains nitrogen or carbon in solution, the instantaneous strain ε_1 is followed by a much smaller strain which gradually tends to a final value (after effect).

If later the tensile stress is removed, the total strain disappears immediately in pure iron; with iron containing nitrogen or carbon the same instantaneous shortening will occur, followed by the gradual disappearance of the extra strain. This extra strain is much smaller than indicated in the figure.

x, y and z sites, i.e. one third in each. However, this doesn't mean that they occupy fixed positions; as will be shown in following chapters they jump from one interstice to another unless the temperature is very low. At room temperature their average time of stay at any one interstice is of the order of magnitude of one second.

When a crystal is elastically stretched in the z direction then there is a preference for the elongated z sites and a shift in equilibrium occurs, such that the number of occupied z sites increases whilst the number of occupied x and y sites decreases[(1)]. This redistribution takes a little while and results in an additional elongation in the z direction, occurring after the instantaneous initial elastic elongation and gradually approaching an end value (Fig. 12). The rate at which this happens is determined by the frequency of jumping of the N and C atoms and consequently by the temperature. This phenomenon is usually called the *elastic after-effect*. A complete ordering of the interstitial atoms remains far from being reached since the tendency of the entropy to a maximum opposes it. For a more detailed discussion of this phenomenon and its many applications in metal research the reader's attention is drawn to Chapters 3 and 4.

Distortion caused by internal stresses

A familiar source of internal stresses are the dislocations, well-known one-dimensional lattice imperfections giving rise to the plastic deformability of metals. They are shown in Figs. 13 and 14, representing two extremes: the screw dislocation and the edge dislocation. They can be regarded as originating as follows (although it cannot be realized in this simplified way in practice): cut a crystal in the direction parallel to a slip plane and displace one part by sliding it along the other part. The inside edge of the cut is the dislocation line. It corresponds to a screw dislocation when the slip has taken place in a direction parallel to this line (Fig. 13), but to an edge dislocation when it has taken place in a direction at right angles to the line (Fig. 14). In both cases it is assumed that slip has occurred in such a way that at great distance from the dislocation line the lattice is undisturbed. One could regard both figures as instantaneous pictures of a dislocation on its way from one side of the crystal to the other (from front to back in Fig. 13 and from right to left in Fig. 14).

From this it follows that when an edge dislocation exists, one part of the crystal is compressed compared to the other part. Fig. 15 shows a cross-section of the crystal of Fig. 14, the atoms being represented by circles and assuming the crystal has a simple cubic structure. The broken line represents

1. J. L. Snoek, *Physica*, 's *Grav.* **8**, 711 (1941); **9**, 862 (1942).

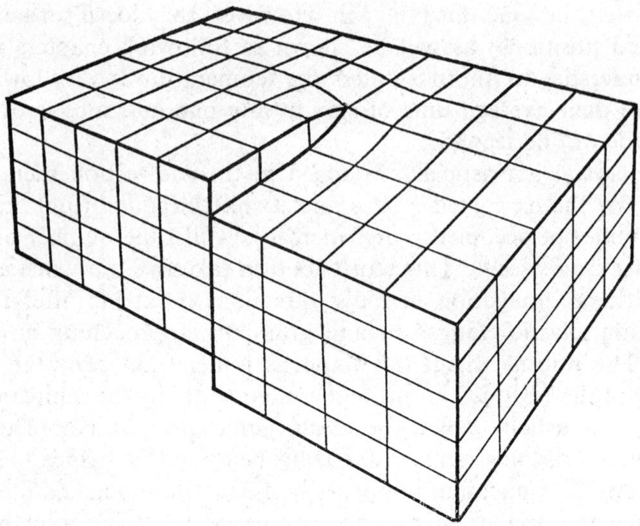

FIG. 13. Symbolized representation of a screw dislocation at right angles to the top face of the crystal. The Burgers vector is parallel to the dislocation line.

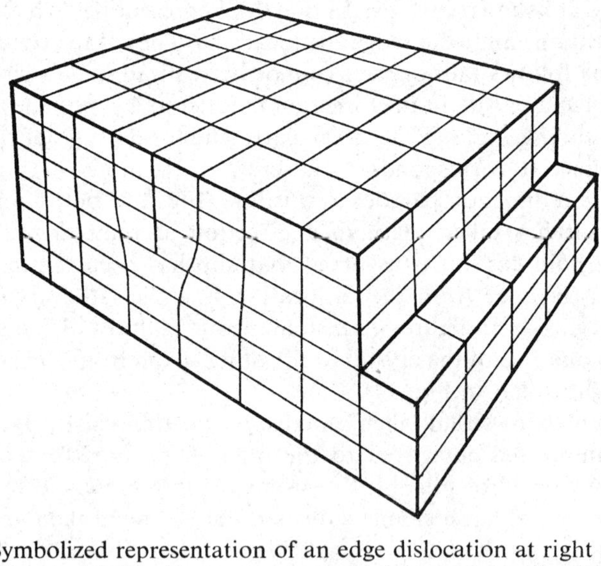

FIG. 14. Symbolized representation of an edge dislocation at right angles to the front face of the crystal. The Burgers vector is at right angles to the dislocation line.

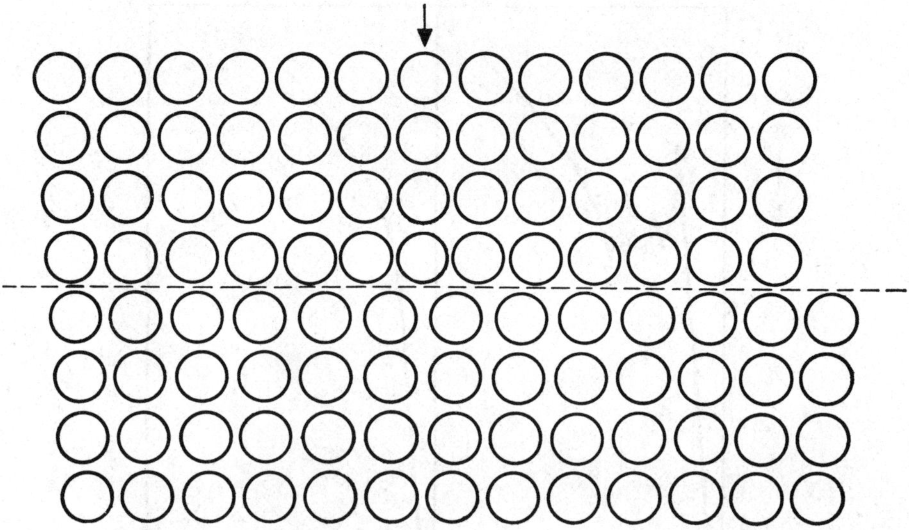

FIG. 15. Symbolized representation of an edge dislocation in a simple cubic lattice of metal atoms. The dislocation can also be seen as produced by a forced introduction of an extra half-plane of atoms (indicated by the arrow) in the upper half of the crystal. The broken line indicates the slip plane. The figure corresponds to a cross-section of Fig. 14.

the plane in which shearing can occur. The edge dislocation is at right angles to the plane of the drawing; its presence causes the lattice to be compressed at one side of the slip plane and elongated at the other side. The distortion diminishes rapidly with increasing distance from the imperfection. The introduction of a nitrogen or carbon atom into an enlarged interstice of the elongated region will require less deformation energy than the introduction into a normal interstice. Dissolved N or C atoms that reach the elongated region through diffusion will be unable to leave it again at low temperatures (e.g. room temperature). If a sufficient number of N or C atoms are present in the lattice then these will form chains or clouds of interstitial atoms spreading over the whole length of any edge dislocation.

The formation of these chains or clouds greatly affects the plastic properties of a metal, because a detaching of the dislocations from the atom clouds must precede plastic deformation. This separation brings the metal into a state of higher energy since after it has taken place the dissolved atoms are in normal interstices. The initiation of the deformation therefore requires an extra stress which, once the dislocations have broken away, is no longer needed. In this way we can understand the occurrence of an upper and lower yield point in the stress–strain curve of mild steel (Fig. 16a). It also

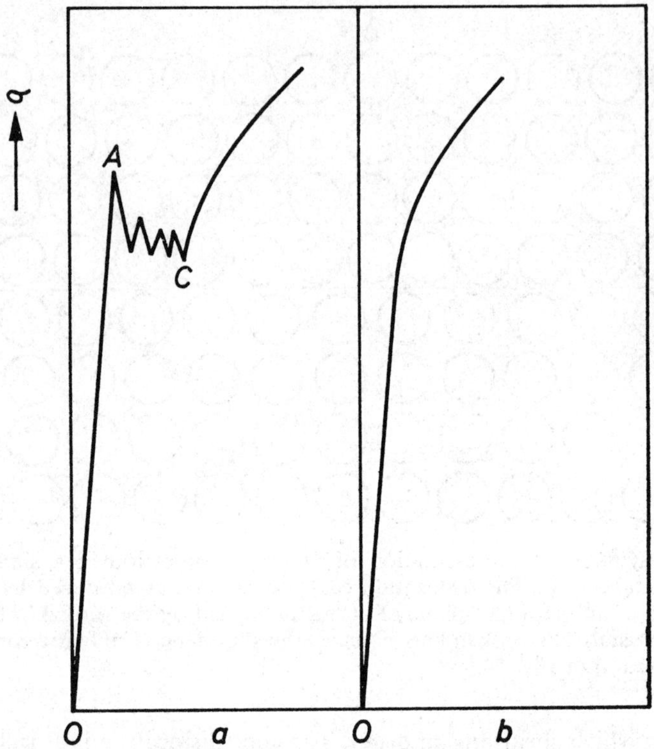

FIG. 16. Stress–strain graphs of mild steel: (*a*) the steel shows a sharp upper yield point at A; (*b*) after plastic deformation to C and reapplying the load, the metal shows no distinct yield point.

explains why immediately after a slight plastic deformation—that is when the existing dislocations are still free—no distinct yield point occurs (Fig. 16b). If the metal is then left alone for some considerable time, the nitrogen and carbon atoms again diffuse to the dislocations, so that the upper and lower yield points return and the metal becomes more resistant to deformation (harder). This spontaneous process is known as *strain ageing*. It is primarily due to nitrogen atoms, the solubility of nitrogen in iron being much greater than that of carbon.

The above theory of the occurrence of an upper and lower yield point and of the strain ageing originates from Cottrell[(1)]. Our own experiments[(2)] have shown that at not too low a temperature strain ageing even occurs

1. A. H. COTTRELL, *Prog. Metal Phys.* **1,** 77 (1949).
2. J. D. FAST, *Revue Métall. Paris* **47,** 779 (1950).

when the iron contains nitrogen or carbon virtually only in the form of a precipitate of iron nitride or iron carbide and none in solution. In this case nitrogen or carbon atoms diffuse from the precipitate to the dislocations, implying the entire or partial solution of the precipitate. One can conclude from these experiments that nitrogen and carbon atoms are more strongly bound in a dislocation than in an iron nitride or iron carbide crystal. This is the more remarkable since in normal interstices they are less strongly bound than in the crystals mentioned. The influence of the size factor and the corresponding deformation energy is very evident from these facts. When the nitrogen or carbon is present in steel in the form of a very stable nitride or carbide then it does not age after deformation because nitrogen and carbon are, in these compounds (e.g. TiN or TiC), more strongly bound than in dislocations. It is therefore possible to produce a non-ageing steel by the addition of a sufficient quantity of Ti, Zr, V, Nb or Ta. Since the ageing is mainly due to nitrogen and since the above-mentioned elements have a greater affinity to nitrogen than to carbon, satisfactory results are obtained with relatively small additions. Since aluminium has only a small affinity for carbon it is often used in practice because it has a great affinity for nitrogen.

In the above only the interaction between interstitial atoms and edge dislocations, whose strain fields possess both hydrostatic and shear components, was discussed. The strain field around a screw dislocation is symmetrical, having only shear components. In fcc and hcp metals therefore, interstitial atoms have a greater preference for edge dislocations than for screw dislocations. In bcc metals interstitial atoms cause, as was shown above, both volume expansion and tetragonal deformation, and will consequently be attracted strongly by both edge and screw dislocations. In bcc metals the mechanical properties are therefore more adversely affected by interstitial atoms than in fcc and hcp metals.

1.5. Electronic interaction

The previous sections illustrated the great influence of the size factor on the solubility of a gas in a metal; they could therefore create the wrong impression that this factor is the determining one. In point of fact it is only a factor inhibiting atoms going into solution because strain energy has to be provided in order to introduce foreign atoms into the interstices. A second opposing factor is the dissociation energy required to split the molecules into atoms before these can be absorbed by the metal. As illustration we consider the dissolving of a nitrogen molecule in an arbitrary metal. We imagine this process to proceed in two stages: (1) the molecule is split into atoms at some distance from the metal; (2) the atoms are introduced into the interstices of the metal:

$$N_2 \rightarrow 2N \qquad (1.5.1)$$

$$2N \rightarrow 2[N] \qquad (1.5.2)$$

The total energy required for the dissolving process is the sum of the energies corresponding to the processes (1.5.1) and (1.5.2):

$$E_{sol} = E_1 + E_2$$

E_1 already amounts to at least $1{\cdot}2 \times 10^{-18}$ joule per molecule (7·4 eV)[1]. Therefore, if E_2 represented only the strain energy discussed above, positive like E_1, then nitrogen would not dissolve in any metal to a notable extent. Solubilities of any importance can only be expected if E_2 is negative and, as far as its absolute value is concerned, of the same order of magnitude as E_1. This can only be so when a favourable electronic interaction occurs between the metal and the nitrogen atoms. The importance of this interaction is demonstrated by the fact that helium, due to its very stable configuration of two paired electrons, does not dissolve to a notable extent in any metal notwithstanding the facts that helium atoms are smaller than oxygen, nitrogen and carbon atoms and that no dissociation of molecules need precede the dissolving process[2].

That a strong electronic interaction is a prerequisite for the solution of a gas in a metal is in accordance with modern theory of metals. This theory, which is further discussed in Sections 1.7 and 1.8, describes a metal as a three-dimensional lattice of positive ions, floating freely as it were, in a dense electron 'gas' filling the space between the ions. The mutual electrostatic repulsion of the ions causes them to stay as much apart as possible and their distances adjust themselves such that the pressure of the electron gas, the electrostatic attraction between ions and electrons and the mutual repulsion of the ions compensate each other. This results in the ions arranging themselves in a regular lattice. It is now possible to transfer the picture of the symmetrically arranged rigid spheres, as discussed in preceding sections, to the ions but in this case it is no longer necessary that the spheres touch each other[3]. In the alkali metals the ions are relatively far apart, though in many other metals they are closer together. Fig. 17 shows the positions of the ions in the electron gas in the most closely packed lattice planes for the metals sodium and copper[4]. In sodium approximately 70% of the metal volume is taken up by

1. In the literature, values of $1{\cdot}2 \times 10^{-18}$, $1{\cdot}4 \times 10^{-18}$, $1{\cdot}6 \times 10^{-18}$ J (7·4, 8·6 and 9·8 eV respectively) have been reported.

2. According to investigations by A. van Wieringen helium dissolves to a measurable extent in the non-metallic elements germanium and silicon which crystallize with the loosely packed diamond structure. See his paper in the conference book *La Diffusion dans les Métaux* (editors: J. D. Fast, H. G. van Bueren and J. Philibert), Eindhoven, 1957.

3. However, when an atom or ion is regarded as a positive nucleus surrounded by a cloud of electrons, the notion of touching becomes very vague.

4. W. Shockley, *J. appl. Phys.* **10**, 543 (1939).

FIG. 17. Arrangement of ions in the closest packed lattice planes of sodium {110}, on the left, and copper {111}, on the right, according to Shockley. The ion boundaries are idealized; in reality the electron density falls gradually from a high value inside the ion to a much lower constant value outside it.

the electron gas. In copper, however, the ions are so close together that a mutual penetration of their electronic charge clouds occurs.

The spaces between the metal atoms can therefore, in many cases, be larger than would be expected from the primitive model of the touching rigid spheres. On the other hand, these spaces are not empty, as in the model, because they are filled with electrons. Consequently, the introduction of an atom (even a very small atom) into an interstice implies a strong electronic interaction with the metal.

Somewhat artificially one can separate from the electronic interaction energy, E_2, a part corresponding to the strain energy, discussed in the previous sections. Several investigators have tried to calculate this strain energy starting from the continuum theory of elasticity, by using the values of Young's modulus in different crystallographic directions. However, the application of a typical macroscopic theory to processes on a scale of atoms is a doubtful procedure. The separation of a strain energy component from the total electronic interaction energy is in itself somewhat suspect since the two parts are interdependent. The part that remains after the separation is strongly dependent on the interatomic distances and thus on the strain energy.

It has already been stated (in Section 1.1) that the behaviour of hydrogen in various metals is exceptional. This does not only apply to its large diffusion coefficient but also to the influence of the size of the interstices on its solubility. Striking in this respect is the fact that, as opposed to nitrogen and oxygen, the solubility of hydrogen in the bcc forms of titanium and zirconium is much greater than in the hcp forms of these metals. The electronic interaction is apparently more favourable when the hydrogen occupies the small interstices of the bcc metal than when it is in the larger interstices of the hcp form.

1.6. Chemical factors

Chemical affinity

Instead of stating that the electronic interaction is the primary factor determining the degree of solubility of a gas in a metal, we could as well say that the chemical affinity between metal and gas is of primary importance. In doing so we concern ourselves no longer with the solution of separate molecules but move to thermodynamic considerations which always relate to macroscopic quantities. In equations (1.5.1) and (1.5.2) the chemical symbols then no longer refer to one molecule or atom but to a gram-molecule or gram-atom.

The concept of 'chemical affinity' has a wider meaning than 'electronic interaction', since it does not only encompass changes in energy accompanying a reaction at constant volume but also changes in entropy. When the pressure is kept constant instead of the volume then the affinity is determined not by the changes in energy and entropy but by the changes in enthalpy, H, and entropy, S. In this connection we may remind the reader of the equation $\Delta G = \Delta H - T\Delta S$, where G represents the free energy (Gibbs free energy) which has a minimum value in the equilibrium state.

Since the enthalpy and entropy are functions characteristic of the state of a system the enthalpy or entropy of solution of molecular nitrogen is given by the sum of the enthalpies or entropies corresponding to equations (1.5.1) and (1.5.2):

$$\Delta H_{sol} = \Delta H_1 + \Delta H_2 \tag{1.6.1}$$

$$\Delta S_{sol} = \Delta S_1 + \Delta S_2 \tag{1.6.2}$$

Fig. 18*a* shows the enthalpy diagram corresponding to equation (1.6.1) for the case in which the total enthalpy of solution, ΔH_{sol}, is negative (liberation

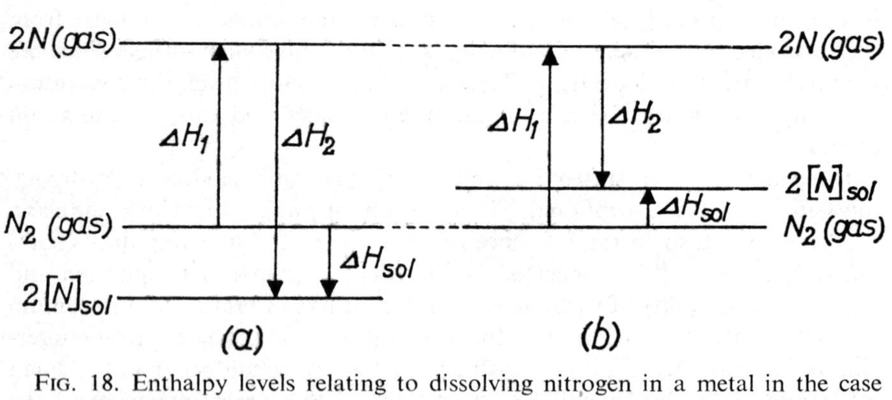

FIG. 18. Enthalpy levels relating to dissolving nitrogen in a metal in the case where the enthalpy of solution, ΔH_{sol}, is negative (*a*) and in the case where it is positive (*b*).

of heat during the dissolving of N_2 at constant temperature and pressure), Fig. 18*b* corresponds to the case in which the total enthalpy of solution is positive.

The change in entropy, ΔS_{sol}, that occurs during the dissolving of a gas of normal atmospheric pressure in a metal is negative, unless the solution is very dilute. The partial entropy of the gas in the metal can roughly be written as the sum of two terms: one is a *configuration term* determined by the number of ways in which the gas atoms can be distributed over the interstices under consideration; the other is a *vibration term* taking account of the vibrations of the gas atoms in their interstices, and also of the changes in the vibrations of the metal atoms resulting from the presence of the interstitial atoms. According to an investigation by Eichenauer *et al.*[1], hydrogen and deuterium dissolved in single crystals of nickel and copper give a contribution to the specific heat which corresponds to what would be expected if they behaved as harmonic oscillators without seriously affecting the vibrations of the metal atoms. In other words, the vibrations of hydrogen and deuterium in these solutions can be defined by means of the well-known Einstein function containing only one vibration frequency. According to Thomasch[2] this also applies to hydrogen and deuterium in the delta phases of the Zr–H and Zr–D systems at the much larger concentrations H/Zr and D/Zr = 1·58 and, according to Bergsma and Goedkoop[3], also to hydrogen in palladium at the concentration H/Pd = 0·63.

The statement that the existence of chemical affinity between metal and gas is a necessary requirement for the formation of interstitial solutions may of course not be reversed into the statement that a gas will dissolve interstitially in a metal when the chemical affinity between the two is sufficiently great. Chlorine has a great affinity for many metals but does not dissolve in any solid metal to a notable extent, probably because the size factor is too unfavourable[4]. Another opposing factor is the great stability of metal chlorides.

Stability of a coexisting phase

As indicated by equation (1.6.1) the enthalpy of solution, ΔH_{sol}, and consequently the solubility of an element in a metal, depends to a large extent on the stability of the coexisting phase acting as a source of the

1. W. Eichenauer, W. Löser and H. Witte, *Z. Metallk.* **56,** 287 (1965).
2. W. J. Thomasch, *Phys. Rev.* **123,** 510 (1961).
3. J. Bergsma and J. A. Goedkoop, *Physica, 's Grav.* **26,** 744 (1960).
4. In some *liquid* metals the solubility of chlorine is very great. As an extreme example it can be mentioned that Cs and CsCl form an uninterrupted series of liquid solutions. See: M. A. Bredig, H. R. Bronstein and W. T. Smith, *J. Am. Chem. Soc.* **77,** 1454 (1955).

interstitial atoms. In the case of Fig. 18, this stability is indicated by ΔH_1. When the coexisting phase is not N_2 of, for instance, normal atmospheric pressure but a nitride of low stability, then ΔH_1 is much smaller and greater solubilities are found. Thus the solubility of nitrogen in α iron in equilibrium with Fe_8N or Fe_4N is much greater than in equilibrium with N_2 at normal atmospheric pressure.

The effect of the stability of the coexisting phase also shows very clearly when comparing the solubilities of carbon, nitrogen and oxygen in iron. Although the sizes of the atoms C, N and O decrease in this order the solubility of oxygen in both forms of iron in the presence of 'FeO' or Fe_3O_4 as second phase is much smaller than that of nitrogen in the presence of Fe_8N or Fe_4N, or of carbon with Fe_3C or graphite as second phase. This is, without doubt, largely due to the fact that iron oxides have greater stability than iron nitrides and carbides.

Solubility of oxygen in copper, silver and gold

The previous paragraphs mentioned two factors which, in addition to the solution entropy, determine the interstitial solubility of an element in a metal: the stability of the phase acting as a source of the interstitial atoms and the total electronic interaction energy (or enthalpy) between the element in solution and the metal. From the latter the part determined by the size factor was separated in a rather artificial manner. The literature on solutions of gases in metals shows several examples of overestimating the effect of the size factor (the strain energy) and underestimating the effect of the stability of the coexisting phase. For instance, in a paper by well-known researchers in this field: 'The structural reason for the very low solubility of oxygen in iron (and in copper) as contrasted with its measurable solubility in silver and very large solubility in zirconium, titanium and vanadium, is almost certainly that oxygen is too large to be accommodated interstitially in the iron lattice.' The incorrectness of this reasoning has already been indicated by Meijering(1). In the first place, the authors of the cited sentence did not take into account the fact that vanadium has a bcc lattice and consequently smaller interstices than Cu, Ag, Zr, Ti and fcc Fe (see Table 1). A more serious error, however, is the complete neglecting of the chemical factor and of the influence of the coexisting phase. In order to demonstrate this we will consider the solubility of oxygen in copper, silver and gold.

According to Table 1 gold and silver have interstices of equal size but, in agreement with their different affinities for oxygen, the solubility of

1. J. L. Meijering, *Acta Metall.* **3**, 157 (1955).

oxygen in silver is relatively great, whereas that in gold is negligible. At 770°C silver in equilibrium with oxygen of normal atmospheric pressure contains 0·0032% O[1], whereas at this temperature copper in equilibrium with Cu_2O contains 0·0020% O[2]. These solubilities are of the same order of magnitude, but obviously only solubilities at one and the same oxygen pressure (i.e. one and the same chemical potential of the oxygen) can be compared. The dissociation pressure at 770°C of Cu_2O according to the reaction

$$2Cu_2O \rightleftharpoons 4Cu + O_2$$

has only a value of about 10^{-5} N m^{-2} (10^{-10} atm). According to the $\sqrt{p}$ rule, then, at 770°C and a given low oxygen pressure, copper will dissolve about 10^5 times as much oxygen as silver does. If a eutectic Ag–Cu alloy at this or any other temperature contains a quantity of oxygen in solution then practically all of it will be in the copper-rich phase despite its smaller lattice interstices.

In agreement with the decreasing affinity for oxygen, the solubility of this gas in the sequence copper, silver, gold thus decreases very rapidly. This demonstrates again that in general the chemical factor is of much greater importance than the size factor. The effect of the latter only becomes clear when the chemical factor remains constant or nearly constant, as is the case when the solubilities of a particular gas in different crystal modifications of one and the same metal are compared. Even then any quantitative consideration based on the continuum theory of elasticity must be regarded with some suspicion.

1.7. Brillouin zones and energy bands

Various scientists have looked for a connection between the solubility of gases (especially hydrogen) in transition metals and the occupation of their Brillouin zones with electrons. Before discussing this we would like to remind the reader of the significance of the concepts of Brillouin zone and energy band. To this end we will consider firstly a gas of free electrons; following this, account will be taken of the fact that the conduction electrons in a metal move in the periodic potential field originating from the metal ions.

1. W. Eichenauer and G. Müller, *Z. Metallk.* **53**, 321 (1962).
2. F. Bouillon *et al.*, *Acta Metall.* **10**, 647 (1962).

The free electron gas

Let us consider an electron moving freely in a cubic box with impenetrable walls. According to quantum mechanics the particle cannot assume any arbitrary energy. The stationary states of the electron permitted by the theory can be visualized geometrically in momentum space by means of the coordinates p_x, p_y and p_z. If this space is divided into many cells of size h^3/V, where h represents Planck's constant and V the volume of the cube, then each of these cells corresponds to two possible states of the electron, differing only in their spin direction. The energy of the particle and its momentum are connected by the quadratic relation $E = p^2/2m$ (see Fig. 19). Cells near to the origin correspond to low energies, those further away to larger energies of the electron.

If the cubic box contains not one but N electrons, then according to Pauli's exclusion principle all must be in different quantum states. This can be expressed more simply by stating that each cell in momentum space can

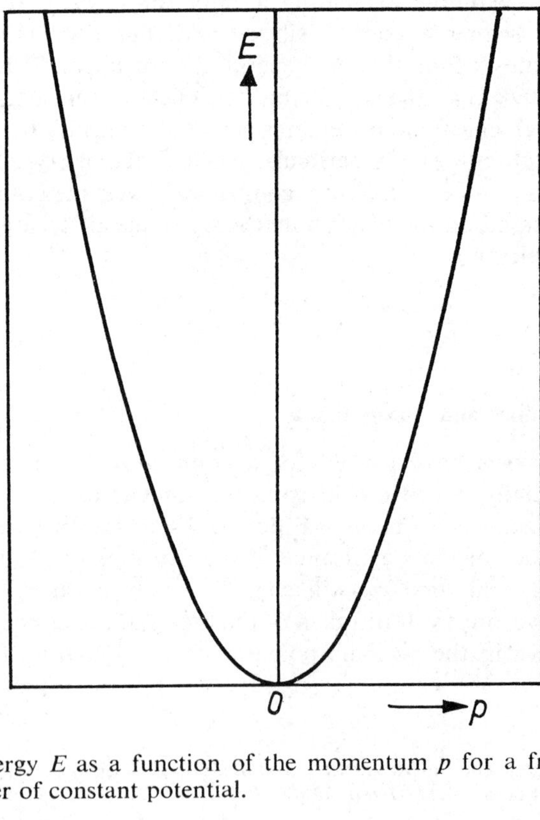

FIG. 19. Energy E as a function of the momentum p for a free electron in a closed container of constant potential.

'contain' at most two electrons (with opposite spin). At the absolute zero of temperature the energy of the electron gas is a minimum. All cells in a sphere in momentum space are then fully occupied by electrons whilst outside it all cells are empty. The surface separating the full from the empty states is called the Fermi surface. The distribution of the electrons over momentum space dictated by the Pauli principle means that at 0 K their average momentum is already very large. In connection with considerations in Section 1.8 we are particularly interested in the maximum momentum of the electrons at 0 K, i.e. in their momentum p_F at the Fermi level. This can be obtained straight away from the facts that the volume of the Fermi sphere is $4\pi p_F^3/3$ and that it contains $N/2$ cells of volume h^3/V:

$$\frac{4\pi p_F^3}{3} = \frac{Nh^3}{2V} \tag{1.7.1}$$

The energy E_F at the Fermi level is given by

$$E_F = \frac{p_F^2}{2m} \tag{1.7.2}$$

and hence from (1.7.1)

$$E_F = \frac{h^2}{8m}\left(\frac{3N}{\pi V}\right)^{2/3} \tag{1.7.3}$$

Brillouin zones

According to the theory and confirmed by the well-known experiments of Davisson and Germer the behaviour of a beam of electrons is determined by mathematical relations almost identical to those determining the propagation of electromagnetic radiation. The effective wavelength of a beam of electrons of uniform velocity is given by De Broglie's relation

$$\lambda = h/p \tag{1.7.4}$$

Like X-rays, electron beams of suitable velocity are reflected by the lattice planes of a crystal when they strike these at certain well-defined angles. The angles are given by Bragg's expression

$$n\lambda = 2d \sin \theta \tag{1.7.5}$$

where θ is the angle of incidence, d the spacing of the lattice planes concerned and n an integer. Combining (1.7.4) and (1.7.5) gives us the expression for electrons:

$$p \sin \theta = nh/2d \tag{1.7.6}$$

Here $p \sin \theta$ represents the component of the momentum at right angles to the reflecting atom planes. A 'monochromatic' electron beam will therefore only be reflected by a group of lattice planes when the normal component of the momentum of each separate electron has certain discrete values ($h/2d$, $2h/2d$, $3h/2d$, . . .) which depend only on the distance between the reflecting planes.

Let us now consider a crystal being bombarded by electrons. When their momenta satisfy, both as regards direction and magnitude, equation (1.7.6) for one of the possible groups of lattice planes, then they will not be absorbed by the crystal but reflected. From this, we can draw the important conclusion that electrons with those particular momenta cannot be present in the crystal at all. The freedom of a 'free' electron is thus severely restricted in that it cannot possess all the momenta which are allowed for the empty box mentioned earlier. Indicating the permissible values of the momentum by vectors in momentum space, then according to the above, each lattice plane in the crystal excludes all vectors having a normal component of certain definite values. This means that in momentum space planes can be delineated corresponding to prohibited values of the momentum of the electrons. Thus momentum space can be divided in regions (Brillouin zones) whose extent and shape are completely determined by the crystal structure. Fig. 20 shows the first Brillouin zone for the face-centred cubic lattice.

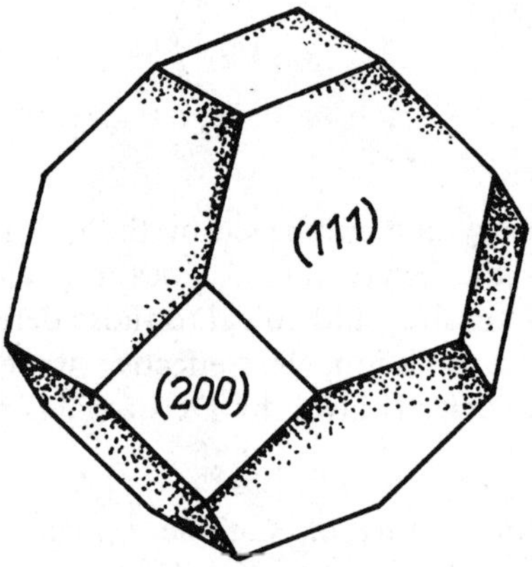

FIG. 20. First Brillouin zone of the fcc lattice. The boundary planes are of the (111) and (200) types.

It can be shown that the existence of the Brillouin zones has important consequences for the permitted energies of the electrons: if, for a given direction in a crystal, the energy of an electron is plotted as a function of its momentum then a prohibited value corresponds to a discontinuity in the energy (Fig. 21). The concept of momentum as denoted by the expression $E = p^2/2m$ loses its significance in these regions. It is therefore preferable to use the language of wave mechanics and to refer to prohibited values of the wave number k, which corresponds to the number of wave crests over

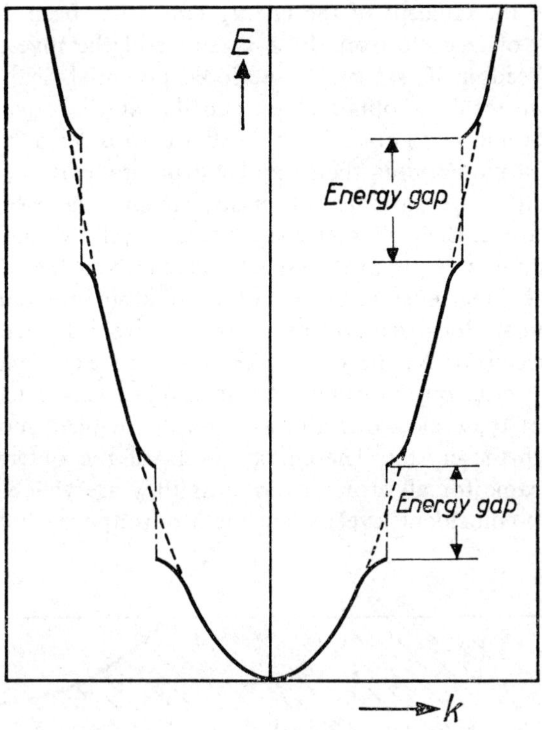

FIG. 21. The continuous line on the graph indicates the energy E as a function of the wave number k for an electron moving in a particular direction through a crystal. The dot-dash lines indicate the critical values of the wave number of this electron. Each of these values corresponds to a region of forbidden energies.

a distance 2π. Where meaningful the relation between p and k is given by the expression

$$k = 2\pi/\lambda = 2\pi p/h \tag{1.7.7}$$

Energy bands

Often, only the energies of the electrons are taken into account disregarding their directions of movement. This leads to a bunching of permissible energies in 'energy bands'. A band contains all the energies occurring in the corresponding Brillouin zone. The zones in momentum space (k space) are completely separate but the energy bands corresponding to the zones can overlap. Overlap between the first and second bands occurs when the second Brillouin zone contains momenta (k values) corresponding to energies smaller than the largest energies occurring in the first band.

In the above, the concept of the energy bands has been developed from the assumption of free electrons in a crystal and the investigation of the limits to this freedom as set by the periodic potential field in which they move. A similar result is obtained by starting at the other extreme, that of completely bound electrons. To this end we consider a crystal in which the atoms, whilst maintaining their regular structural pattern, are sufficiently far apart for their mutual interaction to vanish. The electrons of each separate atom are at their characteristic energy levels; we now imagine the interatomic distances to be progressively decreased to the normal value in the crystal. The innermost electrons of each atom are hardly disturbed by this; they remain localized and retain their sharply defined energy levels. However, the electrons on the outside of the atoms experience strongly the influence of the electrons of neighbouring atoms. This causes a complete delocalization of these electrons and as a result they can move from atom to atom through the crystal. The energy levels of the outermost electrons which are the same for all atoms as long as they are widely separated are replaced by non-coincident levels when the atoms approach each other, i.e.

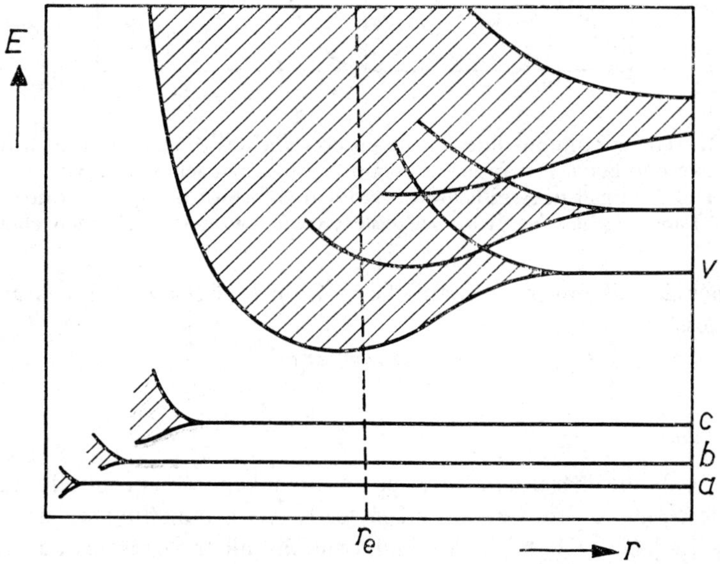

FIG. 22. When atoms approach each other to form a crystal the energy levels of the atoms change into energy bands. The broken line indicates the equilibrium distance, r_e, of the atoms in the crystal. The three lower levels (a, b and c) refer to inner electrons, level v to the valency level of a free atom. In free atoms the higher levels are unoccupied.

an energy level changes into an energy band which becomes wider the closer the atoms approach each other (Fig. 22). The number of energy levels (or, more strictly, quantum states) in each band corresponds to the number of atoms in the crystal. The levels are so close together that within a band the energy can almost be considered as being a continuous variable.

Each discrete level can, according to the Pauli principle, accommodate at the most two electrons with opposite spin. At zero temperature on the Kelvin scale the energy must be a minimum, therefore up to a certain value all levels will be occupied by two electrons whilst above that value all levels will be empty. This filling of levels up to a certain limit can result in all levels in a band being fully occupied whilst the next band in the energy spectrum is completely empty. (In this case it is usual to name the top full band *the* full band and the bottom empty one *the* empty band). It is also possible that the uppermost band containing electrons is only partly filled; this is the case with metals whilst the former holds for insulators.

Energy bands in metals

Typical examples of metals are the alkali metals (Li, Na, K, Rb and Cs) which have one valency electron per atom. The energy band containing the valency electrons is only half filled, because a band in a crystal of N atoms has N levels and thus has room for $2N$ electrons. Divalent elements such as the alkaline-earth metals (Be, Mg, Ca, Sr and Ba) would be insulators if their energy bands were separated by wide prohibited regions. They are metals because their valency band (the outer s band) strongly overlaps higher bands corresponding to unoccupied levels in the free atoms. Fig. 22 shows such an overlapping of energy bands. For most metals it is not possible to derive the populations of the overlapping bands from those of the corresponding shells of the free atoms. We shall clarify this matter by discussing the populations of the bands in some transition metals. This will serve as an introduction to a number of the following sections in which the solubility of hydrogen in two transition metals (palladium and nickel) will be discussed.

The transition metals are characterized by an incomplete inner electron shell inside a completely or partially filled outer shell (Fig. 23). The first series of transition metals, which includes nickel, is characterized by an incomplete 3d shell inside the 4s shell. The table on page 35 gives the occupation of these shells for the free atoms of the last three elements of the series (iron, cobalt, nickel) as well as for copper, the first element in which the 3d shell is fully occupied. The table also gives the spin orientation of the 3d electrons for the ground state of the free atoms.

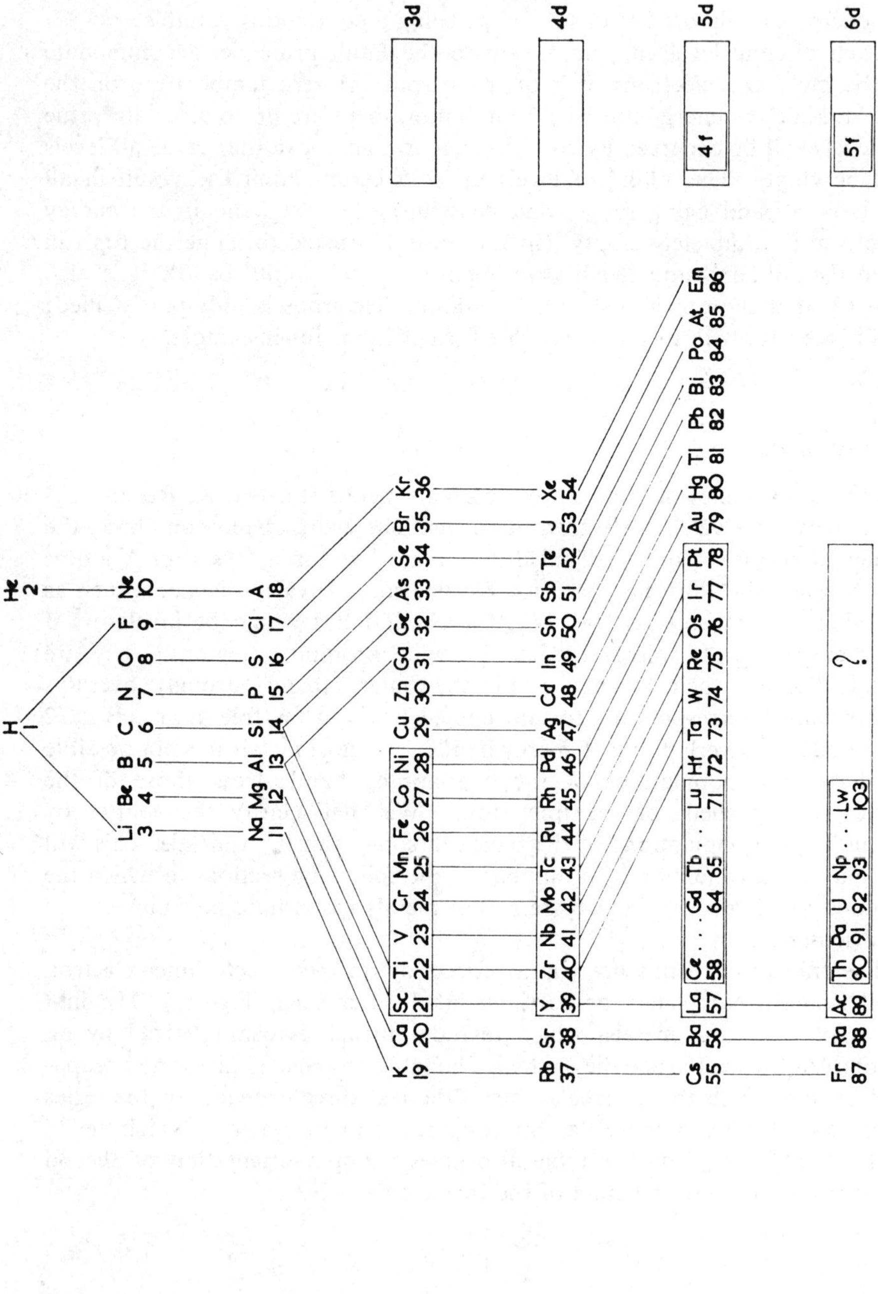

FIG. 23. Periodic table of the elements; the transition metals are shown in boxes. In Sc, Ti, . . ., Ni the 3d level is only partly filled; in Y, Zr, . . ., Pd this applies to the 4d level.

Electron configuration and spin orientation of the 3d *electrons in free atoms of elements* 26 *to* 29

Element	Configuration	3d spin orientation				
Fe (26)	$3d^6 4s^2$	↑↓	↑	↑	↑	↑
Co (27)	$3d^7 4s^2$	↑↓	↑↓	↑	↑	↑
Ni (28)	$3d^8 4s^2$	↑↓	↑↓	↑↓	↑	↑
Cu (29)	$3d^{10} 4s$	↑↓	↑↓	↑↓	↑↓	↑↓

In agreement with this table copper atoms in the solid metal have no resulting magnetic moment, i.e. copper is diamagnetic. When a diamagnetic material is brought into a magnetic field, the movements of its electrons change such that they cause a small induced magnetization in a direction opposite to that of the field. A diamagnetic material has therefore a (very small) negative susceptibility and it will be repelled by a magnet.

The atomic magnetic moments of the ferromagnetic metals iron, cobalt and nickel cannot be derived from that of their free atoms. If it is assumed that both 4s electrons in these metals act as conduction electrons whilst the configuration and spin orientation of the 3d electrons is the same as in the free atoms, then, according to the above table, atomic magnetic moments of 4, 3 and 2 Bohr magnetons, respectively, may be expected. In actual fact the outer electrons of iron, cobalt and nickel are distributed in accordance with a complicated equilibrium over the overlapping 3d and 4s bands. Measurement of saturation magnetizations at low temperatures gives experimental values of 2·22, 1·74 and 0·60 Bohr magnetons per atom of Fe, Co or Ni. Theory is not yet sufficiently advanced to calculate these values.

For the following sections it is of particular importance that, according to the above, there are 0·60 electron 'holes' per atom(1) in the 3d band of nickel. Until quite recently it had been generally accepted from a series of diverse experiments that the same is true for the 4d band of palladium. However, measurements of the de Haas–van Alphen effect(2) indicate that palladium has only 0·36 holes per atom in its 4d band.

1. Strictly speaking, this value should be corrected for a small contribution to the magnetization from orbital electronic motion. After this correction has been applied, the number of holes in the 3d band of nickel is found to be slightly less than 0·60 per atom.
2. J. J. Vuillemin, *Phys. Rev.* **144**, 396 (1966).

1.8. Fermi surface and electronic specific heat

The Fermi surface

In the previous section we concerned ourselves with the behaviour of conduction electrons in metals and we discussed how these electrons form a kind of gas obeying Pauli's exclusion principle. Because they obey this principle they will, at absolute zero, fill all available quantum states in momentum space up to states corresponding to the so-called Fermi energy. The surface of constant energy inside which all states are full and outside which all are empty at 0 K, is called the Fermi surface. For a free electron gas this surface is a perfect sphere and the energy corresponding to it is given by equation (1.7.3).

We also discussed the fact that the regular arrangement of the ions in a metal creates gaps in the range of energies allowed to the conduction electrons. The positions of the energy gaps are defined by the Brillouin zones which are polyhedra in momentum space (k space). Their shapes depend only on the crystal structure of the metal. For the alkali metals (bcc) the first Brillouin zone is a dodecahedron only half filled with electrons. For these metals the Fermi surface, enclosing the occupied states in the zone, is not far from spherical; it lies well within the zone. In terms of energy bands the Brillouin zone under discussion corresponds to the s band of an alkali metal because it contains the outer s electrons. It is half filled, i.e. filled up to the Fermi level. For the alkali metals the energy of this level can, with reasonably close approximation, be calculated by means of equation (1.7.3). In this way we find a value of about $6{\cdot}4\times10^{-19}$ J (4 eV).

The fcc metals, copper, silver and gold, also have one 'free' electron per atom and have their first Brillouin zones (Fig. 20) half filled. The Fermi surface of copper has been investigated by several methods. It could be shown that the sphere has been pulled out along the $\langle 111\rangle$ directions to such an extent that it meets the hexagonal faces of the zone over a substantial area. The Fermi surface of many metals has a shape that is much more complex than that of copper[1,2]. To understand many of the properties of metals the knowledge of their Fermi surface and of the density of electronic states at this surface (see below) is as important as the knowledge of their Brillouin zones and energy bands.

The electronic specific heat

The electrons contribute only little to the specific heat of a metal. This is due to the high value of the Fermi energy which, as mentioned above, is

1. J. M. Ziman, *Electrons in Metals (a short guide to the Fermi surface)*, Taylor & Francis, London (1964).
2. W. A. Harrison and M. B. Webb (editors), *The Fermi Surface*, Wiley, New York (1960).

of the order of 6×10^{-19} J. If we heat a metal from 0 K to the temperature T, an equilibrium between lattice and electrons will be reached by exchange of energy. This exchange takes place in quantities of approximately kT at a time. However, at room temperature kT is only $4{\cdot}8 \times 10^{-21}$ J (0·03 eV), i.e. less than 1 % of E_F. Only those electrons which are so near to the Fermi level that thermal agitation can bring them into unoccupied states, can take part in the exchange. More than 99 % of the electrons cannot take up thermal energy because the levels above them are already occupied and cannot accommodate any more electrons as a consequence of the Pauli principle.

The fraction of the electrons which can be affected by heating are those in a narrow layer near the Fermi level. This fraction is of the order of magnitude of kT/E_F. Since each of these electrons gains an energy of about kT, the total *thermal* energy of the electrons is

$$\Delta U \approx \frac{kT}{E_F} N_0 kT \quad \text{per mole}$$

where N_0 is Avogadro's number. The electronic heat capacity per mole therefore equals

$$C_{\text{elec.}} = \frac{\partial(\Delta U)}{\partial T} \approx \frac{2N_0 k^2 T}{E_F} = \gamma T \tag{1.8.1}$$

The electronic heat capacity is thus in simple cases proportional to the absolute temperature. However, at low temperatures the *lattice* heat capacity, according to Debye's theory, is proportional to T^3. The total heat capacity of a metal at low temperatures can therefore be written as

$$C = \gamma T + \beta T^3 \tag{1.8.2}$$

This equation indicates that in metals the electronic term will always be dominant at low enough temperatures. The values of the two coefficients γ and β can be determined by plotting C/T against T^2:

$$C/T = \gamma + \beta T^2 \tag{1.8.3}$$

If equation (1.8.3) is satisfied we obtain a straight line whose intercept with the vertical axis $T = 0$ gives us the value of γ, while the slope of the line gives β and therefore the value of the Debye temperature Θ according to the equation

$$\beta = \frac{1{\cdot}944}{\Theta^3} \text{ kJ mole}^{-1} \text{ deg}^{-4} \tag{1.8.4}$$

Fig. 24 shows C/T as a function of T^2 for copper over the region 0 to 4K[1,2].

1. J. A. Rayne, *Aust. J. Phys.* **9**, 189 (1956).
2. K. G. Ramanathan and T. M. Srinivasan, *J. Scient. Ind. Res.* **16B**, 277 (1957).

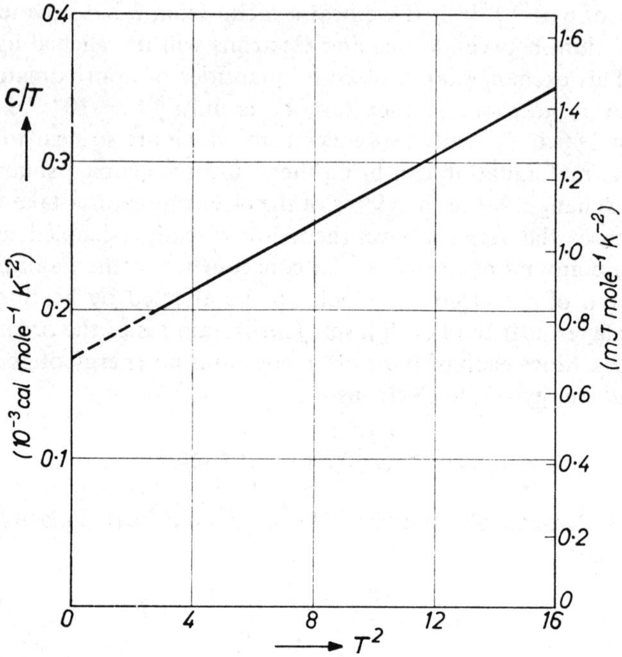

FIG. 24. C/T in 10^{-3} cal mole^{-1} K^{-2} and mJ mole^{-1} K^{-2} as a function of T^2 (K^2) for pure copper.

The electronic specific heat of transition metals

In transition metals, such as palladium and nickel, neither the s band nor the d band is full according to Section 1.7. In consequence both overlapping bands contribute to the electrical and thermal properties of the metal. However, their contributions are quite different. The mutual interaction of the outer s electrons is large so that they combine to form a rather broad band; on the other hand, the d electrons are nearer to the atomic nuclei and therefore the interaction with their partners on neighbouring atoms is much weaker. The d band is thus rather narrow and since it can accommodate ten electrons per atom, the density of electronic states, i.e. twice the number of energy levels per unit energy range, is very high inside the band. This means that in transition metals there are many more electrons near the Fermi level than in non-transition metals. In other words, in transition metals the number of electrons which contribute to the specific heat, i.e. the electrons within a range of about kT from E_F, is abnormally large. This results in large values for γ. For nickel and palladium γ values of

7·3 and 9·4 mJ mole^{-1} deg^{-2} ($1{\cdot}74\times10^{-3}$ and $2{\cdot}25\times10^{-3}$ cal mole^{-1} deg^{-2}, respectively) have been found. The corresponding values for their neighbours in the periodic table, i.e. for copper and silver, are 0·71 and 0·63 mJ mole^{-1} deg^{-2} ($1{\cdot}7\times10^{-4}$ and $1{\cdot}5\times10^{-4}$ cal mole^{-1} deg^{-2}, respectively). It will be remembered that the d bands of the latter elements are completely filled.

We have discussed the electronic specific heat of transition metals for the same reason as we discussed their magnetic properties in the previous section. Both change when the transition metals absorb hydrogen and from the nature of the changes many people have thought that conclusions could be drawn about the state of ionization of the absorbed gas, particularly that of hydrogen in palladium and nickel. In these considerations a connection is sought between the change in the magnetic properties and in the electronic specific heat on the one hand and the occupation by electrons of the d band on the other. For further details we refer the reader to the following sections, especially to Sections 1.12 and 1.13.

1.9. Hydrogen in palladium

The system Pd–H

Palladium belongs to the metals capable of absorbing large amounts of hydrogen. Below about 300°C two solid phases are in equilibrium over a large region of concentrations. The oldest systematic experiments on the equilibrium between palladium and gaseous hydrogen are probably those by Troost and Hautefeuille[1]. The conclusion suggested by their results was that palladium can take up small quantities of hydrogen in solid solution and that, when a limiting concentration is exceeded, the hydride Pd_2H is formed as a coexisting phase. Accurate measurements led Hoitsema[2] in 1895 to the deviating conclusion that phase separation occurs in the Pd–H system, i.e. that two series of analogous solid solutions occur, separated by an interval of immiscibility. X-ray investigations[3–7] have confirmed this and proved the existence of two solid phases (α and β) with the fcc structure of palladium. Fig. 25 gives the lattice parameters of the coexistent fcc phases. Fig. 26 shows the equilibrium diagram of the Pd–H system, derived from

1. L. Troost and P. Hautefeuille, *Annls. Chim. phys.* **2,** 279 (1874).
2. C. Hoitsema, *Z. phys. Chem.* **17,** 1 (1895).
3. L. W. McKeehan, *Phys. Rev.* **21,** 334 (1923).
4. J. O. Linde and G. Borelius, *Annln Phys.* **84,** 747 (1927).
5. F. Krüger and G. Gehm, *Annln Phys.* **16,** 174 (1933).
6. E. A. Owen and E. St J. Williams, *Proc. phys. Soc.* **56,** 52 (1944).
7. A. J. Maeland and T. R. P. Gibb, *J. phys. Chem., Ithaca* **65,** 1270 (1961).

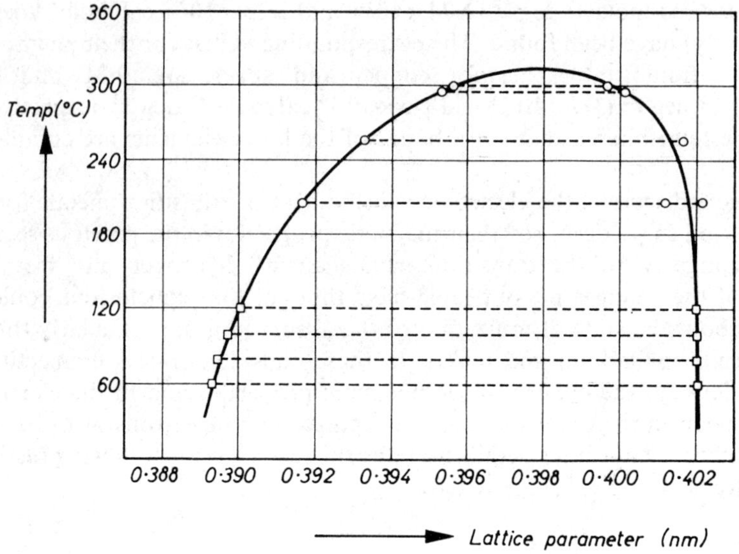

FIG. 25. Lattice parameters of the coexistent fcc phases α and β in the palladium–hydrogen system (squares after Owen and Williams, circles after Maeland and Gibb).

measurements of H_2 equilibrium pressures at various concentrations and temperatures by Hoitsema, *loc. cit.*, Linde and Borelius, *loc. cit.*, and many others[1–6].

Fig. 27 showing the hydrogen content of palladium at 50°C as a function of the hydrogen pressure may serve as an illustration of the type of measurement leading to the construction of the equilibrium diagram (Fig. 26). The percentage hydrogen at first rises with increasing pressure until at the prevailing temperature the α phase is saturated with hydrogen (point A). In this region of concentrations the lattice parameter of the fcc phase increases from 0·3890 nm for pure palladium to 0·3894 nm for the saturated α solution[7]. When the hydrogen content increases further the β phase comes into coexistence with the α phase. According to Gibbs'

1. H. BRÜNING and A. SIEVERTS, *Z. phys. Chem.* **A 163,** 409 (1932).
2. L. J. GILLESPIE and L. S. GALSTAUN, *J. Am. chem. Soc.* **58,** 2565 (1936).
3. P. L. LEVINE and K. E. WEALE, *Trans. Faraday Soc.* **56,** 357 (1960).
4. E. WICKE and G. H. NERNST, *Ber. Bunsenges.* **68,** 224 (1964).
5. An extensive survey of the older literature on the Pd–H system can be found in D. P. SMITH's book: *Hydrogen in Metals*, University Press, Chicago (1948).
6. A good summary of the more recent state of knowledge of the Pd–H system is provided by F. A. LEWIS, *The Palladium–Hydrogen System*, Academic Press, New York (1967).
7. P. C. ABEN and W. G. BURGERS, *Trans. Faraday Soc.* **58,** 1989 (1962).

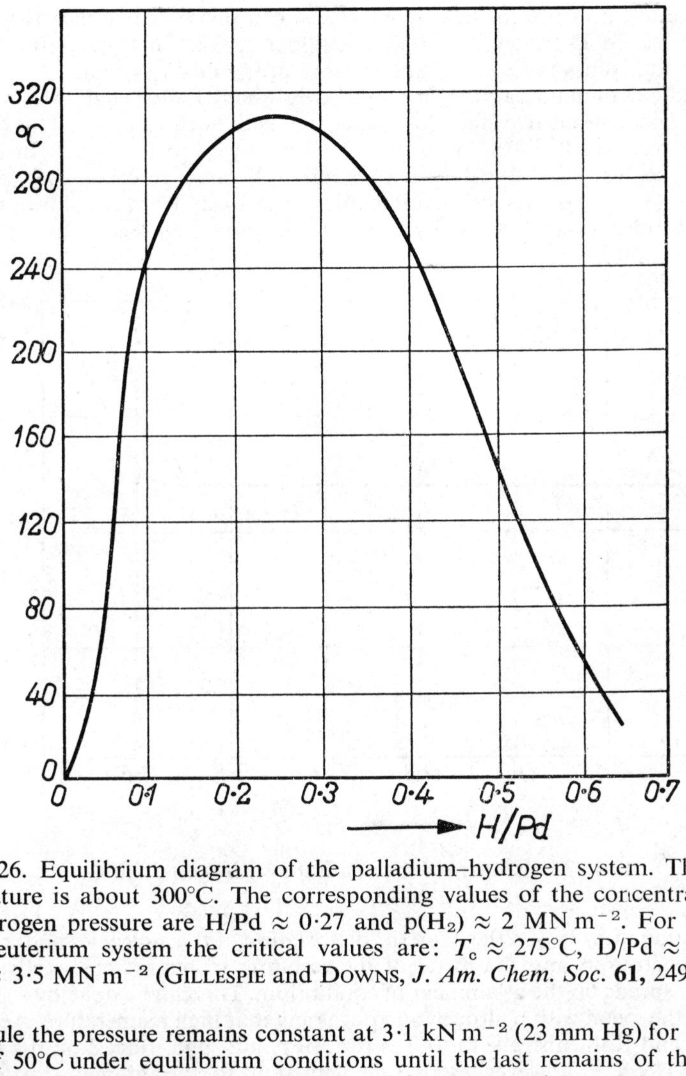

FIG. 26. Equilibrium diagram of the palladium–hydrogen system. The critical temperature is about 300°C. The corresponding values of the concentration and the hydrogen pressure are H/Pd ≈ 0·27 and $p(H_2)$ ≈ 2 MN m^{-2}. For the palladium–deuterium system the critical values are: T_c ≈ 275°C, D/Pd ≈ 0·25 and $p(D_2)$ ≈ 3·5 MN m^{-2} (GILLESPIE and DOWNS, *J. Am. Chem. Soc.* **61**, 2496 (1939)).

phase rule the pressure remains constant at 3·1 kN m^{-2} (23 mm Hg) for a temperature of 50°C under equilibrium conditions until the last remains of the α phase have changed to the β phase (point B). Obviously A and B correspond to two points of the curve in Fig. 26. At 50°C the coexisting α and β phases according to Wicke and Nernst, *loc. cit.*, have the compositions 0·012 and 0·599 H/Pd and according to Aben and Burgers, *loc. cit.*, the lattice parameters 0·3894 and 0·4025 nm. Further increase of the hydrogen content corresponds to an increase of the lattice parameter of the β phase and an increase of the hydrogen pressure. At a pressure of 100 kN m^{-2} (760 mm Hg) the atom ratio H/Pd reaches the value 0·67.

The palladium–deuterium system shows the same picture as the palladium–hydrogen system (see Wicke and Nernst, *loc. cit.*). However, at a given temperature the equilibrium pressure in the region of α and β phase coexistence is markedly higher for the Pd–D system. At 50°C it is about 13·3 kN m^{-2} (100 mm Hg), i.e. more than four times the equilibrium pressure of the Pd–H system.

It should be noted that at temperatures below about 150°C, equilibrium between palladium and gaseous hydrogen (or deuterium) is difficult to attain. This difficulty is not due to a small diffusion coefficient of hydrogen in the metal. In the next chapter we shall see that this diffusivity is relatively large. In cases where only one solid phase (α or β) is present the difficulties arise solely from the slowness of the surface reactions which necessarily precede the diffusion process. In the case of

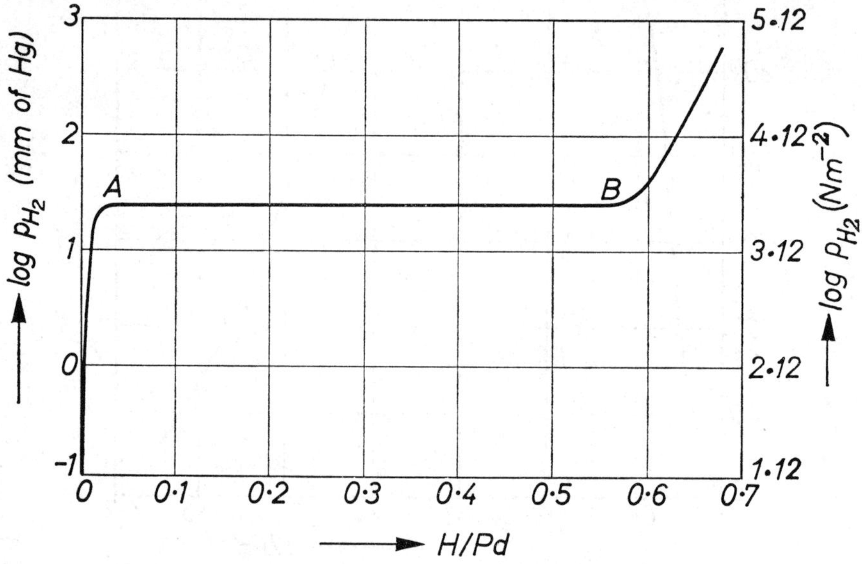

FIG. 27. Isotherm for palladium–hydrogen at 50°C.

the absorption of hydrogen these are the dissociation of the molecules on the surface and the penetration into the metal of the resulting atoms or ions. Activation of the surface speeds up the attainment of equilibrium. This can be done by repeatedly saturating the metal with hydrogen and degassing it at high temperature. According to Krause and Kahlenberg(1) this gives rise to 'points' and 'ridges' on the surface; Smith and Derge(2), however, ascribe the activation to 'intragranular fissures' and 'rifts'. Also the grain boundaries act as active sites: the smaller the grains, the higher the rate of absorption of hydrogen by palladium(3). The so-called 'transfer catalysts' are particularly active(4, 5). They are powders of certain hydrides (e.g.

1. W. KRAUSE and L. KAHLENBERG, *Trans. electrochem. Soc.* **68,** 449 (1935).
2. D. P. SMITH and G. J. DERGE, *Trans. electrochem. Soc.* **66,** 253 (1934).
3. T. J. TIEDEMA, B. C. DE JONG and W. G. BURGERS, *Proc. K. ned. Akad. Wet.* **B 63,** 422 (1960).
4. A. KÜSSNER and E. WICKE, *Z. phys. Chem.* **24,** 152 (1960).
5. A. KÜSSNER, *Z. Elektrochem.* **66,** 675 (1962).

UH_3 or UD_3) which are brought into close contact with palladium and accelerate the absorption and desorption of the gas. Finely divided copper powder also acts in this way.

Many investigators have got round the difficulties due to the surface reactions by electrolytically charging the palladium with hydrogen. Equilibrium pressures can then be calculated from the results of electrode potential measurements. During these experiments it was noted that even without the passage of an electric current the metal absorbs hydrogen directly from solutions of H_2 in water containing sulphuric or hydrochloric acid[1,2]. Similar to the absorption from the gas phase, absorption from hydrogen-saturated acidic solutions is accelerated by activation of the palladium surface. In this case activation is produced by electrolytic oxydation and reduction of the surface prior to the absorption of hydrogen from the solution[3-6].

Deviations from the thermodynamic equilibrium which are certainty not only due to the slowness of surface reactions occur in the region of α and β phase coexistence. Activation of the surface does not alter the fact that in this region the apparent equilibrium pressure corresponding to the formation of the β phase is higher than that corresponding to its decomposition (Fig. 28). This 'hysteresis' decreases as the temperature increases and as the palladium is more finely divided. There is as yet no generally accepted explanation of this phenomenon. In order not to interrupt the continuity of the discussion too much we shall not return to this hysteresis effect until Section 1.11.

The thermodynamic interpretation of the phase separation under discussion is given by the fact that the partial heat of solution of hydrogen in the metal decreases (becomes more negative) with increasing concentration. This means that the process of solution becomes more exothermic at rising concentration or, in other words, that it is energetically more favourable for the hydrogen atoms to be concentrated locally than to be distributed evenly throughout the metal. At low temperatures, where the influence of entropy is small, separation of a homogeneous solution into two phases must therefore be expected, one with a low hydrogen concentration and one containing most of the gas. On the other hand, at high temperatures, where the influence of entropy is dominant, the occurrence of a single solution must be expected. This is exactly what has been observed (Fig. 26). For attempts to interpret these phenomena quantitatively with the aid of the methods of statistical thermodynamics we refer to publications of Lacher[7], Libowitz[8],

1. J. P. Hoare and S. Schuldiner, *J. phys. Chem., Ithaca* **61**, 399 (1957).
2. S. Schuldiner, G. W. Castellan and J. P. Hoare, *J. chem. Phys.* **28**, 16, 20 and 22 (1958).
3. T. B. Flanagan and F. A. Lewis, *Trans. Faraday Soc.* **55**, 1400 and 1409 (1959).
4. R. J. Fallon and G. W. Castellan, *J. phys. Chem., Ithaca* **64**, 4 (1960).
5. A. W. Carson, T. B. Flanagan and F. A. Lewis, *Trans. Faraday Soc.* **56**, 363, 371 and 1332 (1960).
6. T. B. Flanagan, *J. phys. Chem., Ithaca* **65**, 280 (1961).
7. J. R. Lacher, *Proc. R. Soc.* **A 161**, 525 (1937).
8. G. G. Libowitz, *J. appl. Phys.* **33**, 399 (1962).

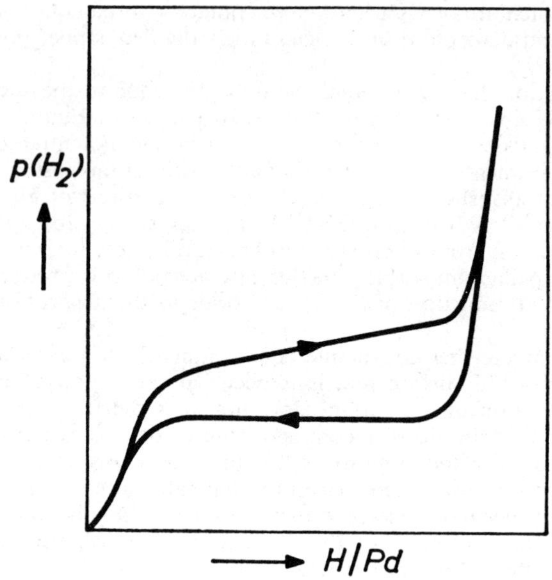

FIG. 28. Diagrammatic representation of the 'hysteresis effect' occurring during the absorption and desorption of hydrogen by palladium at a relatively low temperature, e.g. 100°C. The apparent equilibrium pressure at increasing hydrogen content is higher than that on the release of hydrogen from the system. (See, for example, D. H. EVERETT and P. NORDON, *Proc. R. Soc.* **A 259**, 341 (1960).)

Brodowsky[1] and Simons and Flanagan[2]. Unlike the other investigators, Libowitz in his theoretical treatment does not introduce a direct interaction of interstitial hydrogens but an interaction through attraction between hydrogen vacancies (unoccupied octahedral interstices).

With most other investigators Brodowsky, *loc. cit.*, assumes that hydrogen, dissolved in palladium, exists as protons. He explains their inclination to the formation of clusters by assuming (a) that the expansion of the lattice (Fig. 25) is caused by the vibrations of the protons in their interstices and (b) that the energy required to bring this elastic deformation about is smaller when the protons are close together than when they are far apart. An assumption made by Axelrod and Makrides[3,4] seems more satisfactory to us. They assume the net attracting potential between the protons to be due to attracting interactions via the electrons that provide screening for the protons. Probably we do not modify their opinion too much by adding that

1. H. BRODOWSKY, *Z. phys. Chem.* **44**, 129 (1965).
2. J. W. SIMONS and T. B. FLANAGAN, *Can. J. Chem.* **43**, 1665 (1965).
3. S. D. AXELROD and A. C. MAKRIDES, *J. phys. Chem., Ithaca* **68**, 2154 (1964).
4. A. C. MAKRIDES, *J. phys. Chem., Ithaca* **68**, 2160 (1964).

the dissolved hydrogen, through the formation of clusters, can act more or less as metallic hydrogen where the protons are bound together by shared electron forces. According to calculations by Wigner and Huntington[1] the body-centred cubic form of metallic hydrogen in its pure state is only stable at very high hydrogen pressures, estimated as at least 25 GN m^{-2} (250,000 atm.). However, they consider it possible that a layer-like form has greater stability. Ubbelohde[2] suggested that metallic hydrogen can be stabilized by alloying. In his view the hydrogen-rich phase in the Pd–H system is a typical example of an alloy of hydrogen metal and another metal. It is true that in this phase the proton–proton distance is appreciably larger than in bcc metallic hydrogen of minimum energy, but calculations by Friedel[3] indicate that protons dissolved in palladium or nickel will interact at relatively large distances.

Ebisuzaki and O'Keeffe[4] present arguments which differ from those of most other investigators. They assume that the heat of solution of hydrogen in palladium contains no contributions from hydrogen–hydrogen interactions but depends only on the valence electron : atom ratio, a quantity varying strongly with the composition of the Pd–H alloy. In their theoretical treatment the hydrogen–hydrogen interactions, if any, are weak and repulsive rather than attractive. If so, this would imply that at a sufficiently low temperature there will be an order–disorder transition in a palladium hydride of suitable composition. Ebisuzaki and O'Keeffe are inclined to interpret the specific heat anomaly observed in the β phase at 55 K[5] in these terms.

The positions of the hydrogens in palladium hydride were determined by Worsham *et al.*[6] using neutron diffraction techniques. The diffraction patterns obtained for the composition H/Pd = 0·7 show that the hydrogen atoms are located in the large octahedral interstices of the Pd lattice. The same is true for deuterium atoms in the β phase of the Pd–D system. This means that, in the Pd–H and Pd–D systems, this phase has a structure similar to NaCl, but with a large region of homogeneity where the H or D sites are for a large part unoccupied. From the experiments by Worsham *et al.* it seems reasonable to expect that the upper limit of the existence range of the β phase occurs at the atom ratio H/Pd = 1. The highest value that has so far been reached experimentally is H/Pd = 0·92 at −78°C and 100 MN m^{-2} (1000 atm) H_2[7]. It is also reasonable to assume that the

1. E. Wigner and H. B. Huntington, *J. chem. Phys.* **3**, 764 (1935).
2. A. R. Ubbelohde, *Proc. R. Soc.* **A 159**, 295 and 306 (1937).
3. J. Friedel, *Adv. Phys.* **3**, 446 (1954).
4. Y. Ebisuzaki and M. O'Keeffe, *J. phys. Chem., Ithaca* **72**, 4695 (1968).
5. D. M. Nace and J. G. Aston, *J. Am. Chem. Soc.* **79**, 3623 (1957).
6. J. E. Worsham, M. K. Wilkinson and C. G. Shull, *Physics Chem. Solids* **3**, 303 (1957).
7. P. S. Perminov, A. A. Orlov and A. N. Frumkin, *Dokl. Akad. Nauk S.S.S.R.* **84**, 749 (1952).

interstices occupied by hydrogen in α palladium are of the same kind as those occupied in the β phase. Sköld and Nelin[1] and Maeland[2] have found strong evidence in favour of this assumption.

Influence of hydrogen on the magnetic properties of palladium

It has long been established (Graham[3]) that the paramagnetism of palladium decreases when hydrogen is absorbed. Several workers (in par-

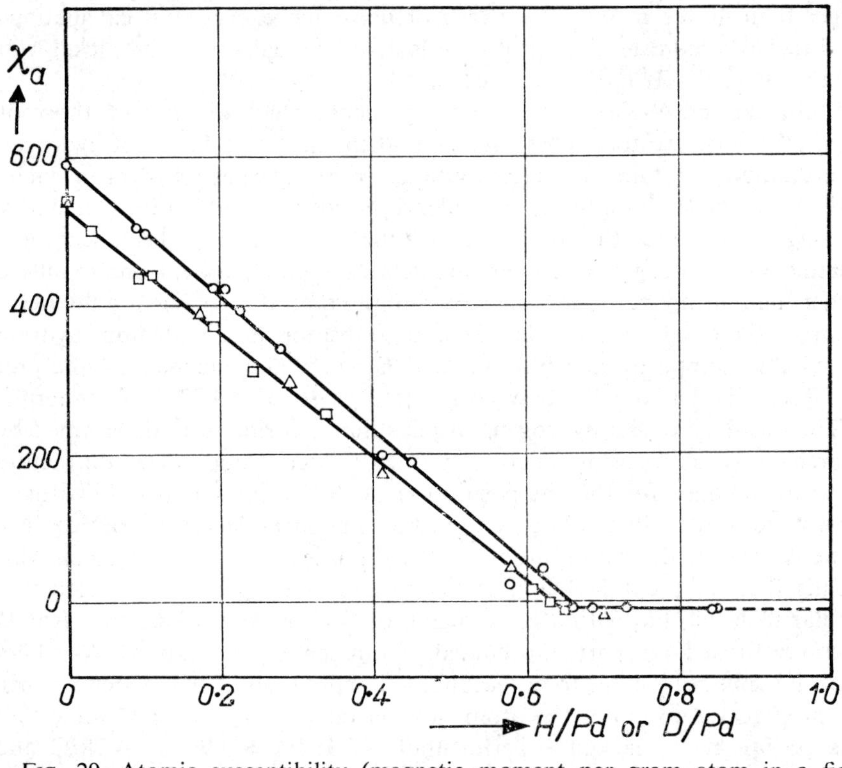

FIG. 29. Atomic susceptibility (magnetic moment per gram atom in a field of 10^{-4} tesla) of Pd–H alloys according to Svensson (circles) and Sieverts (triangles). The susceptibility of Pd–D alloys, according to Sieverts (squares) is also given. The measurements were carried out at room temperature.

1. K. SKÖLD and G. NELIN, *Physics Chem. Solids* **28,** 2369 (1967).
2. A. J. MAELAND, *Can. J. Phys.* **46,** 121 (1968).
3. *The Collected Chemical and Physical Researches of Thomas Graham* (ed. R. A. Smith) Edinburgh (1876).

ticular Svensson[1] and Sieverts and Danz[2]) have carried out accurate measurements of this effect. The paramagnetic susceptibility of palladium decreases linearly with increasing hydrogen content. The straight line representing this decrease (Fig. 29) corresponds approximately to the two-phase (α plus β) region. It changes abruptly into a near horizontal line when a H/Pd ratio of 0·65 is reached. The near horizontal part corresponds to low negative values of the susceptibility indicating that the β phase is diamagnetic at H/Pd ratios greater than 0·65. Sieverts and Danz, *loc. cit.*, have also investigated the effect of deuterium and found that, per atom, it is the same as that of hydrogen (Fig. 29).

Mott and Jones[3] explained these phenomena by assuming that the electrons from the absorbed hydrogen are accommodated in the 4d band of the palladium. As long as this band is not full the overlapping 5s band will be filled to the same level. In view of the fact that the density of states is much smaller in the s band than in the d band, the great majority of the electrons will be accommodated in the latter. As soon as the d band is full, further electrons will have to be accommodated in the s band. The energy of the system and the hydrogen equilibrium pressure will then increase much more rapidly than during the filling of the d band. On the assumption that the number of holes in the d band of pure palladium is about 0·6 per atom, Mott and Jones, *loc cit.*, deduced from the experimental results shown in Fig. 29 that the paramagnetic susceptibility drops to zero when the number of hydrogen atoms taken up equals the number of holes, i.e. they deduced that hydrogen dissolved in palladium is completely ionized. Recent evidence suggests that there are only 0·36 holes per palladium atom in the d band (see Section 1.7). Adherence to the original model of Mott and Jones would lead to the conclusion that hydrogen in palladium is only partly ionized. However, a weak point of their model is the assumption that alloying palladium with hydrogen, or any other element, does not significantly alter the shape of the composite 4d-5s band. At present it seems almost certain that this 'rigid band' approximation is far too crude. If so, it is possible to maintain the model of complete electron donation by assuming that adding hydrogen to palladium changes the band structure in such a way that d band filling occurs at a H/Pd ratio of about 0·6. Strong evidence in favour of this assumption has recently been obtained in susceptibility measurements by Brill and Voitländer[4]. Evidence in favour of the electron donation model has also been obtained by utilizing electronic heat capacity measurements (see Section 1.13).

1. B. Svensson, *Annln Phys.* **18,** 299 (1933).
2. A. Sieverts and W. Danz, *Z. phys. Chem.* **B 38,** 61 (1937).
3. N. F. Mott and H. Jones, *Theory of the Properties of Metals and Alloys*, Oxford University Press (1936), Chapter 6.
4. P. Brill and J. Voitländer, *Z. Naturf.* **24A,** 1 (1969).

One of the objections to the 'proton theory' under discussion is the neglect of the previously mentioned fact that the β phase in the Pd–H system has a structure similar to that of NaCl. Based on this and on other facts, some investigators tend to an approach diametrically opposed to the 'proton theory' by regarding the β phase as a non-stoichiometric compound in which the hydrogen is present as negative ions. The pros and cons of both theories will be discussed in Sections 1.12 and 1.13.

1.10. Hydrogen in palladium alloys

The Pd–Ag system

Palladium and silver form an uninterrupted series of solid solutions. These solutions are not ordered after rapid cooling from a high temperature (e.g., 900°C). Starting with pure palladium, the paramagnetic susceptibility of the non-ordered solutions decreases with increasing silver content and reaches zero at about 60 atom % Ag[1] (Fig. 30). At higher silver contents the susceptibility is slightly negative[2,3]. The alloys Pd–Cu[2] and Pd–Au[4] behave similarly.

The Pd–Ag system is more suitable for checking the theoretical concepts of Mott and Jones, *loc. cit.*, than the earlier discussed Pd–H system, where phase separation occurs and where the hydrogen atoms occupy sites that are unoccupied in pure palladium. Silver, contrary to hydrogen, forms *substitutional* solutions with palladium. In addition, silver contains only one electron per atom more than palladium, being adjacent to it in the periodic table of the elements. Therefore, to a first approximation, if a number of palladium atoms in the lattice are replaced by silver atoms one could regard the system as unaltered except for the fact that an equal number of extra electrons are added and that the lattice parameter is slightly increased (from 0·388 nm for pure palladium to 0·399 nm for the alloy 40% Pd + 60% Ag according to Coles[5]. The attainment of zero susceptibility at 60% Ag agrees very well with the original theory of Mott and Jones, where it was assumed that the number of holes in the 4d band is 0·6 per atom. Since we now know that the number of such holes is only 0·36 per atom in *pure* palladium, the band structure must have changed on alloying with silver. It was shown by Dugdale and Guénault[6] that only a slight

1. F. E. Hoare, J. C. Matthews and J. C. Walling, *Proc. R. Soc.* **A 216,** 502 (1953).
2. B. Svensson, *Annln Phys.* **14,** 699 (1932).
3. J. Wucher, *Annls Phys.* **7,** 317 (1952).
4. E. Vogt, *Annln Phys.* **14,** 1 (1932).
5. B. R. Coles, *J. Inst. Metals* **84,** 346 (1956).
6. J. S. Dugdale and A. M. Guénault, *Phil. Mag.* **13,** 503 (1966).

modification of the Mott–Jones model is needed to make it consistent with all experimental findings.

We should expect the Ni–Cu system to show a similar dependence of susceptibility on composition as the Pd–Ag system, since nickel and copper

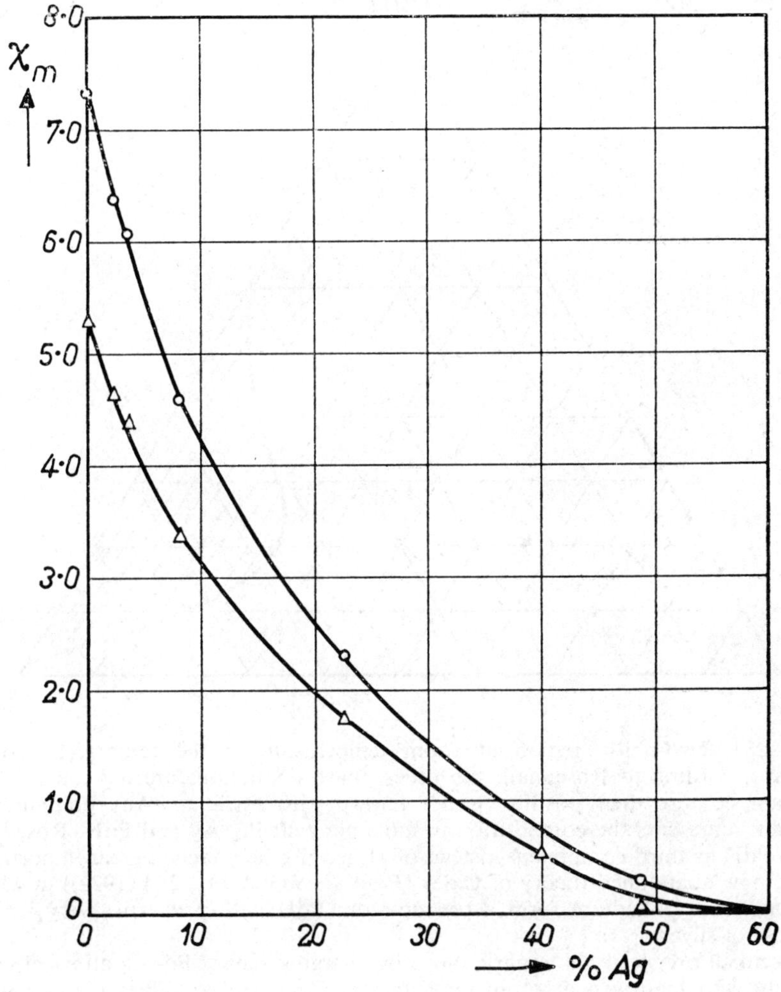

FIG. 30. Mass susceptibility (magnetic moment per gram in a field of 10^{-4} tesla) corrected for core diamagnetism of Pd–Ag alloys at 20 K (circles) and at 290 K (triangles) according to Hoare, Matthews and Walling. The silver content is given in atom per cent which, in this particular case, is almost equal to that in weight per cent.

are also adjacent elements in the periodic table (see Fig. 23). However, in the following section it will be shown that the Ni–Cu system behaves differently.

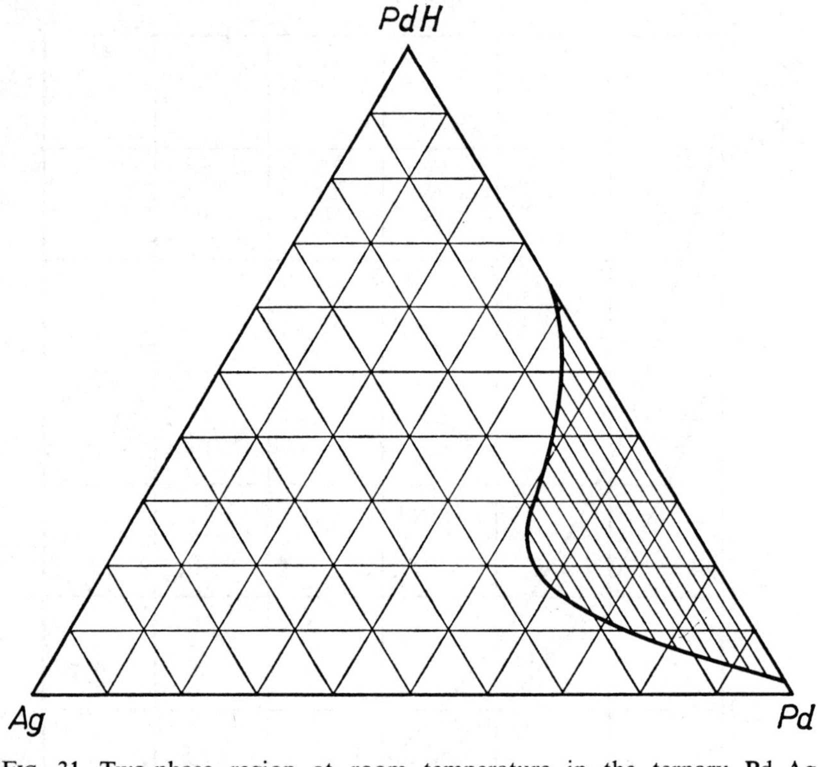

FIG. 31. Two-phase region at room temperature in the ternary Pd–Ag–H system according to Rosenhall. No other phase boundaries are indicated in the diagram because their position is not known with certainty. Any point in the diagram represents the composition in mole per cent Pd, Ag and PdH. Rosenhall chose PdH as third component instead of H because he attached great importance to the now abandoned theory of Oxley (*Proc. R. Soc.* **A 101,** 264 (1922)) in which hydrogen and palladium form a pseudo-atom PdH with properties very similar to those of silver.

Rosenhall prepared his ternary alloys by charging binary Pd–Ag alloys electrolytically with hydrogen at room temperature. Complete equilibria could not be established in this case since neither palladium nor silver diffuse with any noticeable velocity at room temperature. Each of the parallel tie-lines in the two-phase region corresponds to two coexisting phases in which hydrogen has the same chemical potential. This mutual equality does not apply to the chemical potential of Pd or Ag.

The Pd–Ag–H system

Just as in the binary system Pd–H, a two-phase region occurs in the ternary system Pd–Ag–H. One of the interesting properties of the ternary system is the gradual reduction of the miscibility gap with increasing silver content. At room temperature, no phase separation occurs at silver contents in excess of about 25%[(1–4)]. Fig. 31 shows the area of the two-phase region according to Rosenhall, *loc. cit.*, and Fig. 32 the lattice parameters of the ternary phases coexisting at room temperature, as a function of the silver content. With increasing silver content the parameters approach each other and become the same at about 25% silver (Brodowsky and Poeschel, *loc. cit.*).

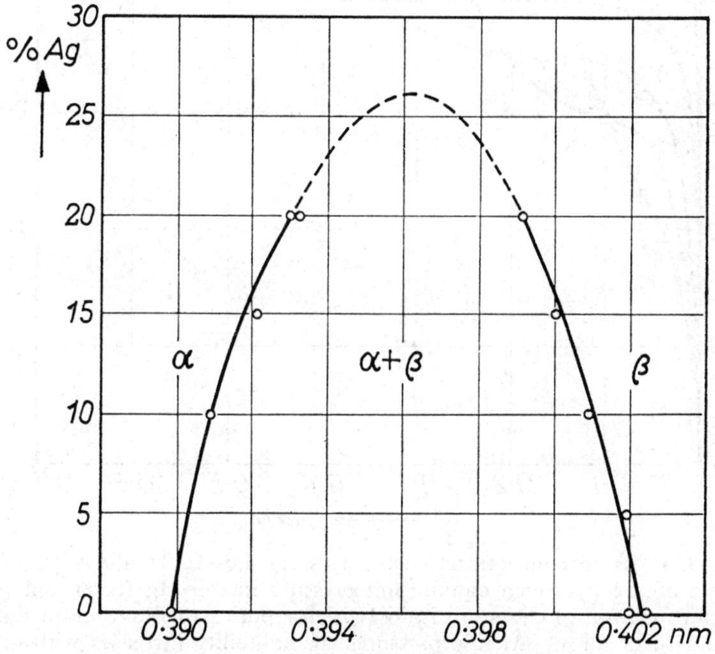

FIG. 32. Lattice parameters at room temperature of the coexistent fcc phases in the palladium–silver–hydrogen system according to Brodowsky and Poeschel. The parameters are given as a function of the silver content in percentages by weight. Similar to Rosenhall's experiments (see description Fig. 31) complete ternary equilibria were not obtained.

1. G. ROSENHALL, *Annln Phys.* **24,** 297 (1935).
2. S. D. AXELROD and A. C. MAKRIDES, *J. phys. Chem., Ithaca* **68,** 2154 (1964).
3. A. C. MAKRIDES, *J. phys. Chem., Ithaca* **68,** 2160 (1964).
4. H. BRODOWSKY and E. POESCHEL, *Z. phys. Chem.* **44,** 143 (1965).

Fig. 33 shows the isotherms at 50°C for pure palladium and for four Pd–Ag alloys containing 10, 20, 30 and 40% silver by weight, respectively. The striking fact here is that the solubility of hydrogen at low hydrogen

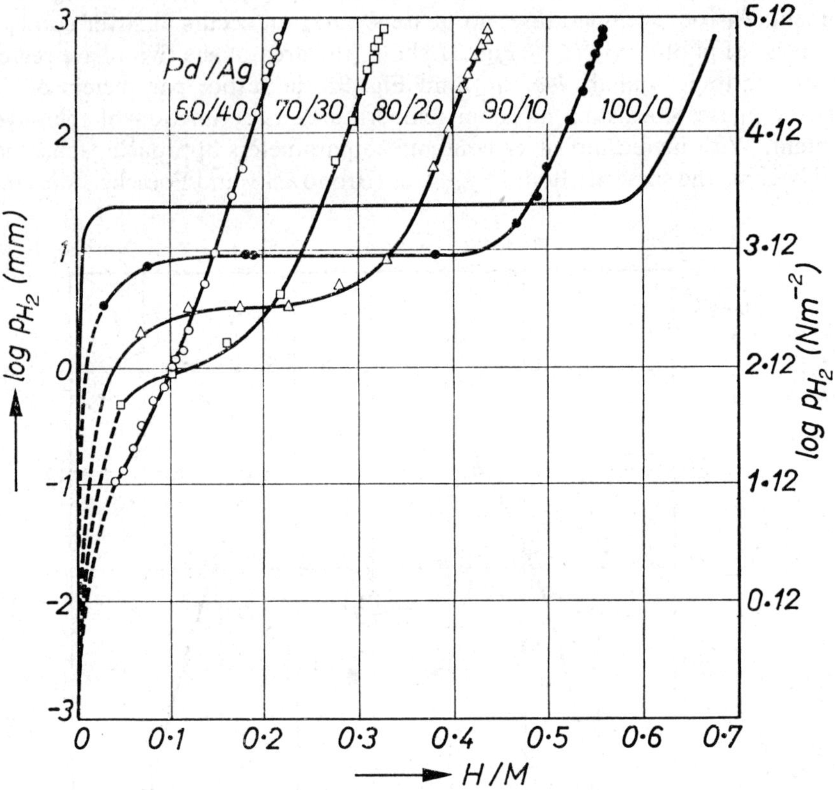

FIG. 33. Pressure–concentration isotherms for Pd–Ag–H alloys at 50°C. The logarithm of the hydrogen equilibrium pressure in mm Hg (torr) and Nm^{-2} is given as a function of the atom ratio H/M for pure palladium and a number of palladium–silver alloys. At low pressures the solubility increases with increasing silver content. At high pressures the reverse occurs. The equilibria were measured by means of the method of direct gas absorption, which was possible by using copper powder as a transfer catalyst (Brodowsky and Poeschel).

pressures increases with increasing silver content whilst at high hydrogen pressures the reverse holds true. In agreement with the behaviour at low hydrogen pressures, and over the investigated region of silver concentrations, Brodowsky and Poeschel found that at a high silver content more heat is released during isothermal solution of the first quantities of hydrogen than at a low silver content[1]. The enthalpy of solution at very low hydrogen pressures is according to Brodowsky and Poeschel $-19{\cdot}0$ kJ mole^{-1} ($-4{\cdot}5$ kcal/mole) for pure palladium, $-39{\cdot}1$ kJ mole^{-1} ($-9{\cdot}34$ kcal/mole) for the alloy 80% Pd+20% Ag, $-46{\cdot}0$ kJ mole^{-1} ($-11{\cdot}0$ kcal/mole) for 70% Pd+30% Ag and $-49{\cdot}0$ kJ mole^{-1} ($-11{\cdot}7$ kcal/mole) for 60% Pd+40% Ag. Makrides, *loc. cit.*, found enthalpies of solution which deviate considerably from these but they cannot fully be compared with the values found by Brodowsky *et al.*, since they were calculated from the temperature coefficient of the plateau potential of time–potential graphs obtained during absorption of hydrogen from aqueous solutions. As has been shown by X-ray measurements the plateau corresponds to the α–β transformation, and therefore Makrides' values for the heats of solution relate to the regions of coexistence of the α and β phases. They do, however, show the same tendency of becoming more negative with increasing silver content as the solution enthalpies of Brodowsky and Poeschel.

It has already been mentioned in the previous section that an elastic distortion of the lattice, resulting from the vibrations of the protons in their interstices, plays an important part in the theoretical considerations of Brodowsky *et al.* According to their theory the influence of silver is due to a reduction of the distortion energy or, in other words, to the increasing compressibility of palladium with increasing silver content. Makrides, *loc. cit.*, on the other hand, attributes the influence of silver to an attractive interaction between silver and hydrogen similar to that between interstitial hydrogen atoms. Values of the partial molar entropy of hydrogen in the alloys at very low hydrogen concentrations would at first sight appear to support Makrides' view. According to Brodowsky and Poeschel, *loc. cit.*, this entropy decreases with increasing silver content from 15 J deg^{-1} ($3{\cdot}5$ cal/degree) per gram atom of hydrogen for pure palladium to $5{\cdot}4$ J deg^{-1} ($1{\cdot}3$ cal/degree) for 80% Pd+20% Ag and to $1{\cdot}3$ J deg^{-1} ($0{\cdot}3$ cal/degree) for 70% Pd+30% Ag. This fall in entropy suggests that in the investigated alloys the interstices are not all identical. When viewed together with the above mentioned values for enthalpy these entropy figures could give the impression that the protons have a preference for octahedral interstices where the number of adjacent silver atoms is greater than that corresponding to the average silver to palladium ratio.

Two facts appear to contradict this view: (a) hydrogen dissolves endothermically in pure silver and only in very small quantities; (b) the decrease

1. Analogous results have been obtained by Allard *et al.* in determinations of the heats of absorption of hydrogen in a series of palladium–gold alloys. Over the investigated region of gold concentrations ($5{\cdot}7$ to $44{\cdot}7$ atom per cent Au) the enthalpy of solution for hydrogen at infinite dilution varies from -25 kJ mole^{-1} ($-6{\cdot}0$ kcal/mole) for $5{\cdot}7$% Au to -39 kJ mole^{-1} ($-9{\cdot}3$ kcal/mole) for $44{\cdot}7$% Au; K. Allard, A. Maeland, J. W. Simons and T. B. Flanagan, *J. phys. Chem., Ithaca* **72,** 136 (1968).

in entropy with increasing silver content is much greater than could be expected if the only cause for the decrease were the preference of hydrogen for certain interstices. It is therefore reasonable to assume that not only the configurational entropy but also the vibrational entropy should be taken into account[1]. The absorption of hydrogen probably causes the vibrational entropy of the alloys to decrease as a consequence of the increase in vibrational frequency of the metal atoms surrounding any hydrogen atom. The results of an investigation by Lieser and Witte[2] do, in our opinion, support this assumption. They report a drop in partial molar entropy of hydrogen in nickel, when zinc or aluminium has been added to it in homogeneous solution; this is similar to the effect discussed here. Comparison of the Ni–Zn–H and Ni–Al–H systems with the Pd–Ag–H system is important because the solubility of hydrogen in silver is much smaller than that in palladium, whilst its solubility in zinc and in aluminium is much smaller than that in nickel.

Brodowsky and Poeschel, *loc. cit.*, call the partial molar entropy of hydrogen in their alloys 'the standard entropy of the dissolved hydrogen'. This could create the wrong impression that the change in entropy of an alloy, which occurs when hydrogen dissolves can be written as the sum of a configurational part and a vibrational part referring only to the hydrogen in solution. In actual fact, changes in the vibrations of the metal atoms can contribute appreciably to the above-mentioned partial molar entropy of hydrogen which, by definition, is the change in entropy of an infinite quantity of the *alloy* when dissolving one mole of hydrogen at constant temperature and pressure.

Besides the favourable effect of silver on the capability of palladium to take hydrogen into solution there is the opposing influence of silver which is noticeable only at high concentrations of hydrogen (cf. Fig. 33). This can best be understood by recognizing that the hydrogen electrons and the 5s electrons of silver are in competition when occupying the holes in the 4d band of palladium. At low hydrogen and silver concentrations they do not interfere with each other in this respect. The original theory of Mott and Jones, *loc. cit.*, would lead to the expectation that the solubility of hydrogen in the alloys approaches zero when the silver content reaches 60%, i.e. the percentage at which all holes in the 4d band of palladium were supposed to be occupied by electrons. However, according to measurements by Rosenhall[3] the 40% Pd + 60% Ag alloy still has a fair capability for dissolving hydrogen. Fig. 34 gives the lattice parameters of a number of

1. Even contributions to the entropy that are neither configurational nor vibrational could be included. Cf. K. M. Myles, *Acta Metall.* **13**, 109 (1965).
2. K. H. Lieser and H. Witte, *Z. Elektrochem.* **61**, 367 (1957).
3. G. Rosenhall, *Annln Phys.* **24**, 297 (1935).

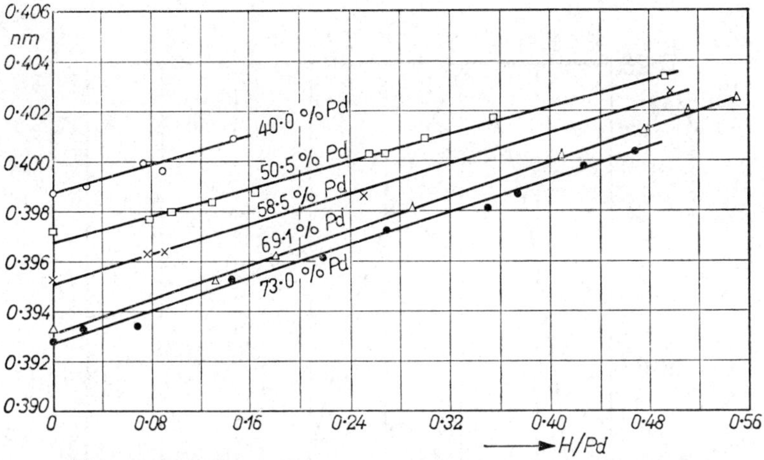

FIG. 34. Lattice parameters at room temperature of a number of single-phase Pd–Ag–H alloys as a function of the atom ratio H/Pd. The alloys were electrolytically charged with hydrogen.

single-phase Pd–Ag–H alloys as a function of the atom ratio H/Pd according to Rosenhall, *loc. cit.*

The foregoing discussion of the influence of silver on the solubility of hydrogen in palladium refers to a temperature of 50°C. As the temperature increases the hydrogen solubility in Pd–Ag alloys decreases. In accordance with this, the favourable influence of silver on the solubility can be noticed at higher hydrogen pressures as the temperature is raised. Fig. 35 shows the results of classical measurements by Sieverts *et al.*(1) on the solubility of hydrogen in Pd–Ag alloys at a hydrogen pressure of 100 kN m^{-2} (1 atm), supplemented by some later measurements by Sieverts and Hagen(2). The two opposing influences of silver are clearly seen in these results also: at first solubility increases with increasing silver content; then the solubility falls when the silver content increases further. Also these measurements show that at the composition 40% Pd + 60% Ag the solubility is still relatively high.

Palladium–silver diffusion cells

In the past few years Pd–Ag alloys containing 25 to 40% silver have found large scale application in diffusion cells for securing very pure hydrogen

1. A. SIEVERTS, E. JURISCH and A. METZ, *Z. anorg. allg. Chem.* **92,** 329 (1915).
2. A. SIEVERTS and H. HAGEN, *Z. phys. Chem.* **A 174,** 247 (1935).

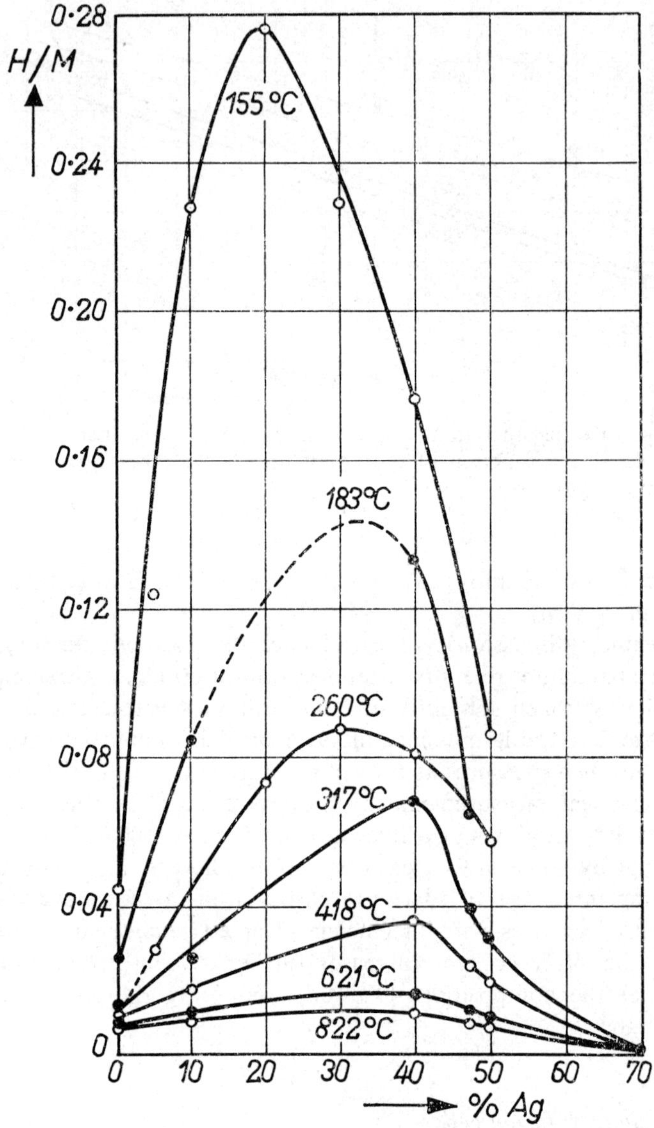

FIG. 35. Solubility of hydrogen at 100 kN m^{-2} in Pd–Ag alloys as a function of the silver content (in atom per cent) at various temperatures. The concentration of the hydrogen is given in atoms per metal atom (data taken from Sieverts *et al.*).

from cracked ammonia or other gas mixtures[1-3] The alloys in question are preferable to pure palladium since no second phase is formed when they take up hydrogen. The α–β phase transition occurring in Pd–H alloys below 310°C makes diffusion membranes of unalloyed palladium dimensionally unstable and liable to severe distortion and failure (cf. Section 1.11, small print).

The Pd–Au–H and Pd–Pt–H systems

Similar to palladium and silver, palladium and gold[4] as well as palladium and platinum[5] form an uninterrupted series of solid solutions.

Two phase regions occur in the ternary Pd–Ag–H, Pd–Au–H and Pd–Pt–H systems at room temperature with boundaries at definite Ag, Au or Pt contents. The graphs giving the lattice parameters at 25°C for the coexisting ternary phases as a function of the gold or platinum content[6,7] show similarity with the graph in Fig. 32 relating to the coexisting α and β phases in the Pd–Ag–H system. However, the graph for the Pd–Pt–H system has a more asymmetric shape than the almost symmetrical graphs for the Pd–Ag–H and Pd–Au–H systems. Two-phase formation at 25°C ceases above about 25 atom % Ag, 17 atom % Au and 14 atom % Pt, respectively. It should be noted that the alloys will not achieve complete equilibrium when charged with hydrogen at room temperature. This is due to the fact that at low temperatures the metal atoms, contrary to the hydrogen atoms, do not diffuse with any noticeable velocity. In all three systems, Pd–Ag–H, Pd–Au–H and Pd–Pt–H, the chemical potential of the hydrogen has the same value in each pair of coexisting α and β phases. This mutual equality does not, however, apply to the chemical potentials, $\mu_{\alpha M}$ and $\mu_{\beta M}$ of each separate metal M.

In the theory of Mott and Jones[8] both gold and silver donate s electrons to the 4d band of palladium whereas platinum does not. Following this theory the Pd–Au–H system would be expected to show greater similarity to the Pd–Ag–H system than the Pd–Pt–H system does. This is indeed so; at high temperatures and with hydrogen of normal atmospheric pressure

1. J. B. Hunter, *Platin. Metals Rev.* **4**, 130 (1960).
2. A. A. Rodina, M. A. Gurevich and N. I. Doronicheva, *Russ. J. phys. Chem.* **41**, 1286 (1967).
3. E. M. Wise, *Palladium*, Academic Press, New York (1968), Chapter 12.
4. M. Hansen and K. Anderko, *Constitution of Binary Alloys*, McGraw-Hill, New York (1958).
5. E. Raab, *J. less-common Metals* **1**, 3 (1959).
6. A. Maeland and T. B. Flanagan, *J. phys. Chem., Ithaca* **69**, 3575 (1965).
7. A. Maeland and T. B. Flanagan, *J. phys. Chem., Ithaca* **68**, 1419 (1964).
8. N. F. Mott and H. Jones, *Theory of the Properties of Metals and Alloys*, Oxford University Press (1936).

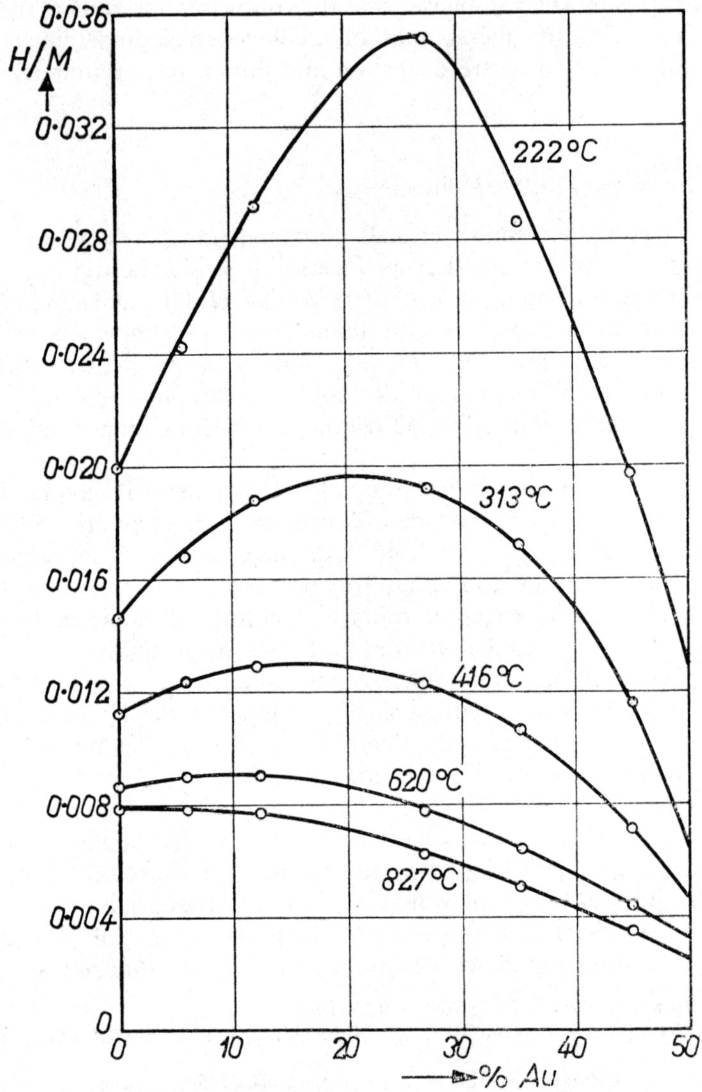

FIG. 36. Solubility of hydrogen at 100 kN m^{-2} in Pd–Au alloys as a function of the gold content (in atom per cent) at various temperatures. The concentration of the hydrogen is given in atoms per metal atom (data taken from Sieverts *et al.*).

an increasing gold content has influence similar to that of silver on the solubility of hydrogen in palladium: the solubility increases to a maximum and then falls to very low values[1] (Fig. 36). On the other hand, in Pd–Pt alloys under similar conditions the solubility of hydrogen decreases continuously with increasing platinum content (Fig. 37).

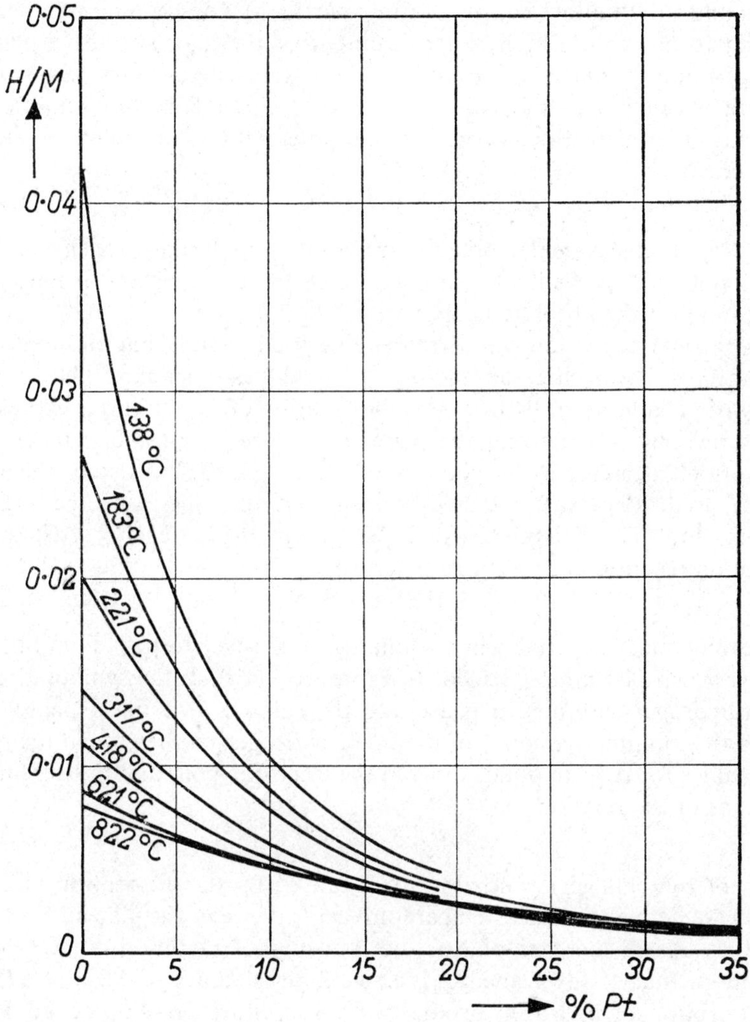

FIG. 37. The solubility of hydrogen at 100 kN m^{-2} in Pd–Pt alloys as a function of the platinum content (in atom per cent) at various temperatures. The concentration of the hydrogen is given in atoms per metal atom (data taken from Sieverts *et al.*).

1. A. SIEVERTS, E. JURISCH and A. METZ, *Z. anorg. allg. Chem.* **92,** 329 (1915).

The difference in influence of gold and platinum also shows in pressure–concentration isotherms at low temperatures (0°–60°C). The course of the absorption of hydrogen by Pd–Au and Pd–Pt alloys from aqueous solutions was followed by measuring the electrode potential of wire specimens[1-3]. As in the case of pure palladium, the potential–time graphs of palladium-rich Pd–Au and Pd–Pt alloys consist of three parts: (a) a descending part corresponding to an increasing hydrogen content of the α phase; (b) a plateau corresponding to the $\alpha \rightarrow \beta$ transformation; (c) a descending part corresponding to an increasing hydrogen content of the β phase. Equilibrium pressures are obtained from the electrode potential measurements by means of the equation

$$-2FE = RT \ln p \qquad (1.10.1)$$

where E is the electrode potential referred to the standard hydrogen electrode as zero potential, p the hydrogen pressure, R the gas constant, F Faraday's constant and T the absolute temperature.

Experiments at the low temperatures under discussion, like those at high temperatures, show that the addition of gold has a similar effect on the affinity of palladium for hydrogen as the addition of silver. This is expressed, for instance, by the fact that the plateau pressure at constant temperature does not only decrease by the addition of silver (Fig. 33) but also by the addition of gold. However, the addition of platinum increases the plateau pressure (Fig. 38). This means that the change in free energy, ΔG, of the absorption reaction

$$H_2 + \alpha \rightarrow \beta \qquad (1.10.2)$$

will become more negative when adding gold or silver to palladium but less negative when platinum is added. In expression (1.10.2) the symbol α refers to the hydrogen-saturated α phase and the symbol β to the β phase containing the minimum amount of dissolved hydrogen at the chosen temperature. Values for ΔG are obtained from the electrode potential measurements by means of the relation

$$\Delta G = -2FE \qquad (1.10.3)$$

Values of the change in enthalpy, ΔH, caused by the absorption reaction (1.10.2) are derived from the temperature dependence of the plateau electrode potentials. As in the case of ΔG, the influence of platinum on ΔH is the opposite of that of silver and gold. Table 2 gives values of ΔG and ΔH for the absorption of hydrogen by the two-phase alloys Pd–Au and Pd–Pt at 25°C and normal atmospheric pressure.

1. A. Maeland and T. B. Flanagan, *J. phys. Chem., Ithaca* **69,** 3575 (1965).
2. A. W. Carson, T. B. Flanagan and F. A. Lewis, *Trans. Faraday Soc.* **56,** 1332 (1960).
3. A. Maeland and T. B. Flanagan, *J. phys. Chem., Ithaca* **68,** 1419 (1964).

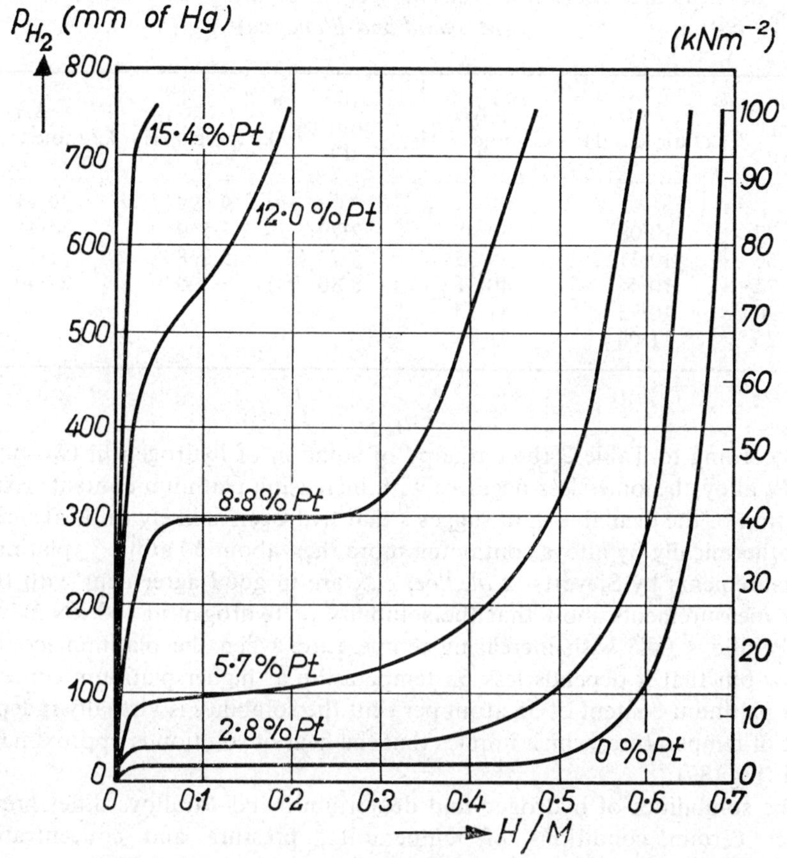

FIG. 38. Pressure–concentration isotherms for Pd–Pt–H alloys at 25°C. The isotherm for pure palladium is given for comparison purposes. The platinum content of the hydrogen-free alloys used as starting material is given in atom per cent. The pressures were obtained indirectly from measurements of electrode potentials with reference to the potential of the standard hydrogen electrode as zero potential. In calculating the equilibrium pressures use was made of the equation $-2FE = RT \ln p$, where E is the electrode potential, p the hydrogen pressure, R the gas constant and F Faraday's constant. The hydrogen content is given as the atom ratio H/M (Carson, Flanagan and Lewis).

TABLE 2

Free energies and enthalpies of solution of hydrogen in Pd–Au and Pd–Pt alloys at 25°C and 100 kNm^{-2} in the region of α and β phase coexistence (Maeland and Flanagan)

Alloy atom % Au	$-\Delta G$ kJ $mole^{-1}$ H_2	$-\Delta H$ kJ $mole^{-1}$ H_2	Alloy atom % Pt	$-\Delta G$ kJ $mole^{-1}$ H_2	$-\Delta H$ kJ $mole^{-1}$ H_2
			0	9·699	39·94
2·77	10·08	39·05	2·79	7·569	34·41
5·66	10·31	39·75	5·73	5·368	31·21
8·70	10·44	40·31	8·80	2·390	27·40
11·90	10·73	41·15			
15·26	11·06	41·15			

According to Table 2 the enthalpy of solution of hydrogen in two-phase Pd–Pt alloys becomes less negative with increasing platinum content. Extrapolation of the available data suggests that hydrogen is likely to be absorbed endothermically by alloys containing more than about 30 atom % platinum. Measurements by Sieverts *et al.*, *loc. cit.*, are in good agreement with this; their measurements show that the solubility of hydrogen at 100 kN m^{-2} in Pd–Pt alloys falls with increasing temperature when the platinum content is low but that it depends less on temperature at higher platinum contents. At a platinum content of 35 atom per cent the solubility is virtually independent of temperature, which implies that the heat of solution is approximately zero (Fig. 39).

The solubilities of hydrogen and deuterium in Pd–Pt alloys differ greatly under certain conditions of temperature, pressure and concentration. However, we cannot discuss this point in great detail here and refer to papers by Flanagan and Maeland[1,2]. Of further interest at this point is the fact that at 25°C and normal atmospheric pressure the equilibrium solubilities in a Pd–Pt alloy containing 8·80 atom per cent of platinum are given by the atom ratios H/Pd = 0·46 and D/Pd = 0·03. The cause of this large difference in gas content lies in the fact that the plateau pressures for the Pd–Pt–D alloys at 25°C are appreciably higher than those for the Pd–Pt–H alloys. For this reason the Pd–8·8% Pt–D alloy is still in the low-content α region whilst the Pd–8·8% Pt–H alloy is in the high-content β region at 25°C and 100 kN m^{-2}.

1. T. B. FLANAGAN, *J. phys. Chem., Ithaca* **67,** 203 (1963).
2. A. MAELAND and T. B. FLANAGAN, *J. phys. Chem., Ithaca* **68,** 1419 (1964).

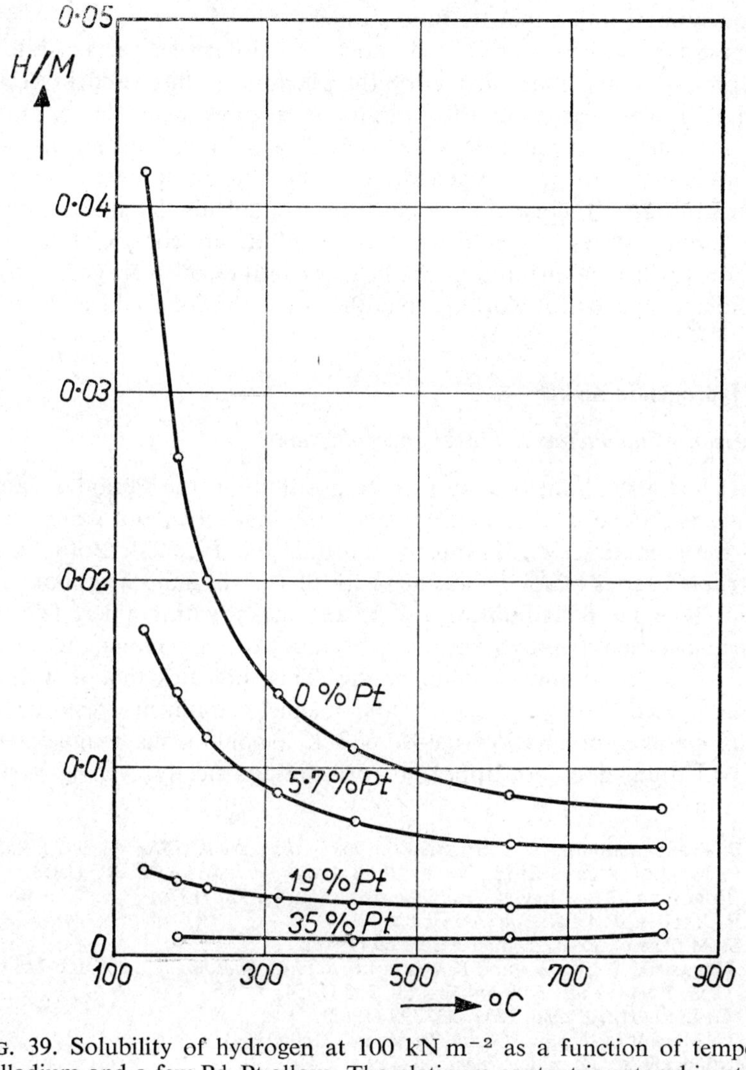

FIG. 39. Solubility of hydrogen at 100 kN m^{-2} as a function of temperature in palladium and a few Pd–Pt alloys. The platinum content is entered in atom per cent. The concentration of hydrogen is given in atoms per metal atom (data taken from Sieverts *et al.*).

For data on the palladium–rhodium–hydrogen and palladium–nickel–hydrogen systems we refer to papers by Russian investigators[1, 2], and for data on the palladium–boron–hydrogen and palladium–silver–boron–hydrogen systems to a paper by Burch and Lewis[3].

In this section we have not entered into the quantitative aspects of attempts to interpret, in terms of atomic theory, the phenomena discussed in Sections 1.9 and 1.10. The reason for this is that the theories available to date still contain too many uncertainties, which is evident when various papers on the subject are compared[4–15]. In general it can be stated that there is at present no real quantitative theory of alloys of transition metals. Even the statement that the unfilled states in the 4d band of palladium are completely occupied by electrons when the metal absorbs large quantities of hydrogen, can only be considered as a useful working hypothesis (cf. Sections 1.12 and 1.13).

1.11. Hydrogen in nickel

Comparison of the elements nickel and palladium

Nickel and palladium occupy similar positions in the Periodic Table of the elements. Nickel is the last element of the first series and palladium the last of the second series of transition metals (cf. Fig. 23). Both form an uninterrupted series of solid solutions with their right hand neighbour in the Periodic Table, i.e. palladium with silver and nickel with copper. The major difference between the two metals is that nickel is ferromagnetic below 632 K, while palladium is paramagnetic. This indicates that in nickel an interaction occurs between the atomic magnetic moments resulting in a mutually parallel orientation below 632 K (spontaneous magnetization), whilst palladium does not appear to order magnetically even at very low temperatures.

1. I. P. Tverdovskii and A. I. Stetsenko, *Dokl. Akad. Nauk SSSR* **84,** 997 (1952).
2. I. P. Tverdovskii and Zh. L. Vert, *Dokl. Akad. Nauk SSSR* **88,** 305 (1953).
3. R. Burch and F. A. Lewis, *Trans. Faraday Soc.* **66,** 727 (1970).
4. J. R. Lacher, *Proc. R. Soc.* **A 161,** 525 (1937).
5. J. S. Anderson, *Proc. R. Soc.* **A 185,** 69 (1946).
6. A. Harasima, T. Tanaka and K. Sakaoku, *J. phys. Soc. Japan* **3,** 208 and 213 (1948).
7. A. L. G. Rees, *Trans. Faraday Soc.* **50,** 335 (1954).
8. G. G. Libowitz, *J. appl. Phys.* **33,** 399 (1962).
9. A. C. Makrides, *J. phys. Chem., Ithaca* **68,** 2160 (1964).
10. H. Brodowsky, *Z. phys. Chem.* **44,** 129 (1965).
11. H. Brodowsky and E. Poeschel, *Z. phys. Chem.* **44,** 143 (1965).
12. J. W. Simons and T. B. Flanagan, *Can. J. Chem.* **43,** 1665 (1965) and *J. phys. Chem., Ithaca* **70,** 3750 (1966).
13. K. Allard, A. Maeland, J. W. Simons and T. B. Flanagan, *J. phys. Chem., Ithaca* **72,** 136 (1968).
14. Y. Ebisuzaki and M. O'Keeffe, *J. phys. Chem., Ithaca* **72,** 4695 (1968).
15. R. Burch, *Trans. Faraday Soc.* **66,** 736 and 749 (1970).

Since pure nickel has about 0·6 holes per atom in its 3d band, the rigid-band model predicts filled d states, and therefore diamagnetism, when the concentration of copper in nickel exceeds about 60 atom %. We have seen that palladium–silver alloys do indeed behave in this way. Nickel–copper alloys, however, behave differently; although, as expected, the ferromagnetism vanishes at approximately 60 atom % copper[1], the alloys that are richer in copper are not diamagnetic but strongly paramagnetic[2]. These facts are exemplified by Figs. 40 and 41 showing the atomic magnetic moment and the Curie temperature of ferromagnetic Ni–Cu alloys and also by Fig. 42 giving the magnetic susceptibility of non-ferromagnetic Ni–Cu alloys. Nickel–copper alloys only become diamagnetic at a copper content exceeding 95%.

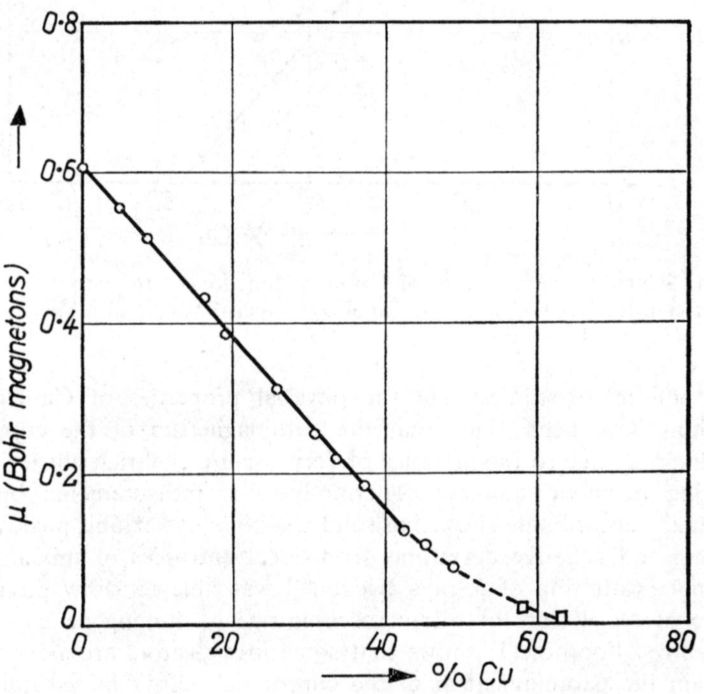

FIG. 40. Atomic magnetic moment in Bohr magnetons as a function of the copper content in atom per cent for ferromagnetic Ni–Cu alloys (Ahern, Martin and Sucksmith).

1. See, for example, S. A. AHERN, M. J. C. MARTIN and W. SUCKSMITH, *Proc. R. Soc.* **A 248,** 145 (1958).
2. See, for example, E. W. PUGH and F. M. RYAN, *Phys. Rev.* **111,** 1038 (1958).

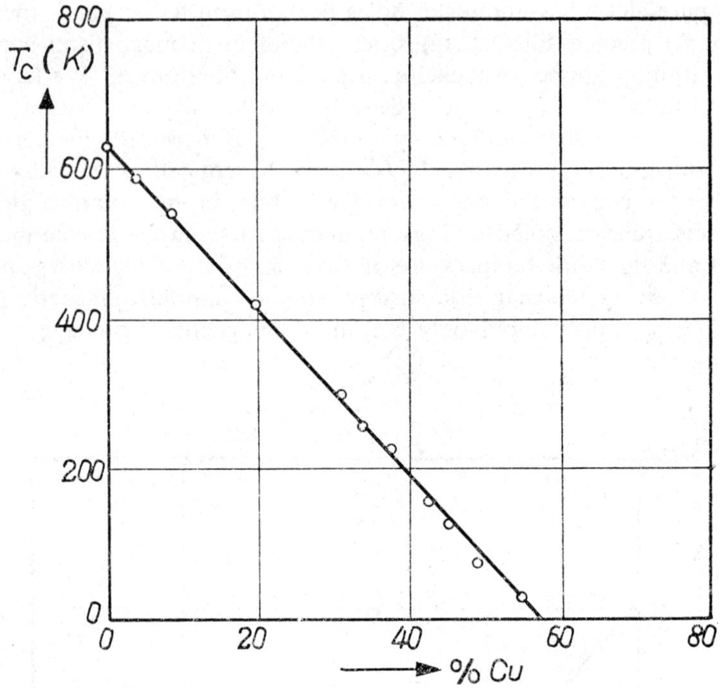

FIG. 41. Curie temperatures in Kelvin as a function of the copper content in atom per cent for ferromagnetic Ni–Cu alloys (Ahern, Martin and Sucksmith).

Based on many studies[1] of the physical properties of Cu–Ni alloys suggestions have been made that the paramagnetism of the copper-rich alloys could be due to the presence of very small nickel-rich clusters with a fluctuating magnetic moment. Thermodynamic measurements by Rapp and Maak[2] accordingly show that solid Cu–Ni alloys exhibit positive heats of mixing and negative deviations from ideal entropies of mixing. Using small-angle scattering of X-rays Kidron[3] was able to show directly the presence of Ni clusters of submicroscopic size in 'homogeneous' copper–nickel alloys. Footnote(1) shows that some investigators are more inclined to explain the paramagnetism of the copper-rich alloys by assuming that there are holes in the common 3d band of homogeneous Ni–Cu alloys almost up to pure copper.

1. P. A. BECK (ed.), *Electronic Structure and Alloy Chemistry of the Transition Elements*, Interscience, New York (1962).
2. R. A. RAPP and F. MAAK, *Acta Metall.* **10,** 63 (1962).
3. A. KIDRON, *Phys. Lett.* **26A,** 593 (1968) and *Phys. Rev. Lett.* **22,** 774 (1969).

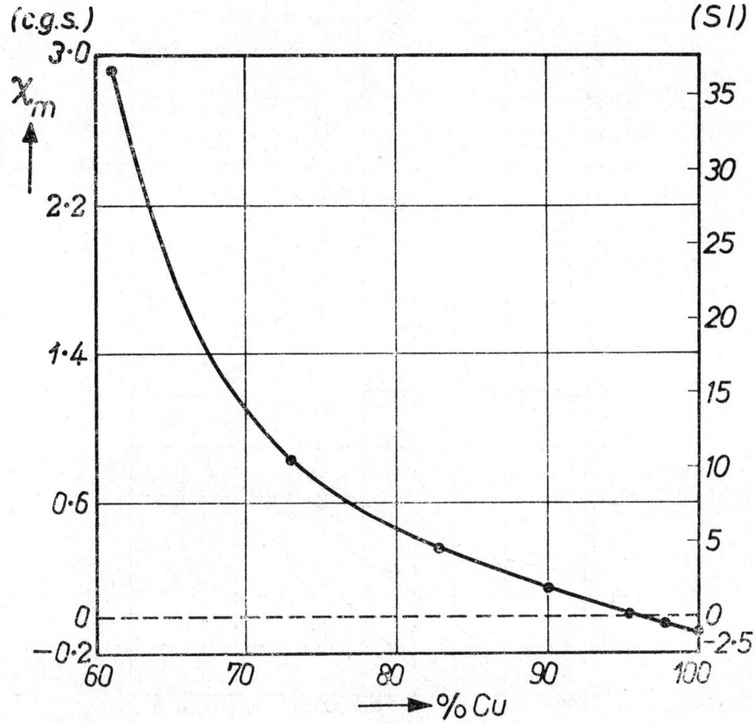

FIG. 42. Mass susceptibility in cgs and SI units as a function of the copper content in atom per cent for non-ferromagnetic Ni–Cu alloys at 295 K (Pugh and Ryan).

The Ni–H system

Hydrogen has a much smaller affinity for nickel than for palladium. It dissolves endothermically in nickel, but exothermically in palladium (see Fig. 43). Notwithstanding this there is great similarity between the systems Pd–H and Ni–H. From investigations by Polish scientists[1,2] the conclusion can be drawn that, not only in the Pd–H system but also in the Ni–H system, two fcc phases with entirely different hydrogen contents can be in equilibrium. So little hydrogen dissolves in the hydrogen-poor α phase that the lattice parameter of the hydrogen-bearing metal is virtually the same as that of pure nickel (0·352 nm). In the hydrogen-rich β phase the atom ratio, H to Ni, is approximately 0·7 at room temperature. Its lattice parameter is 0·372 nm,

1. B. BARANOWSKI and M. SMIALOWSKI, *Physics Chem. Solids* **12,** 206 (1959).
2. A. JANKO, *Bull. Acad. pol. Sci. Sér. Sci. chim.* **8,** 131 (1960).

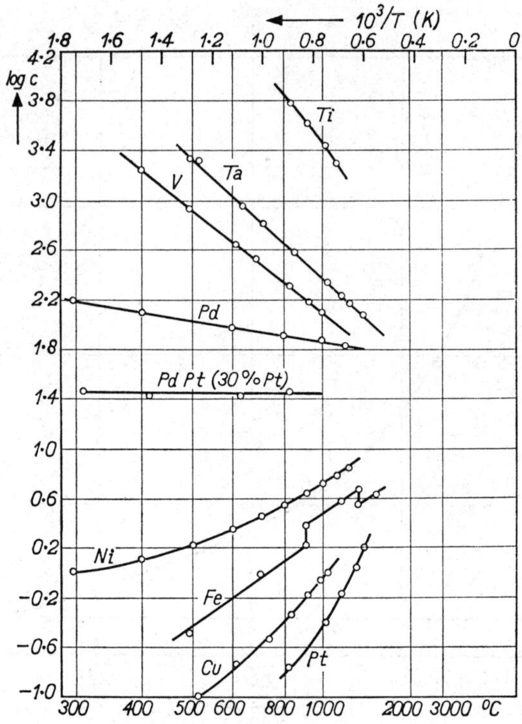

FIG. 43. Logarithm of the solubility of hydrogen in several metals at 100 kNm^{-2} of H_2 as a function of the reciprocal of the absolute temperature. The concentration of hydrogen is given in atoms per 10^4 metal atoms. The data on which the figure is based have mostly been derived from publications by A. Sieverts and co-workers (cf. G. BORELIUS, *Annln Phys.* **83,** 121 (1927) and *Metallwirt., Metallwiss., Metalltech.*, Feb. 1929). The points relating to the Pd–Pt alloy containing 30 wt% (19 atom per cent) Pt are based on the same measurements as the corresponding points for the 19 atom per cent Pt–Pd alloy in Fig. 39.

i.e. about 6% larger than that of pure nickel. These results have been confirmed by Boniszewski and Smith[1]. An investigation of the nickel–deuterium system gave similar results[2]. Neutron diffraction patterns show that the hydrogen atoms occupy the octahedral interstices, as in the case of palladium[3].

1. T. BONISZEWSKI and G. C. SMITH, *Physics Chem. Solids* **21,** 115 (1961).
2. A. STROKA and B. BARANOWSKI, *Bull. Acad. Pol. Sci. Sér. Sci. chim.* **10,** 147 (1962) and **14,** 419 (1966).
3. E. O. WOLLAN, J. W. CABLE and W. C. KOEHLER, *Physics Chem. Solids* **24,** 1141 (1963).

The phase with an atom ratio H/Ni = 0·7 could, at first, only be obtained by charging nickel with hydrogen electrolytically in a very special process. Only in 1966 did Baranowski *et al.* manage to prepare it from nickel and hydrogen under high pressure (see below). At room temperature the nickel hydride under discussion decomposes spontaneously into nickel and hydrogen, even under hydrogen pressures of the order of 100 MN m^{-2}[(1)]. Quantitatively there are therefore great differences between the Pd–H and Ni–H systems. Under equilibrium conditions at room temperature the hydrogen-rich phase of the Pd–H system readily forms at a hydrogen pressure of the order of 100 N m^{-2}. Although not a sufficient number of equilibrium measurements for the Ni–H system is available, it may be expected that this system behaves qualitatively in a manner similar to the Pd–H system for which the equilibrium diagram is shown in Fig. 26.

As in the case of palladium, the thermodynamic explanation of the phenomena under discussion lies in the fact that the partial heat of solution of hydrogen in nickel decreases with increasing hydrogen concentration. This means that the process of solution becomes less endothermic, or even exothermic, with increasing hydrogen concentration. From the atomic viewpoint the further assumptions of dissociation of the dissolved hydrogen into protons and electrons, an occupation of holes in the 3d band of nickel by hydrogen electrons and an attractive interaction between the dissolved protons via their screening atmospheres of electrons, may be made. It should, however, be mentioned that differences between the systems Pd–H and Ni–H may be expected comparable to those observed between the systems Pd–Ag and Ni–Cu (see above).

Only very incomplete data at present exist on the hydrogen content of the hydrogen-poor Ni–H phase (α phase) coexisting with the hydrogen-rich β phase in the two-phase region. Extrapolation of measurements of solubility by Sieverts[(2)], Smittenberg[(3)] and Armbruster[(4)] indicate that nickel in equilibrium with hydrogen of normal atmospheric pressure and at room temperature has a hydrogen content of the order of magnitude H/Ni = 10^{-5}. According to electrochemical measurements[(5)] the hydrogen content of this α phase, in equilibrium with the β phase at 25°C, is of the order of magnitude H/Ni = 10^{-2} or 10^{-1}, i.e. one thousand to ten thousand times higher. This would correspond to an equilibrium pressure in the two-phase region at 25°C of approximately 1 TN m^{-2} H_2 if the square root p rule still applies at very high pressures. In actual fact, the solubility at low concentrations and high pressures is proportional to the square root of the

1. B. Baranowski, *Bull. Acad. Pol. Sci. Sér. Sci. chim.* **10,** 451 (1962).
2. A. Sieverts, *Z. Metallk.* **21,** 37 (1929).
3. J. Smittenberg, *Recl. Trav. chim. Pays-Bas Belg.* **52,** 112, 339 (1933); **53,** 1065 (1934).
4. M. H. Armbruster, *J. Am. chem. Soc.* **65,** 1043 (1943).
5. B. Baranowski and Z. Szklarska-Smialowska, *Electrochim. Acta* **9,** 1497 (1964).

fugacity[1] of the gas; therefore, not the equilibrium pressure but the equilibrium fugacity is of the order of magnitude of 1 TN m^{-2}. At relatively low pressures there is virtually no difference between pressure and fugacity. Unfortunately the relation between these two quantities is not known for very high pressures. According to a rough estimate the equilibrium fugacity of about 1 TN m^{-2} in the α plus β region at 25°C corresponds to an equilibrium pressure of about 1 GN m^{-2}. The experiments, discussed below, show that this value is of the right order of magnitude.

At 25°C, a pressure of 344 MN m^{-2} proved sufficient to stop dissociation of nickel hydride[2]. The preparation of this phase from nickel and gaseous hydrogen at the same temperature required a pressure of approximately 1 GN m^{-2} H_2[3]. Assuming the occurrence of hysteresis, the equilibrium pressure in the two-phase (α plus β) region must therefore lie somewhere between 344 MN m^{-2} and 1 GN m^{-2} at 25°C.

It is to be expected that the nickel–hydrogen system will show hysteresis phenomena similar to those occurring in the palladium–hydrogen system. As shown diagrammatically in Fig. 28 these manifest themselves over the α-plus-β region in higher apparent equilibrium pressures during the absorption of hydrogen than during its release.

Wagner[4] explains the hysteresis with the undeniable fact that the diffusivity of palladium in the Pd–H alloys is negligible compared with that of hydrogen. This means that exchange of palladium atoms between the coexisting α and β phases is virtually impossible. In the two-phase ($\alpha + \beta$) region therefore, the equilibrium condition $\mu_{H,\alpha} = \mu_{H,\beta}$ is satisfied but not the condition $\mu_{Pd,\alpha} = \mu_{Pd,\beta}$. Thus, the β phase only begins to form during the absorption of hydrogen by palladium when the α phase contains more hydrogen than that corresponding to complete equilibrium. In the reverse direction, during the release of hydrogen, formation of the α phase only starts when the β phase contains less hydrogen than that corresponding to the thermodynamic equilibrium between the α and β phases.

It should be noted that the establishment of equilibrium between the α and β phases in a piece of metal would not require a displacement of palladium atoms if both fcc phases had the same lattice parameter. In that case all the palladium atoms could remain in the same position and only diffusion of hydrogen would be necessary. In other words, the large increase in the lattice parameter on formation of the β phase plays an important part in the hysteresis under discussion. The β crystals that are being formed on absorption of hydrogen require a greater volume and are thus under compressive stress from the surrounding crystals. In 1937 Ubbelohde[5] observed that the dissociation pressure of the β phase is increased by this compressive stress. Scholtus and Hall[6] have contributed to this view by

1. Fugacity (P^*) is defined by the equation $\mu = RT \ln P^* + \mu^\circ$. The ratio P^*/P, where P is the actual pressure, approaches unity as P tends to zero, i.e., as the gas approximates more closely to ideal behaviour.
2. B. Baranowski and K. Bochenska, *Z. phys. Chem.* **45**, 140 (1965).
3. B. Baranowski and R. Wisniewski, *Bull. Acad. Pol. Sci. Sér. Sci. chim.* **14**, 273 (1966).
4. C. Wagner, *Z. phys. Chem.* **193**, 386 (1944).
5. A. R. Ubbelohde, *Proc. R. Soc.* **A 159**, 295 (1937).
6. N. A. Scholtus and W. K. Hall, *J. chem. Phys.* **39**, 868 (1963).

pointing out that the internal pressure opposing β formation increases with the amount of hydrogen absorbed until the yield stress of the solid is reached and irreversible plastic deformation starts. The well-known strain hardening causes the yield stress to rise with increasing plastic strain and, in consequence, absorption takes place at increasing hydrogen pressure even after plastic deformation has started.

In contrast to the absorption isotherm, the hydrogen-release isotherm deviates only slightly from the thermodynamic equilibrium curve, which is also indicated by the fact that this isotherm has a virtually horizontal part (Fig. 28). This is to be expected since the release of only a small amount of hydrogen suffices to relieve the β phase from residual elastic stresses. Desorption then takes place at a lower pressure, almost corresponding to equilibrium between the β phase, the gas phase and the newly formed α phase. Severe mechanical distortions and fractures in pieces of palladium after repeated charging and discharging with hydrogen show that plastic deformation does play an important part.

Everett and Nordon[1] made an extensive study of hysteresis in the Pd–H system and analysed their data in terms of the statistical mechanical theory of hydrogen in metals as developed by Lacher[2]. In this theory the interaction energy between hydrogen atoms in the lattice is taken into account, in addition to the energy of absorption of isolated hydrogen atoms. The difference in specific volume between α and β is allowed for by Everett and Nordon, *loc. cit.*, by the logical assumption that the absorption and interaction energies are a function of the lattice parameter, and thus have different values for the α and β phases. On this basis two separate pressure–composition isotherms are derived, one for the absorption and one for the release of hydrogen.

The influence of hydrogen on the magnetic properties of nickel

Hydrogen, as well as copper, can reduce the ferromagnetism of nickel to zero when it is added to the metal in sufficient quantity. Fig. 44 shows the results of measurements carried out by Bauer and Schmidbauer[3] on 6 μm-thick nickel foil. The graphs give the magnetization (on an arbitrary scale) as a function of the magnetic field strength in oersteds and amperes per metre. The measurements were done at field strengths ranging from 220 kA m^{-1} (2770 oersted) to 0 kA m^{-1}. Graph (a) refers to the metal before it is charged with hydrogen, graph (b) to the metal 3·5 minutes after it had been charged up to the maximum atom ratio H/Ni = 0·7. At this concentration the saturation magnetization amounts to only about 1 % of that of the pure metal.

In the course of time the saturation magnetization increases again due to the fact that the absorbed hydrogen leaves the metal spontaneously at room temperature. This return of the ferromagnetism is shown by the graphs

1. D. H. Everett and P. Nordon, *Proc. R. Soc.* **A 259,** 341 (1960).
2. J. R. Lacher, *Proc. R. Soc.* **A 161,** 525 (1937).
3. H. J. Bauer and E. Schmidbauer, *Z. Phys.* **164,** 367 (1961); see also: W. Andrä, *Phys. stat. sol.* **1,** K135 (1961).

between (a) and (b), the time elapsed since the charging being indicated for each graph. After 6·5 hours hardly any further change takes place. As shown in Fig. 44 the metal does not completely return to its original state. The coercivity of the hydrogen-containing nickel is greater than that of the pure metal but does not, after spontaneous degassing, return to the starting value either.

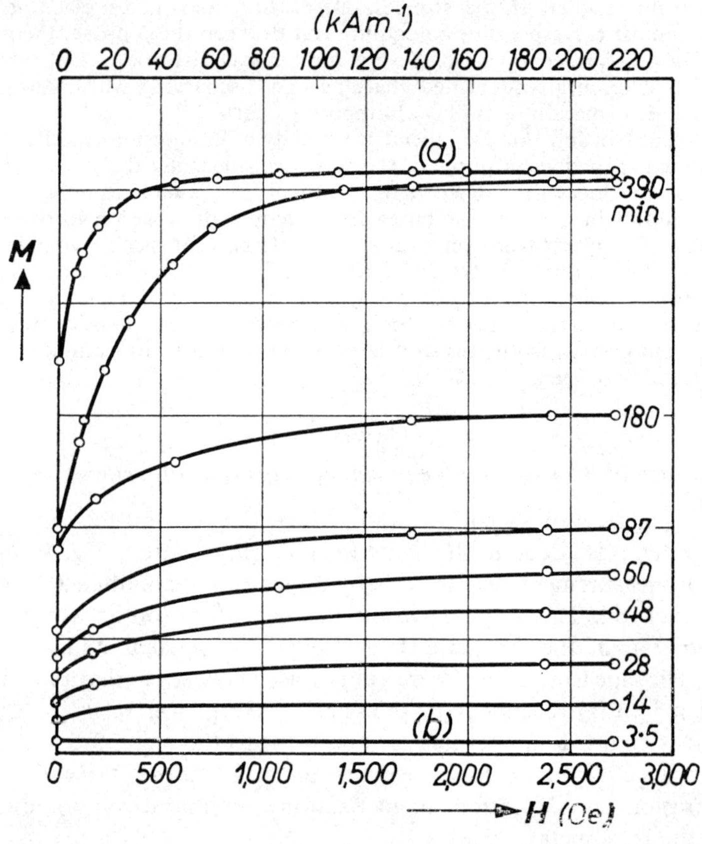

FIG. 44. Magnetization of 6 μm-thick nickel foil measured at decreasing field strength. The figure gives the field strength in oersteds and amperes per metre and the magnetization on an arbitrary scale. Graph (a) refers to the metal before it has been charged with hydrogen, graph (b) to the same nickel sample 3·5 minutes after it had been charged up to the atom ratio H/Ni = 0·7. The other graphs give results of measurements after further time intervals (14, 28, 48, 60, 87, 180 and 390 minutes). The measurements were carried out at 12·5°C (Bauer and Schmidbauer).

The lasting changes in the metal are probably due to the high dissociation pressure of the nickel hydride which makes it possible for molecular hydrogen to precipitate under high pressure at the grain boundaries thus causing permanent damage in the form of micro-cracks and plastic deformation of the metal in the neighbourhood of the cracks[1,2]. These phenomena are similar to the damage that can occur in iron by electrolytically charging it with hydrogen. However, in the case of iron no hydride is formed as an intermediate phase. In this connection it should be noted that Smialowski[3] has suggested that a hydrogen-rich β phase also exists in the iron–hydrogen system. Up to now no evidence supporting this hypothesis has been found.

The loss of ferromagnetism in nickel following the absorption of hydrogen can, according to Bauer and Schmidbauer, *loc. cit.*, be explained by the filling of the 3d band of the metal with hydrogen electrons. The effect could, on the other hand, also be due to the 6% increase in the lattice parameter, resulting in a decrease of the quantum mechanical exchange forces acting between electrons in neighbouring nickel atoms. Recent measurements by Faessler and Schmid[4] of the X-ray absorption spectrum of hydrogen-bearing nickel make the first explanation the more probable. The fine structure of the K absorption edge of nickel hydride, as compared with that of nickel, indicates that electrons of the absorbed hydrogen occupy empty 3d and 4s states of the conduction band of the metal. In this connection measurements of the Mössbauer effect of ^{57}Fe in nickel hydride[5] are also interesting but not conclusive.

1.12. The nature of bonding in transition-metal hydrides

Interstitial solid solutions and compounds

There is still great uncertainty about the nature of bonding in the transition-metal hydrides. In the first tentative theory they were not considered to be real chemical compounds. Their composition and structure were assumed to be governed by geometrical factors and even hydrogen-rich phases were thought to be a kind of solid solution or 'interstitial structure', where neutral hydrogen atoms were located in the interstices of the metal. The borides, carbides and nitrides of many transition metals were also regarded

1. D. J. van Ooyen, *Physics Chem. Solids* **23,** 1173 (1962).
2. B. A. Wilcox and G. C. Smith, *Acta Metall.* **13,** 331 (1965).
3. M. Smialowski, *Hydrogen in Steel*, Pergamon Press, London (1962), p. 57.
4. A. Faessler and R. Schmid, *Z. Phys.* **190,** 10 (1966).
5. G. K. Wertheim and D. N. E. Buchanan, *Physics Chem. Solids* **28,** 225 (1967).

in this way. In this theory the stability of the 'interstitial structures' is completely determined by the ratio of the radii of the atoms[1,2]. This ratio should be such that the parent atoms are nearly in mutual contact and that the atoms of the non-metallic element are distributed over the interstices between them in such a way that they are in contact with their metal neighbours, causing a slight expansion of the metal lattice. Subject to this condition, the non-metal atoms occupy the largest interstices in the structure in which this contact can be attained. Thus, if the metal has a fcc structure, the non-metal atoms, if sufficiently large, will occupy the octahedral interstices. If all these sites are occupied the structure is that of sodium chloride and corresponds to the composition MX. If the non-metal atom is too small for this type of arrangement it occupies the smaller tetrahedral interstices. If all these sites are occupied, the structure is that of fluorite and corresponds to the composition MX_2.

Investigations of crystal structures have proved this theory, where the transition metal hydrides are considered as solid solutions, to be untenable in its original form. The investigations showed that it is necessary to distinguish between a metal containing hydrogen in solution and the phase, much richer in hydrogen, which is formed after the solubility limit has been exceeded. The former is usually indicated as primary solution and the latter as hydride. The old theory discussed above is undermined by the fact that the metal sub-lattice in most hydrides is different from the lattice of the parent metal. Exceptions to this are the systems Pd–H and Nb–H, where at not too low temperatures primary solution and hydride have the same structure, fcc in the former and bcc in the latter case. With Gibbs' phase rule in mind each of these systems must be regarded as having two different solid phases, but in actual fact they form only one phase which is unstable over a certain range of compositions. With rising temperature this region narrows, and above a critical temperature (300°C in the case of Pd–H and 140°C for Nb–H) no phase separation occurs. We must therefore assume that in neither of the two systems is there an essential difference between the nature of the chemical bond in the 'solid solution' and in the 'hydride'.

The case of the intermediate phases rich in 'interstitials', such as the hydrides under discussion, is far removed from the idea that their stability is determined by size factors only and that there is no particular chemical bonding between the 'interstitials' and the metal. Surprisingly enough many recent papers on primary solid solutions are still consciously or subconsciously based on the assumption

1. G. HÄGG, *Z. phys. Chem.* **B 6,** 221 (1929); **B 7,** 339 (1930); **B 8,** 445 (1930); **B 12** 33 (1931).
2. R. C. EVANS, *An Introduction to Crystal Chemistry*, Cambridge University Press (1939).

that their stability is determined by size factors only. However, as already discussed in Section 1.6, the chemical factor plays an important part in primary solid solutions also. In the case of palladium, nickel and niobium there is therefore no reason to assume that there is a difference in the nature of the solid solution and that of the hydride.

Proton theory and anion theory

In a previous section (Section 1.9) it was stated that, even in the simple case of the Pd–H system two strongly opposed opinions exist on the nature of the chemical bonding. Whilst most scientists assume that the hydrogen is present in the metal as protons, Gibb[(1)] and Libowitz[(2)], in particular, defend the theory that palladium hydride is a non-stoichiometric compound in which the hydrogen is present as negative ions, i.e. in the same state as in the saline hydrides of the alkali and alkaline-earth metals. The arguments in favour of the 'proton theory' derive mainly from the study of the palladium–hydrogen system whilst those quoted in defence of the 'anion theory' are largely derived from the properties of the hydrides of other transition metals, including the lanthanon and actinon series. It is therefore possible that the two theories do not exclude each other but apply to different groups of 'metallic hydrides'. In addition, these theories refer to idealized models. The real chemical bonding in the compounds under discussion may be expected to consist of a mixture of metallic and ionic bonds, and perhaps even covalent bonds, the degree of ionic character varying from hydride to hydride.

The arguments discussed in previous sections in favour of the proton theory are: (a) the paramagnetism of palladium and the ferromagnetism of nickel both vanish during the absorption of about 0·65 hydrogen atoms per metal atom; (b) the diffusivity of hydrogen in palladium and in many other metals is abnormally high; (c) under the influence of an electric potential gradient, dissolved hydrogen migrates towards the negative pole of a palladium, iron or tantalum wire.

Some of the arguments put forward in favour of the anion theory are: (a) there is close similarity in crystal structure between the hydrides and the halides of the transition metals (this applies even to palladium hydride whose composition, at low temperatures and high hydrogen pressures, approaches PdH and whose crystal structure is the same as that of the alkali-metal halides and hydrides); (b) the interatomic distances in the hydrides of most transition metals agree well with the calculated values, taking the hydrogen atoms in these compounds as negative ions with radius 0·127 nm; (c) assuming

1. T. R. P. Gibb, *Prog. inorg. Chem.* **3,** 315 (1962).
2. G. G. Libowitz, *The Solid-State Chemistry of Binary Metal Hydrides*, Benjamin, New York (1965).

ionic-type bonding, the lattice energies of a number of metallic hydrides, as determined from experimental data with the aid of the Born–Haber cycle, are in fair agreement with the calculated values; (d) in a number of transition metals hydrogen and oxygen can be present together in solid solution; cases are also known where hydrogen and oxygen coexist in metallic hydrides. In view of the fact that it is generally accepted that oxygen, in metallic solutions and compounds, has a negative charge this should, according to the supporters of the anion theory, also apply to the coexisting hydrogen. Points (b) and (c) will be discussed in greater detail below whilst all four points will be critically reviewed in Section 1.13.

Interatomic distances in metallic hydrides

The presence of negative hydrogen ions in the hydrides of many transition metals is suggested by the relatively large specific volume of these compounds. In order to illustrate this with an example, Table 3 gives the value of the shortest distance between titanium atoms in hexagonal titanium as well as in TiN, TiC and the hydride phase of the composition TiH[1]. This table shows hydrogen to cause a greater expansion than nitrogen or carbon. The figures in the table are mutually comparable because the coordination number in the lattice or sub-lattice of the titanium atoms is twelve in each of the four phases.

TABLE 3

Shortest distance between titanium atoms in Ti(hcp), TiN, TiC and TiH

Phase	αTi	TiN	TiC	TiH
d(Ti–Ti) in nm	0·289	0·300	0·305	0·311

The previously mentioned radius of 0·127 nm for the negative hydrogen ion, which Libowitz, *loc. cit.*, assumed in his calculation of the interatomic distances in the hydrides of the transition metals, is somewhat smaller than that of the hydrogen ion in the alkali or alkaline-earth metals. This appears reasonable since the electropositivities of the latter are known to be higher than those of the transition metals. In order to get the calculated and measured values to agree, it was necessary to ascribe to the metal ions (excepting Eu and Yb) smaller radii and hence higher positive charges

1. The existence range of cubic titanium hydride extends from TiH to TiH_2 (idealized formulae). The figures in Table 3 are taken from the book by W. B. PEARSON, *Handbook of Lattice Spacings and Structures of Metals and Alloys*, Pergamon Press, London (1958).

than are needed for the formation of H^- ions (Table 4). Gibb and Libowitz, *loc. cit.*, assume that the extra released electrons take part in metallic bonding. This would imply that the chemical bonding in the metallic hydrides consists of a mixture of ionic and metallic bonds. The model offers an explanation for the high electrical conductivities of the metallic dihydrides mentioned in Table 4. Exceptions to the rule are EuH_2 and YbH_2, the cations of which

TABLE 4[1]

Interatomic distances M–H in metallic hydrides according to Libowitz, loc. cit., *assuming a hydrogen anion radius of* 0·127 *nm and cation oxidation states as indicated in the table*

Hydride	Cation radius	Cation oxidation state	Coordination number	Coordination number correction	M–H distance, nm	
					Calc.	Obs.
ScH_2	0·068	+3	8	0·08	0·203	0·207
TiH_2	0·060	+4	8	0·08	0·195	0·193
VH_2	(0·053)	+5	8	0·08	0·188	0·185
YH_2	0·088	+3	8	0·08	0·223	0·225
ZrH_2	0·077	+4	8	0·08	0·212	0·209
NbH_2	0·067	+5	8	0·08	0·202	0·197
LaH_2	0·104	+3	8	0·08	0·239	0·245
CeH_2	0·102	+3	8	0·08	0·237	0·242
PrH_2	0·100	+3	8	0·08	0·235	0·239
NdH_2	0·099	+3	8	0·08	0·234	0·237
SmH_2	0·097	+3	8	0·08	0·232	0·233
EuH_2	0·109	+2	7	0·04	0·240	0·245
GdH_2	0·094	+3	8	0·08	0·229	0·230
TbH_2	0·092	+3	8	0·08	0·227	0·227
DyH_2	0·091	+3	8	0·08	0·226	0·225
HoH_2	0·089	+3	8	0·08	0·224	0·224
ErH_2	0·087	+3	8	0·08	0·222	0·222
TmH_2	0·086	+3	8	0·08	0·221	0·220
YbH_2	0·093	+2	7	0·04	0·224	0·231
LuH_2	0·084	+3	8	0·08	0·219	0·218
HfH_2	0·077	+4	8	0·08	0·212	0·205
AcH_2	0·111	+3	8	0·08	0·246	0·246
ThH_2	0·099	+4	8	0·08	0·234	0·238
Th_4H_{15}	0·099	+4	9	0·11	0·237	0·229
PaH_3	0·090	+5	12	0·19	0·236	0·232
UH_3	0·083	+6	12	0·19	0·229	0·232
PuH_2	0·100	+3	8	0·08	0·235	0·232

1. The cation radii used in the table are those given by ZACHARIASEN in *The Actinide Elements* (editors Seaborg and Katz), Nat. Nucl. Energy Series IV, 14A (1954). Since these radii refer to cations having six anions as nearest neighbours, it was necessary to apply corrections for coordination numbers other than six.

are assumed to have a charge of +2 because the structure and properties of these two hydrides largely resemble those of the alkaline-earth dihydrides. They are the most nearly ionic of all lanthanon dihydrides. This is related to the fact that the 4f shells of Eu and Yb have an extra degree of stability by being half filled and completely filled, respectively (Eu: $4f^7 5s^2 5p^6 6s^2$ and Yb: $4f^{14} 5s^2 5p^6 6s^2$).

Lattice energies

The lattice energy U_l of an ionic crystal is the energy required to separate the ions from their normal distance in the crystal to an infinite distance apart. It can be calculated in a simple way if it is assumed that the only forces holding the crystal together are the Coulomb electrostatic forces and the short-range repulsive forces between the ions, which prevent the crystal from collapsing under the net attraction of the Coulomb forces. The resulting relation is known as the Born–Landé equation:

$$U_l = -\frac{N_0 A Z_1 Z_2 e^2}{r_0}\left(1-\frac{1}{n}\right) \qquad (1.12.1)$$

where U_l refers to one mole of the ionic compound, N_0 is Avogadro's number, e the charge of an electron, Z_1 and Z_2 the charges on cation and anion (+2 and −1 in the case of an ionic hydride MH_2), A the Madelung constant (a factor arising when summing the electrostatic potential energy of two ions over all interionic distances existing in the crystal), r_0 the shortest equilibrium distance between centres of oppositely charged ions in the normal crystal lattice and n the exponent of r in the expression for the repulsive energy. The value of A is given by the geometry of the crystal lattice; that of n can be inferred from measurements of the compressibility of the crystal.

In the case of an ionic hydride, MH_2, equation (1.12.1) represents the increase in internal energy accompanying the reaction

$$MH_2(s) \rightarrow M^{2+}(g) + 2H^-(g) \qquad (1.12.2)$$

at 0 K, and in close approximation also at 25°C.

U_l can also be determined experimentally by means of the method of Born and Haber (often referred to as the Born–Haber cycle) since the reaction (1.12.2) to which the value of U_l applies can in principle be achieved in a number of simple stages:

$$MH_2(s) \rightarrow M(s) + H_2(g) \qquad (1.12.3)$$

$$M(s) \rightarrow M(g) \qquad (1.12.4)$$

$$M(g) \rightarrow M^{2+}(g) + 2e^- \qquad (1.12.5)$$

$$H_2(g) \rightarrow 2H(g) \qquad (1.12.6)$$

$$2H(g) + 2e^- \rightarrow 2H^-(g) \qquad (1.12.7)$$

$$\overline{MH_2(s) \rightarrow M^{2+}(g) + 2H^-(g)}^{\;+} \qquad (1.12.2)$$

The first step represents the reverse of the formation of solid MH_2 from solid M and gaseous H_2; the change in energy accompanying this step is given by the negative of the standard energy of formation ΔU_f° of MH_2[1]. The second step corresponds to the sublimation of the metal and the third step to its ionization. The fourth step represents the dissociation of H_2 into atoms and the fifth step the formation of hydrogen anions out of atoms. Summation of these steps gives reaction (1.12.2) and thus

$$U_l = -\Delta U_f^\circ + U_{subl} + I_1 + I_2 + D - 2E \qquad (1.12.8)$$

where U_l is the lattice energy, ΔU_f° the standard energy of formation of MH_2, U_{subl} the molar sublimation energy of the metal, D the molar dissociation energy of H_2, I_1 and I_2 the first and second ionization energies for N_0 metal atoms, and E the electron affinity for N_0 hydrogen atoms.

Gibb, *loc. cit.*, has shown that the experimental and calculated lattice energies show a very satisfactory agreement for the saline hydrides and a fair agreement for those metallic hydrides where sufficient information is available to make the comparison. This is shown by Table 5 for some metallic hydrides.

TABLE 5

Lattice energies of four metallic hydrides according to Gibb and Libowitz, loc. cit.

Hydride	U_l(calc.) MJ mole^{-1}	U_l(exp.) MJ mole^{-1}
TiH_2	2·86	2·94
ZrH_2	2·80	2·92
LaH_2	2·38	2·54
GdH_2	2·49	2·59

1.13. Confrontation of proton theory and anion theory

Both theories discussed in Section 1.12 account for the metallic properties of the hydrides under discussion, the proton theory by describing them as intermediate phases of transition metals and *metallic* hydrogen[2,3], the anion theory by allowing for residual metallic bonding by those valency

1. This change in energy can be derived from the standard enthalpy of formation with the aid of the relation $\Delta U_f^\circ = \Delta H_f^\circ - \Delta(pV) \approx \Delta H_f^\circ - RT$.
2. A. R. UBBELOHDE, *Proc. R. Soc.* **A 159**, 295 (1937).
3. I. ISENBERG, *Phys. Rev.* **79**, 736 (1950).

electrons of the metal that remain after each hydrogen atom has acquired one electron. In the latter theory the purely ionic hydrides are those where the number of valency electrons of the metal equals the number of H^- ions in the hydride. It offers a natural explanation for the gradual disappearance of the metallic properties with increasing hydrogen content in the case of those lanthanon hydrides that can be hydrided to the point where all valency electrons are used up, i.e. over the composition range LnH_2 to LnH_3.

Proton theory

None of the arguments put forward in support of the two theories constitute absolute proof. Any of them can be countered by other arguments. The supporters of the proton theory point out the high diffusivity and the migration of hydrogen towards the negative electrode in palladium–hydrogen and a number of other metal–hydrogen systems. Gibb, *loc. cit.*, counters this with the remark that this need not be in contradiction with the anion theory. He assumes that H^+ occurs as an excited state of the hydrogen present as anions in the solid solutions and hydrides under discussion. In view of the fact that the proton is about 10^5 times smaller than the H^- ion its mobility must be vastly greater. Even its presence in only minute quantities would suffice to explain the large diffusivity of hydrogen and its migration towards the cathode under the influence of an electric potential field.

The 'protonists' derive one of their arguments from the dependence of the paramagnetic susceptibility of palladium on its hydrogen content (Section 1.9). Gibb *et al.*[1,2] offer several reasons against the value of this argument. (1) The two-phase nature of the Pd–H system at room temperature is often ignored. It is true that the small, negative value of the susceptibility of β palladium agrees with the electron donation model, but the linear decrease of the susceptibility accompanying the increase of the H/Pd ratio from zero to about 0·6 (Fig. 29) is merely the result of the fraction of β palladium changing from zero to one. (2) Volume effects have not been taken into account sufficiently. Thermal expansion of pure palladium lowers its susceptibility. The mere expansion of the lattice due to hydrogen uptake beyond the H/Pd ratio of 0·6 might in itself explain the loss of paramagnetism. (3) The susceptibility measurements (Fig. 29) all refer to the absorption of hydrogen. In Gibb's own measurements hydrogen was released from very finely divided palladium hydride samples. One sample was diamagnetic, having an initial hydrogen content corresponding to H/Pd $\approx$ 0·7. It released hydrogen at room temperature as soon as the hydrogen pressure dropped below 2·4kN m^{-2} (18 mm of Hg). The loss of hydrogen did not

1. *Engelhard Industries Techn. Bull.* **7,** 28 (1966).
2. T. R. P. Gibb, J. MacMillan and R. J. Roy, *J. phys. Chem., Ithaca* **70,** 3024 (1966).

change the susceptibility of the sample until a hydrogen content of about H/Pd = 0·3 was reached. With further release of hydrogen the susceptibility increased rapidly. Also, from palladium hydride powder that was slightly paramagnetic due to a lower hydrogen content, an appreciable quantity of hydrogen could be removed without much change in its susceptibility.

It is not easy to account for these phenomena since it can be expected that the removal of hydrogen from palladium hydride at a relatively low temperature will leave a material containing many lattice imperfections. In line with this, palladium obtained by decomposition of the hydride has a *higher* residual electrical resistivity than annealed palladium at 4·2 K[1]. However, the imperfections are not the only difference between the two kinds of palladium as is shown by the fact that the phonon contribution to the resistivity (the resistivity due to lattice vibrations) is *smaller* for the palladium obtained by hydrogen release than for annealed palladium.

Specific heat measurements at low temperatures

Specific heat measurements at low temperatures have been carried out on a number of palladium hydride specimens by Mackliet and Schindler[2]. According to their interpretation, the results of the measurements offer direct support for the proton theory. They define the specific heat of a H–Pd specimen as its heat capacity, divided by the number of moles of palladium in the specimen, regardless of the hydrogen content. The temperature dependence of the specific heat of each of the four specimens with H/Pd atom ratios of 0·57, 0·70, 0·81 and 0·88 can be accurately expressed by equation (1.8.2) in the temperature range 1·2 to 4·2 K. The Debye temperatures of the four samples were found to be virtually equal to that of pure palladium (about 275 K). This is shown by the fact that the graphs C/T as a function of T^2 are parallel straight lines in the region 1·2 to 4·2 K. On the other hand, the value of γ for the three samples richest in hydrogen, amounts to about 1·5 mJ mole^{-1} deg^{-2} ($3{\cdot}6 \times 10^{-4}$ cal/mole deg^{-2}) as compared to 9·41 mJ mole^{-1} deg^{-2} ($22{\cdot}5 \times 10^{-4}$ cal/mole deg^{-2}) for pure palladium.

Mackliet and Schindler draw the conclusion that the density of electronic states, at the Fermi level, in the three hydrides is roughly one-sixth that of pure palladium and virtually independent of composition, at least over the range H/Pd = 0·70 to 0·88. This is in keeping with the expectation that the 4d band is completely filled in all three hydrogen-rich alloys. The authors feel that their results offer direct evidence in favour of the electron donation model, i.e. in favour of the proton theory.

1. R. J. Smith, A. I. Schindler and E. W. Kammer, *Phys. Rev.* **127,** 179 (1962).
2. C. A. Mackliet and A. I. Schindler, *Phys. Rev.* **146,** 463 (1966).

Evidence in favour of the proton theory has also been obtained by Phillips and Kimball[1] in measurements of the Mössbauer effect of ^{57}Fe in Pd–Fe–H alloys and by Zamir[2] in nuclear magnetic resonance studies of vanadium–hydrogen and niobium–hydrogen alloys.

Ebisuzaki and O'Keeffe[3] have recently given a theoretical description, in terms of the proton theory, of the behaviour of hydrogen in transition metals. They conclude that the heat of solution of hydrogen is mainly determined by the energy of interaction between the dissolved proton and the electrons screening it. In their opinion the donation of hydrogen electrons to the d band in both primary solutions and hydrides is not incompatible with a density of electrons (mainly s electrons) in the vicinity of the proton that is similar to that in a free hydrogen atom or even a hydrogen anion (H^-).

Anion theory

Supporters of the anion theory, as mentioned in Section 1.12, draw attention to the close similarity in crystal structure between the hydrides and the halides of many transition metals. This is a reasonable argument but not conclusive proof that the nature of the bonding is the same in both categories of compounds. Even less convincing is the argument that the observed interatomic distances in many transition-metal hydrides agree well with the values calculated after ascribing reasonable values to the radii of the positive metal ions and the negative hydrogen ion. The similarity indicates that the anion theory is not incompatible with the data, but Mackay[4] correctly points out that for many hydrides an equally good agreement is obtained by using the tabulated covalent radii in the calculation. Also, the fair agreement which exists, for a number of transition-metal hydrides, between the value of the lattice energy as calculated on the basis of ionic-type bonding and the value as determined from experimental data, only proves that the anion theory is not incompatible with the facts. Libowitz, *loc. cit.*, himself mentions that the equations (1.12.1) and (1.12.8) frequently give good agreement, not only for ionic compounds but also for compounds with a low degree of ionic bonding. The reader should note that Gibb and Libowitz ascribe to the ions of Ti, Zr, La and Gd the charges $+4$ or $+3$ when calculating the interatomic distances in the dihydrides (Table 4), but the charge $+2$ when calculating the lattice energies of the same hydrides (Table 5).

Of the arguments in support of the anion theory, listed in Section 1.12, one remains to be discussed: the fact that hydrogen and oxygen can coexist

1. W. C. Phillips and C. W. Kimball, *Phys. Rev.* **165**, 401 (1968).
2. D. Zamir, *Phys. Rev.* **140**, A271 (1965).
3. Y. Ebisuzaki and M. O'Keeffe, *Prog. Sol. State Chem.* **4**, 187 (1967).
4. K. M. Mackay, *Hydrogen Compounds of the Metallic Elements*, E. & F. N. Spon, London (1966).

in a number of transition metals and transition-metal hydrides[1-4]. It is therefore possible that hydrogen and oxygen both carry a negative charge in these phases. In view of indications (see Section 1.14) that oxygen acts as an electron acceptor in some transition metals and as an electron donor in others, no arguments can be derived from the above-mentioned coexistence to support the sole right of existence of one of the competing theories.

Summarizing, it can be stated that both theories must be regarded as useful simplifications. There are cases where the proton theory can be used with advantage and others where the anion theory is more suitable. In general the proton theory is more applicable to hydrides of the transition metals at the end of the 3d, 4d and 5d Groups, especially those of nickel and palladium (cf. Fig. 23) whilst the anion theory offers advantages in the discussion of hydrides of the transition metals at the beginning of the long Periods, including lanthanon and actinon hydrides. This agrees with the increasing electronegativity of the transition metals from left to right in the Periodic Table of the elements. Although no great importance can be attached to quantitative values of the electronegativity[5], we give in Table 6 the values for the 3d, 4d and 5d transition metals according to Allred and

TABLE 6

Electronegativity values for the 3d, 4d and 5d transition metals (Little and Jones)

Sc 1·20	Ti 1·32	V 1·45	Cr 1·56	Mn 1·60	Fe 1·64	Co 1·70	Ni 1·75
Y 1·11	Zr 1·22	Nb 1·23	Mo 1·30	Tc 1·36	Ru 1·42	Rh 1·45	Pd 1·35
*	Hf 1·23	Ta 1·33	W 1·40	Re 1·46	Os 1·52	Ir 1·55	Pt 1·44

The asterisk indicates the lanthanons, La, Ce, Pr, Nd, Pm, Sm, Eu, Gd, Tb, Dy, Ho, Er, Tm, Yb and Lu, whose electronegativities all lie between 1·01 and 1·14 units on this scale.

1. M. N. A. HALL, S. L. H. MARTIN and A. L. G. REES, *Trans. Faraday Soc.* **41**, 306 (1945) and **50**, 343 (1954).
2. K. P. SINGH and J. GORDON PARR, *Trans. Faraday Soc.* **59**, 2248 (1963).
3. M. T. HEPWORTH and R. SCHUHMANN, *Trans. AIME* **224**, 928 (1962).
4. R. K. EDWARDS and P. LEVESQUE, *J. Am. Chem. Soc.* **77**, 1312 (1955).
5. H. O. PRITCHARD and H. A. SKINNER, *Chem. Rev.* **55**, 745 (1955).

Rochow[1] and Little and Jones[2]. The values are given on the Pauling scale[3], on which hydrogen has a value of 2·1 units.

1.14. Oxygen in transition metals and in transition-metal alloys

In the previous section we have seen that only a few concrete statements can be made on the state of ionization of hydrogen in transition metals and in transition-metal hydrides. Our knowledge of the state of ionization of oxygen in transition metals is even smaller. Below we shall discuss a few interesting data that have become available in recent years and largely refer to the behaviour of oxygen in niobium and a few niobium alloys.

Solubility of oxygen in niobium alloys

The solubility of oxygen in niobium, an element of Group VA, is relatively high and amounts to 2·0 atom per cent at 1000°C[4]. On the other hand, the solubility of oxygen in molybdenum (a Group VIA element) at this same temperature is very low[5]. An investigation on the solubility of oxygen in Nb–Mo alloys is therefore of interest, particularly since Nb and Mo are neighbours in the Periodic Table and form an uninterrupted series of bcc solid solutions. According to an investigation by Bryant, *loc. cit.*, the solubility of oxygen in Nb–Mo alloys falls with increasing Mo content and approaches zero at 75 atom per cent molybdenum. Bryant also investigated the solubility at 1000°C of oxygen in bcc alloys of niobium with the Group VIIA and Group VIIIA transition metals rhenium and ruthenium. The solubility of oxygen in niobium is reduced by both metals and reaches a very low value at 35 atom per cent Re and at 25 atom per cent Ru. Assuming that the numbers of valence electrons per atom are 5, 6, 7 and 8 in Nb, Mo, Re and Ru, respectively, the point of effectively zero oxygen solubility is reached at an electron to atom ratio of about 5·7 in all three cases. A comparison of the solubilities is meaningful because the affinity of niobium for oxygen is much greater than that of Mo, Re and Ru; it can therefore be assumed that in all three cases the coexisting oxide phase is virtually pure NbO.

Jones *et al.*[6–8] found that the paramagnetic susceptibility of Nb–Mo

1. A. L. Allred and E. G. Rochow, *J. inorg. nucl. Chem.* **5**, 264 (1958).
2. E. J. Little and M. M. Jones, *J. chem. Educ.* **37**, 231 (1960).
3. L. Pauling, *The Nature of the Chemical Bond*, Cornell University Press, New York (1945).
4. R. T. Bryant, *J. less-common Metals* **4**, 62 (1962).
5. W. E. Few and G. K. Manning, *Trans. AIME* **194**, 271 (1952).
6. D. W. Jones, N. Pessall and A. D. McQuillan, *Phil. Mag.* **6**, 455 (1961).
7. D. W. Jones and A. D. McQuillan, *Physics Chem. Solids* **23**, 1441 (1962).
8. D. W. Jones, *Phil. Mag.* **9**, 709 (1964).

alloys shows a minimum at an electron concentration of approximately 5·7 electrons per atom. The same was found for bcc Ti–Mo alloys. The heat of solution of hydrogen in alloys of Nb–Mo, Nb–Re and Ti–Mo was shown to fall to zero at about 5·6 electrons per atom. In all the binary transition-metal alloys mentioned (Nb–Mo, Nb–Re, Nb–Ru and Ti–Mo) a change in electronic structure apparently occurs at the same critical electron concentration. This is also indicated by measurements of the Hall coefficient of Ti–Mo alloys at room temperature(1). This coefficient is positive throughout and has a sharp maximum at about 5·6 electrons per atom.

From all the facts mentioned, Jones *et al.*, *loc. cit.*, and Bryant, *loc. cit.*, draw the conclusion that the d band in the binary alloys under discussion is probably split into overlapping sub-bands and that the lower sub-band is filled at an electron concentration of about 5·6. Both hydrogen and oxygen appear here as donors of electrons, i.e. the dissolved oxygen appears to be positively charged in these alloys and not negatively as commonly supposed.

Influence of oxygen on the superconducting critical temperature of niobium

Data on the electronic structures of the niobium alloys mentioned can, in principle, also be derived from measurements of the superconducting critical temperature T_c, i.e. the temperature below which the electrical resistivity vanishes. Theory(2) indicates that T_c is closely related to the density of electronic states at the Fermi surface and thus to the electronic specific heat coefficient γ.

The critical temperature of niobium is high, 9·5 K, that of molybdenum low, 0·92 K. Starting from pure niobium the critical temperature of Nb–Mo alloys decreases with increasing molybdenum content, reaching a minimum value of approximately 0·02 K at the electron concentration of 5·6 electrons per atom, i.e. at the same critical concentration as mentioned before. The addition of Re or Ru to niobium also reduces T_c, but at equal atom fractions the effect on T_c increases in the order Mo, Re, Ru(3–6). Plotting T_c as a function of the atom fractions of Mo, Re and Ru gives a different graph for each of the solutes. However, when plotting T_c as a function of the electron to atom ratio in the substitutional solid solutions based on niobium, all measurement results can be represented by one single graph(7). In the

1. N. V. Grum-Grzhimailo and V. G. Gromova, *Zh. neorg. Khim.* **2,** 2426 (1957).
2. J. Bardeen, L. N. Cooper and J. R. Schrieffer, *Phys. Rev.* **108,** 1175 (1957).
3. T. H. Geballe, *Rev. Mod. Phys.* **36,** 134 (1964).
4. R. A. Hein, J. W. Gibson and R. D. Blaugher, *Rev. Mod. Phys.* **36,** 149 (1964).
5. W. DeSorbo, *Phys. Rev.* **130,** 2177 (1963) and **140,** A914 (1965).
6. R. O. Davies (ed.), *Proc. 8th Intern. Conf. Low Temp. Phys.*, Butterworths, London (1963).
7. K. M. Ralls and J. Wulff, *J. less-common Metals* **11,** 127 (1966).

calculation of the electron to atom ratio it was again assumed that Mo, Re and Ru contribute 6, 7 or 8 electrons per solute atom to the niobium-based solid solutions, i.e. 1, 2 or 3 excess electrons per solute atom.

Based on the data on bcc niobium alloys discussed above Ralls and Wulff, *loc. cit.*, constructed six graphs giving T_c as a function of the atom fraction for solutes contributing one to six excess electrons per atom to the solid solution. Points which relate to relatively dilute Nb–H and Nb–O interstitial solutions fall on the graphs corresponding to 1 and 4 excess electrons per solute atom, respectively. From this they draw the conclusion that, in relatively dilute solutions, each hydrogen atom donates one electron and each oxygen atom four electrons to the metal.

Whilst it is interesting to note that Ralls and Wulff, *loc. cit.*, as well as Bryant, *loc. cit.*, arrive at a positive charge for oxygen dissolved in niobium, it should be borne in mind that all conclusions on the state of ionization of interstitial solutes in metals are still very uncertain. Based on his measurements on niobium containing hydrogen, nitrogen or oxygen Wiseman[1] arrived at the conclusion that the reduction in critical temperature of Nb caused by these solutes is determined solely by the lattice expansion of the metal and not by the nature of the solute itself. By plotting the change in critical temperature as a function of the increase in volume of the unit cell of niobium he obtained one single graph and observed no T_c characteristics that are dependent upon the specific element beyond the effect that each has on the distortion of the lattice.

Earlier, Seraphim *et al.*[2] proved that the critical temperature of superconductivity in tantalum also decreases with increasing concentration of hydrogen, nitrogen or oxygen. According to their results the magnitude of the effect depends on the change in residual resistivity and by this standard is virtually independent of the nature of the interstitial impurity.

Electrotransport of interstitial solutes

It has been known for a long time that oxygen, dissolved in a wire of zirconium or titanium migrates towards the anode when a d.c. current is passed through which heats the specimen to a relatively high temperature[3,4]. On the other hand, Rudman[5] found that oxygen dissolved in niobium moves towards the cathode.

In the past it would have been concluded from these results that oxygen

1. C. D. Wiseman, *J. appl. Phys.* **37**, 3599 (1966).
2. D. P. Seraphim, D. T. Novick and J. I. Budnick, *Acta Metall.* **9**, 446 (1961).
3. J. H. de Boer and J. D. Fast, *Recl Trav. chim. Pays-Bas Belg.* **59**, 161 (1940).
4. F. Claisse and H. P. Koenig, *Acta Metall.* **4**, 650 (1956).
5. P. S. Rudman, *Electrotransport Seminar*, 94th Annual AIME Meeting, Chicago, 1965.

carries a negative charge in Ti and Zr but a positive one in Nb. Today it is accepted that no firm conclusions can be drawn from these experiments since the direct influence of the electric field cannot be separated from the influence of the transfer of momentum from the electrons or the holes to the solute atoms. For a summary of the experimental and theoretical aspects of electrotransport in solid metals and alloys we refer the reader to publications by Heumann[(1)], Verhoeven[(2)], Oriani and Gonzalez[(3)] and Pratt[(8)].

To eliminate as far as possible the influence of momentum transfer, Miller and Gurov[(4, 5)] proposed that electrotransport be measured in Hall fields, i.e. in fields perpendicular to the motion of the current carriers. Bibby and Youdelis[(6)] succeeded in measuring Hall-field electrotransport of carbon and nitrogen in α iron using a phase-boundary migration technique. The Hall-field charges were found to be close to the 'metallic' valences of $+4$ and $+5$ for carbon and nitrogen respectively. On the other hand, nitrogen exhibits a strong apparent negative charge in direct-field electrotransport measurements[(7)], suggesting momentum transfer from the conduction electrons to prevail over the influence of the coulombic field.

1. Th. Heumann, *The Physical Chemistry of Metallic Solutions and Intermetallic Compounds, Vol.* 1, *Paper* 2*C*, Her Majesty's Stationery Office, London (1959).
2. J. D. Verhoeven, *Metall. Rev.* **8,** 311 (1963).
3. R. A. Oriani and O. D. Gonzalez, *Trans. AIME* **239**, 1041 (1961).
4. Yu. G. Miller, *Soviet Phys. Solid St.* **3,** 1728 (1962).
5. Yu. G. Miller and K. P. Gurov, *Soviet Phys. Solid St.* **3,** 2096 (1962).
6. M. J. Bibby and W. V. Youdelis, *Can. J. Phys.* **44,** 2363 (1966).
7. M. J. Bibby, L. C. Hutchinson and W. V. Youdelis, *Can. J. Phys.* **44,** 2375 (1966).
8. J. N. Pratt, *Electrotransport in Metals and Alloys*, Trans. Tech. Publications, Riehen, Switzerland (1973).

2

DIFFUSION IN INTERSTITIAL ALLOYS

2.1. Introduction

The diffusion of atoms plays an important part in many processes that can take place in solid metals and alloys. By diffusion we understand here the phenomenon where atoms move through the crystal lattice by jumping from one available site to another. For instance, in iron at room temperature an interstitial nitrogen atom jumps on average once every second from one interstice to another. On the other hand, the jump frequency of atoms that are substitutionally dissolved in a high-melting metal, as for instance the jump frequency of nickel atoms in iron, is negligibly small at room temperature. With rising temperature the rate of diffusion increases rapidly in all metals and alloys.

In the simplest cases the net result of the diffusion is that existing differences in concentration are reduced and finally disappear altogether. Examples of metallurgical processes where diffusion is essential are: (a) surface hardening of certain steels by inward diffusion of nitrogen or carbon (case hardening); (b) formation of alloys from mixtures of metal powders; (c) homogenization of segregated alloys; (d) precipitation hardening; (e) ageing of steel; (f) external and internal oxidation of alloys; (g) permeation of gases through metal walls.

In view of the fact that gases dissolve interstitially in metals, we will restrict ourselves in this book almost entirely to a discussion of diffusion in interstitial alloys. For the time being, however, the difference between interstitial and substitutional diffusion can be disregarded, since we will begin by discussing diffusion without reference to the atomic structure of matter, i.e. by discussing only its macroscopic aspects (Sections 2.2 to 2.6). The atomic processes involved in diffusion will be discussed from Section 2.7 onwards.

2.2. Steady-state diffusion

In the first attempts to treat diffusion problems quantitatively a parallel was drawn with the transport of heat. When the ends of a (say cylindrical) rod are kept at the constant temperatures T_1 and T_2 ($T_1 > T_2$), then, under steady-state conditions, the amount of heat flowing through a transverse cross-section of A m^2 in the time t is given by the equation

$$q = kA\frac{T_1 - T_2}{l}t \qquad (2.2.1)$$

where l is the length of the rod and k a material constant, the *thermal conductivity* of the substance in question. This equation is due to Fourier.

In analogy with equation (2.2.1) Fick, in the middle of the nineteenth century, proposed an equation for unidirectional diffusion, now known as *Fick's first law*. In this equation the flow of heat is replaced by a flow of matter and the temperatures T_1 and T_2 by the concentrations c_1 and c_2 of the diffusing substance ($c_1 > c_2$); this results in the equation

$$m = DA\frac{c_1 - c_2}{l}t = -DA\frac{c_2 - c_1}{l}t \qquad (2.2.2)$$

or, in differential form:

$$dm = -DA\frac{\partial c}{\partial x}dt \qquad (2.2.3)$$

The negative sign indicates that the diffusion takes place in the direction of decreasing concentration. In the two equations above, where c can for instance be expressed in kg/m^3, m or dm represents the quantity of substance diffusing in time t or dt through A m^2 of the transverse cross-section under consideration. Just as $(T_2 - T_1)/l$ is known as the temperature gradient along the rod, $(c_2 - c_1)/l$ or $\partial c/\partial x$ is called the concentration gradient. The situation described by equation (2.2.2) may for instance be visualized by the flow of nitrogen through an iron rod where the constant nitrogen concentrations c_1 and c_2 are maintained at its two ends. The quantity D is called the *diffusion coefficient* or *diffusivity* of nitrogen in iron.

The significance of D becomes clear if we write equation (2.2.2) in the form

$$D = -\frac{m}{A\dfrac{c_2 - c_1}{l}t} \qquad (2.2.4)$$

Thus D indicates the amount of dissolved substance diffusing in unit time through a plane of unit area when the concentration gradient in the direction perpendicular to the plane equals -1. From equation (2.2.4) it follows that D has the dimension [L^{2}T^{-1}]; it is usually given in terms of square millimetres (or square centimetres) per second.

In the form given above, Fick's law is only applicable to diffusion problems where the concentration in the flow region does not change with time, i.e. to cases of steady-state flow. The concentration of the diffusing substance then varies linearly with distance in the flow direction (Fig. 45), unless D

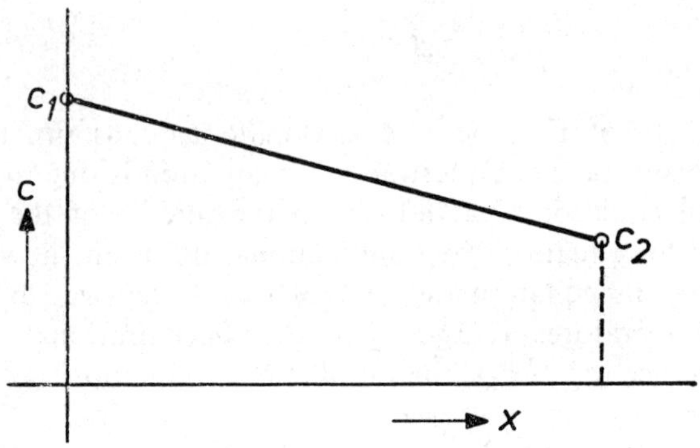

FIG. 45. In steady-state diffusion through a membrane the concentration of the diffusing substance falls linearly with distance if the diffusivity is independent of concentration.

is a function of the concentration. For the present we will regard the diffusion coefficient as a quantity which is independent of concentration, and which thus has a constant value at constant temperature and pressure. Later on we shall see that this is a reasonable approximation only when the differences in concentration over the diffusion region are relatively small.

In the experimental determination of diffusivities, steady-state diffusion measurements can be used only in exceptional cases. In the framework of this book these exceptions are, however, of particular interest since they are those cases where the diffusing solute can be introduced at one side of a membrane, and removed from the other side, as a gas. An example is the diffusion of oxygen through a silver membrane or, in general, the diffusion of a diatomic gas through a metal membrane in which it has a measurable solubility. The determination of the diffusivity of the gas in the metal by this permeation method is only possible at temperatures and membrane thicknesses at which the diffusion of the gas through the metal takes much more time than the entry and exit through the boundary faces. Keeping the gas pressures on the two sides of the membrane at the constant values p_1 and p_2 ($p_1 > p_2$), the concentrations c_1 and c_2 in the surface layers will then virtually be equal to the equilibrium concentrations at the pressures p_1 and p_2 and the temperature T involved. In view of the fact that gases dissolve in metals as atoms, the solubility of diatomic gases at constant temperature is proportional to the square root of the pressure:

$$c = K\sqrt{p} \tag{2.2.5}$$

Therefore, equation (2.2.2) takes the form

$$m = DKA\,\frac{\sqrt{p_1}-\sqrt{p_2}}{d}\,t \qquad (2.2.6)$$

where m is the amount of gas passing through a membrane of thickness d in the time t, and K the solubility of the gas in the metal at $p = 1$. A permeability experiment of this type therefore does not give the desired value of the diffusivity, but the product of diffusivity and solubility. If the solubility of the gas is known as a function of the temperature, then the value of D at various temperatures can be derived from permeability measurements.

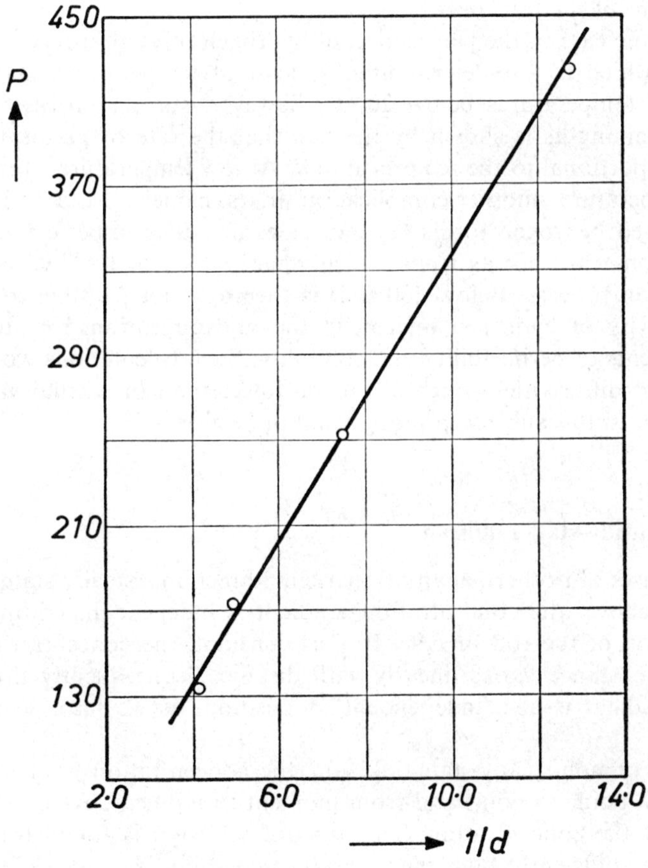

FIG. 46. Rate of permeation P of oxygen (cm^3 per m^2 per hour) through silver discs at 580°C, plotted against the inverse of the thickness. The actual values plotted (indicated by the small circles) are for thicknesses of 79, 135, 205, and 248 μm (Johnson and Larose).

It can be seen from equation (2.2.6) that when the membrane thickness is halved, the rate of permeation will be doubled. Fig. 46 gives the rate of permeation of oxygen through silver membranes at 580°C plotted against the reciprocal of the thickness for thicknesses varying from 80 to 250 μm[1]. Since the graph in Fig. 46 is a straight line the diffusion must determine the rate of permeation in this case, so that equation (2.2.6) is applicable. A different relation between rate of permeation and wall thickness would have been found if the rate had been determined also by a surface process. In the extreme case where one of the surface processes (for instance the dissociation of molecules into atoms at the entry surface) is very much slower than the diffusion in the metal, the rate of permeation will be virtually independent of the thickness.

Also in the case of the permeation of hydrogen through iron walls it is the diffusion which is rate-determining, at least at temperatures above about 300°C. At temperatures below 200°C, however, surface processes become rate-determining as is shown by the fact that the rate of permeation is no longer proportional to the reciprocal of d. At low temperatures, for example room temperature, another complication arises, namely the fact that part of the dissolved hydrogen forms H_2 molecules in lattice imperfections. These imperfections thus act as traps which remove part of the hydrogen from the diffusion process. Below 200°C it is therefore not possible to calculate the diffusivity of hydrogen in iron in the above-mentioned manner from measurements of permeability and solubility. Such calculations would result in apparent diffusivities which are much lower than the actual values. We shall return to this subject in more detail in Section 2.13.

2.3. Non-steady-state diffusion

In the cases of isothermal and isobaric unidirectional steady-state diffusion discussed above, the concentration gradient is independent of time in any cross-section of the rod or disc. If D is constant, the concentration of the diffusing substance varies linearly with distance; consequently the concentration gradient is also independent of position, as we have seen already in Fig. 45.

In cases of non-steady-state diffusion, the concentration gradient usually varies from point to point and from moment to moment, even if D is independent of the concentration. This type of diffusion is found for instance when, at a sufficiently high temperature, two cylinders with different concentrations of the diffusing substance are brought into intimate contact

1. F. M. G. Johnson and P. Larose, *J. Am. chem. Soc.* **46,** 1377 (1924) and **49,** 312 (1927).

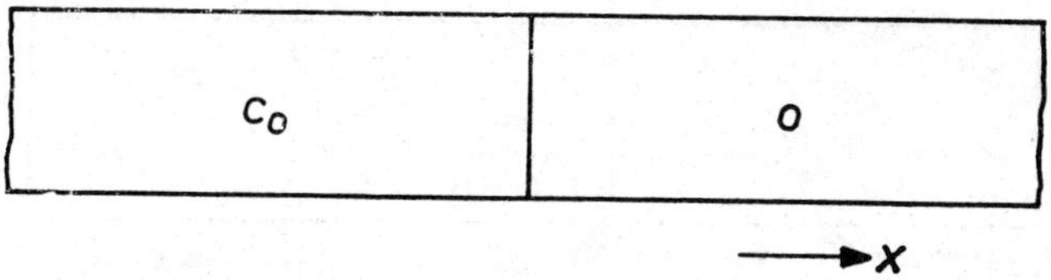

FIG. 47. Welded specimen as is often used in the investigation of diffusion in metals. It consists of two solutions of the diffusing substance, e.g. solutions of nitrogen in iron. At time $t = 0$ the concentrations in the two parts are constant but unequal (for instance c_0 and 0).

with each other, e.g. by electric butt-welding; the situation is sketched in Fig. 47. In the figure the initial concentrations are c_0 and zero and the concentration as a function of distance in the direction of the diffusion thus has a discontinuity at time $t_1 = 0$ (curve 1 in Fig. 48). If the diffusivity is independent of the concentration, then the concentrations at times t_2 and t_3

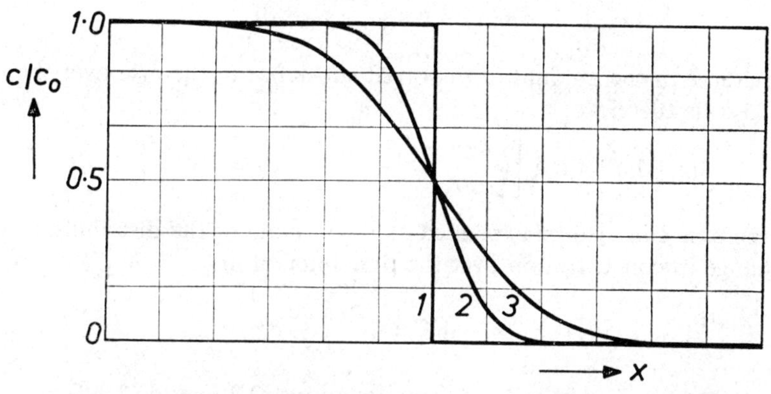

FIG. 48. The concentration as a function of x in the cylinder of Fig. 47 after different time intervals. The discontinuous curve marked 1 corresponds to the initial conditions, curves 2 and 3 represent the situation at the times t_2 and t_3 ($t_3 > t_2$) when the diffusivity is independent of concentration.

($t_3 > t_2$) are given by curves 2 and 3. Unlike the case with steady-state diffusion, the amount of substance passing through a cross-section at right angles to the x axis (the axis of the cylinder) now varies with x.

In order to find the relation between the rate of change of the concentration at x and the change in the concentration gradient in the x direction, let us consider a disc of thickness dx at right angles to the x axis (Fig. 49).

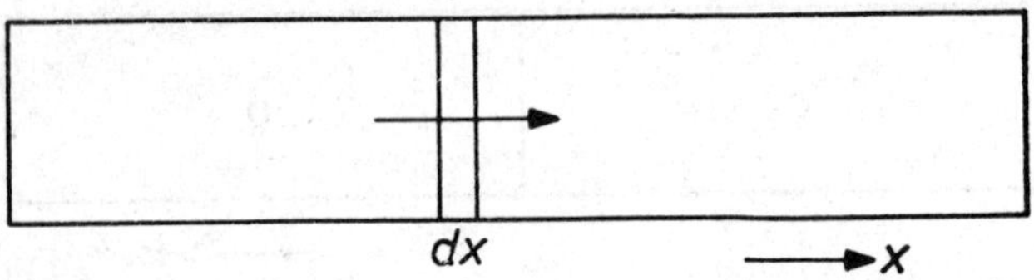

FIG. 49. In non-steady-state diffusion in the x direction the quantity of diffusing substance entering, during a given time interval, at one side of a disc of thickness dx is different from the quantity leaving it at the other end during that same time interval.

The amount of the dissolved substance entering this disc from the left in time dt is given by

$$\mathrm{d}m = -DA\left(\frac{\partial c}{\partial x}\right)_x \mathrm{d}t \tag{2.3.1}$$

Since $\partial c/\partial x$ varies with x, the amount leaving the disc in the same time on the right-hand side will be slightly different:

$$\mathrm{d}m' = -DA\left(\frac{\partial c}{\partial x}\right)_{x+\mathrm{d}x} \mathrm{d}t \tag{2.3.2}$$

The increase in the amount of dissolved substance in the disc over the time interval dt is therefore

$$\mathrm{d}m - \mathrm{d}m' = DA\left[\left(\frac{\partial c}{\partial x}\right)_{x+\mathrm{d}x} - \left(\frac{\partial c}{\partial x}\right)_x\right]\mathrm{d}t = DA\frac{\partial^2 c}{\partial x^2}\,\mathrm{d}x\,\mathrm{d}t \tag{2.3.3}$$

This increase can also be expressed as the volume of the disc multiplied by the change in concentration over the time interval dt:

$$\mathrm{d}m - \mathrm{d}m' = A\,\mathrm{d}x\,\frac{\partial c}{\partial t}\,\mathrm{d}t \tag{2.3.4}$$

The relationship sought follows from equations (2.3.3) and (2.3.4):

$$\frac{\partial c}{\partial t} = D\frac{\partial^2 c}{\partial x^2} \tag{2.3.5}$$

This equation is often referred to as Fick's second law, whilst equation (2.2.3), as mentioned above, represents Fick's first law. They can be seen as two forms of one and the same law, since one can be derived from the other. In Section 2.5 we will see which form equation (2.3.5) assumes when D is dependent on the concentration. For the time being, however, we continue to regard D as independent of concentration.

In order to solve the partial differential equation (2.3.5) it is necessary to know the initial and boundary conditions. If the cylinders depicted in Fig. 47

are so long that the concentrations at their far ends do not change during the diffusion experiment, then the solution of (2.3.5) is given by the equation

$$c = \frac{c_0}{2}\left[1 - \frac{2}{\sqrt{\pi}}\int_0^{\omega} \exp(-y^2)\,dy\right] \tag{2.3.6}$$

where c is a function of x and t. The second term inside the brackets is the Gaussian error function. In this equation ω is the value of the introduced auxiliary variable $y = x/2\sqrt{Dt}$ for certain values of x and t. The temperature is assumed constant resulting in D (for the purpose of this section) having a constant value for a given system.

For each pair of values of D and t, equation (2.3.6) gives the concentration as a function of x, i.e. as a function of the distance from the interface. The concentration curves in Fig. 48 were obtained by means of equation (2.3.6); the concentration at the interface itself remains constant at $c_0/2$. The penetration curve has at all times precisely the same shape; i.e. any one curve at time t_n can be obtained from any other by uniformly stretching or contracting the horizontal axis. One single curve is obtained when c/c_0 is plotted against $x/\sqrt{Dt}$.

The Gaussian error function is often indicated as erf ω. Equation (2.3.6) then reduces to

$$c = \frac{c_0}{2}(1 - \operatorname{erf}\omega) \tag{2.3.7}$$

where

$$\omega = \frac{x}{2\sqrt{Dt}} \tag{2.3.8}$$

The diffusivity D for a particular case is found by determining, after a suitably chosen time t, the concentration c at a certain distance from the interface; equation (2.3.7) then gives the value of erf ω. The corresponding value of the upper limit of the integral, i.e. the corresponding value of ω, can now be looked up in a table (see Table 7). Since x and t are both known, D can then be found from equation (2.3.8). As an illustration, let us consider the case where at a given temperature and after one hundred hours the concentration of the dissolved substance at 1 mm distance from the interface in the right-hand cylinder is 25% of the initial concentration c_0 on the left; thus $c/c_0 = 0{\cdot}25$, $x = 1$ mm, $t = 100 \times 60 \times 60$ s. Substituting c/c_0 in equation (2.3.7) gives a value of 0·50 for erf ω and hence from Table 7 $\omega = 0{\cdot}48$. Substituting this value for ω as well as the values for x and t in equation (2.3.8) gives $D = 3\ (\mu\text{m})^2\ \text{s}^{-1}$.

A less accurate method of determining D by means of two cylinders welded together as shown in Fig. 47 is to observe the 'penetration depth' of the dissolved substance after a time t. Let the arrival of this substance at a point x

TABLE 7

$$\operatorname{erf}\omega = \frac{2}{\sqrt{\pi}}\int_0^{\omega}\exp(-y^2)\,dy$$

ω	erf ω	ω	erf ω	ω	erf ω
0·02	0·02256	0·52	0·53790	1·02	0·85084
0·04	0·04511	0·54	0·55494	1·04	0·85865
0·06	0·06762	0·56	0·57162	1·06	0·86614
0·08	0·09008	0·58	0·58792	1·08	0·87333
0·10	0·11246	0·60	0·60386	1·10	0·88021
0·12	0·13476	0·62	0·61941	1·12	0·88679
0·14	0·15695	0·64	0·63459	1·14	0·89308
0·16	0·17901	0·66	0·64938	1·16	0·89910
0·18	0·20093	0·68	0·66378	1·18	0·90484
0·20	0·22270	0·70	0·67780	1·20	0·91031
0·22	0·24430	0·72	0·69143	1·22	0·91553
0·24	0·26570	0·74	0·70468	1·24	0·92051
0·26	0·28690	0·76	0·71754	1·26	0·92524
0·28	0·30788	0·78	0·73001	1·28	0·92973
0·30	0·32863	0·80	0·74210	1·30	0·93401
0·32	0·34913	0·82	0·75381	1·32	0·93807
0·34	0·36936	0·84	0·76514	1·34	0·94191
0·36	0·38933	0·86	0·77610	1·36	0·94556
0·38	0·40901	0·88	0·78669	1·38	0·94902
0·40	0·42839	0·90	0·79691	1·40	0·95229
0.42	0·44747	0·92	0·80677	1·42	0·95538
0·44	0·46623	0·94	0·81627	1·44	0·95830
0·46	0·48466	0·96	0·82542	1·46	0·96105
0·48	0·50275	0·98	0·83423	1·48	0·96365
0·50	0·52050	1·00	0·84270	1·50	0·96611

of the right-hand cylinder be detected (e.g. by means of a microscope) when $c = 0{\cdot}1c_0$. According to equation (2.3.7) this will occur when erf $\omega = 0{\cdot}8$ and hence from Table 7:

$$\omega = \frac{x}{2\sqrt{Dt}} = 0.9$$

or

$$D \approx \frac{x^2}{3t} \tag{2.3.9}$$

If the arrival of the dissolved substance at point x can only be detected when $c = 0{\cdot}2c_0$, then the denominator in expression (2.3.9) is $1{\cdot}5t$ instead of $3t$. It is often difficult to ascertain which value of concentration corresponds to an observed 'penetration depth' x. In that case a rough estimate of D is obtained by using the equation

$$D \approx \frac{x^2}{2t} \tag{2.3.10}$$

2.4. Measurement of diffusivities of gases in metals

In Section 2.2 we saw that diffusivities of gases in metals can be calculated from measurements of permeability and solubility if diffusion determines the rate of the process of permeation, i.e. if diffusion is much slower than the surface reactions. Diffusivities of gases in metals can be determined in a more direct manner by making use of non-steady-state methods. It is possible for instance to determine the diffusivity of nitrogen in iron by means of the welded cylinder method indicated by Fig. 47. If diffusion in the metal is rate-determining, then the required result can be obtained in a simpler manner by measurement of the rate of absorption or desorption of the gas by or from cylinders or spheres of the metal concerned. When using cylinders these should for preference have a ratio of length to diameter, sufficiently large for diffusion to take place virtually only in a radial direction. Solutions for Fick's equation (2.3.5) or its extended form for two or three dimensions, for these and several other cases are known and can be found in various books on diffusion[(1)]. Below we consider the case for a cylinder[(2)].

A cylinder (e.g. a circular cross-section wire) with a uniform initial concentration c_i of the dissolved gas is assumed to be kept at a constant temperature, while the external pressure of the gas is also held constant. The average concentration $\bar{c}$ of the gas in the metal as a function of time is then given by

$$\frac{\bar{c}-c_e}{c_i-c_e} = \sum_{v=1}^{v=\infty} \frac{4}{\xi_v^2} \exp\left(-\xi_v^2 Dt/r^2\right) \tag{2.4.1}$$

where c_e is the final equilibrium concentration corresponding to the external gas pressure and r the radius of the cylinder. The values of ξ_v are the roots of the equation $J_0(x) = 0$, where $J_0(x)$ is the Bessel function of zero order ($\xi_v = 2{\cdot}405, 5{\cdot}520, 8{\cdot}654, \ldots$).

Equation (2.4.1) applies to both $c_i > c_e$ and $c_i < c_e$, i.e. to both desorption and absorption of the gas. For large values of t the sum in equation (2.4.1) may be replaced by its first term with good approximation

1. See e.g., W. Jost, *Diffusion in Solids, Liquids, Gases*, Academic Press, New York (1952).
2. H. Dünwald and C. Wagner, *Z. phys. Chem.* **B 24,** 53 (1934).

$$\log \frac{\bar{c}-c_e}{c_i-c_e} \approx -0{\cdot}160-0{\cdot}4343\,\frac{t}{t'} \tag{2.4.2}$$

where t' is a constant with the dimension of time:

$$t' = \frac{r^2}{5{\cdot}784D} \tag{2.4.3}$$

Equation (2.4.2) gives satisfactory results for $t > t'/3$ or in other words for

$$\log \frac{\bar{c}-c_e}{c_i-c_e} < -0{\cdot}3 \tag{2.4.4}$$

If the diffusion is indeed rate-determining, then it follows from equations (2.4.2) and (2.4.4) that a straight line will be obtained when plotting values of $\log (\bar{c}-c_e)/(c_i-c_e) < -0{\cdot}3$ against time. This straight line, extrapolated to zero time, should have an intercept of $-0{\cdot}16$ on the axis of ordinates. If, on the other hand, a reaction at the gas–metal interface determines the rate, no straight line satisfying this condition will be obtained. In the first case (diffusion rate-determining), t' is given by the slope of the straight line according to equation (2.4.2) and D is given by (2.4.3).

As an illustration of the determination of the diffusivity of a gas in a metal according to the described method, let us consider that of nitrogen in iron at 500°C. Pure iron wires (diameter 700 μm) were charged with 0·026 wt % nitrogen by heating them for a few hours at 950°C in a gas flow of the composition 99 volume % nitrogen: 1 volume % hydrogen(1). In order to determine the diffusivity, the nitrogen-bearing wires were then heated to 500°C in a flow of hydrogen(2). Each wire was quenched in water and subsequently homogenized after its preselected period of heating. Since in these desorption experiments the final equilibrium concentration c_e is zero, we only need to know the ratio $\bar{c}/c_i$ as a function of time in order to be able to find the value of D with the aid of equations (2.4.2) to (2.4.4). The ratio $\bar{c}/c_i$ was determined by measurement of the internal friction before and after the degassing process (see Chapter 3). This method yields results that are more accurate than those obtained through determining the nitrogen contents by means of chemical analysis, e.g. by the micro-Kjeldahl method.

Fig. 50 shows that the graph $\log \bar{c}/c_i$ against time is a straight line through the point (0, $-0{\cdot}16$), indicating that at 500°C diffusion is rate-determining for the release of nitrogen from iron in an atmosphere of hydrogen. The value for D, derived from Fig. 50 with the aid of equations (2.4.2) and (2.4.3) is $D(\text{N}, 500°\text{C}) = 3{\cdot}8\ (\mu\text{m})^2\ \text{s}^{-1}$.

1. J. D. Fast and M. B. Verrijp, *J. Iron Steel Inst.* **180**, 337 (1955).
2. J. D. Fast and M. B. Verrijp, *J. Iron Steel Inst.* **176**, 24 (1954).

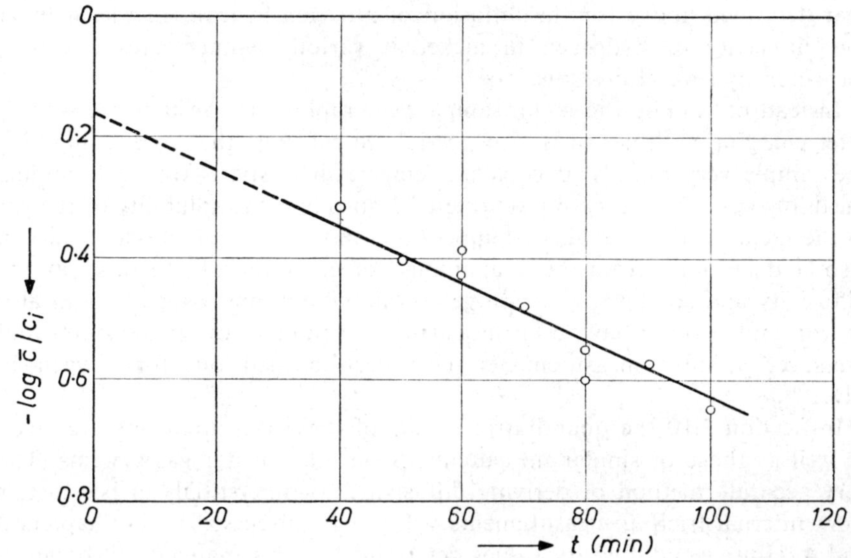

FIG. 50. Release of nitrogen from nitrogen-bearing iron wires in an atmosphere of hydrogen at 500°C. The time is plotted horizontally in minutes, and the logarithm of the ratio of the average nitrogen concentration $\bar{c}$ and the initial concentration c_1 is plotted vertically. Each point relates to an individual wire.

For a sphere the equations corresponding to (2.4.2) and (2.4.3) take the form:

$$\log \frac{\bar{c} - c_e}{c_1 - c_e} \approx -0{\cdot}216 - 0{\cdot}4343 \frac{t}{t'} \tag{2.4.5}$$

and

$$t' = \frac{r^2}{\pi^2 D} \tag{2.4.6}$$

In the example given above, the rate of gas release is derived from measurements of physical properties or from chemical analysis of metal samples. In many cases it is more convenient to determine the gas release in a more direct manner, as for instance by measuring the increase of the pressure in a container of known volume and temperature in which the released gas is collected by means of a pump. Following this method, a cylindrical or spherical metal body is first equilibrated at constant temperature with gas of a certain pressure. After this has been done there are several possibilities for the subsequent steps. Some investigators cool the metal body very rapidly at constant gas pressure. The desorption is then carried out in an evacuated container, at a temperature which is usually different from the charging temperature. The diffusivity of the gas in the metal is derived from the rate of increase in pressure in the collecting vessel in a way similar to that described above for the diffusion of nitrogen in iron. In this manner the diffusivity of hydrogen in nickel at various temperatures has been measured by several investigators[1–3].

Instead of cooling the metal sample very rapidly at constant pressure p_0 after charging with gas, it is also possible to evacuate the vessel containing the sample very rapidly at constant temperature. Apart from determining the diffusivity of the gas from its rate of desorption, the solubility of the gas in the metal at the prevailing temperature and the initial pressure p_0 can also be determined from the total amount of gas given off. In this way the diffusivity and solubility of hydrogen in aluminium and copper[4] and also in iron and silver[5] have been measured at a number of temperatures and pressures. Similar measurements have been carried out for oxygen in silver[6].

In Section 2.10 the quantitative results of these measurements are given as well as those of similar measurements on other metal–gas systems. The very accurate method of deriving diffusivities of interstitials in bcc metals from internal friction measurements will not be discussed until Chapters 3 and 4. However, diffusivity values determined in this manner will be given in Section 2.10 in addition to those determined in other ways.

2.5. Dependence of diffusivity on concentration

Steady-state measurements

So far it has been assumed that the diffusivity is independent of the concentration. This assumption is only justified if the differences in concentration over the diffusion region are small. This condition is satisfied in the examples of diffusion of gases in metals discussed in the previous section. If the concentration differences are fairly large, the dependence of D on the concentration must be taken into account. The following example of steady-state diffusion may illustrate this. Fick's first law for unidirectional diffusion (equation (2.2.3)) can be written in the form

$$J = -D \frac{\partial c}{\partial x} \tag{2.5.1}$$

where J is called the flux, i.e. the mass of diffusing substance passing per unit time through unit area of a plane perpendicular to the direction of the

1. G. Euringer, *Z. Phys.* **96,** 37 (1935).
2. M. L. Hill and E. W. Johnson, *Acta Metall.* **3,** 566 (1955).
3. A. G. Edwards, *Br. J. appl. Phys.* **8,** 406 (1957).
4. W. Eichenauer and A. Pebler, *Z. Metallk.* **48,** 373 (1957).
5. W. Eichenauer, H. Künzig and A. Pebler, *Z. Metallk.* **49,** 220 (1958).
6. W. Eichenauer and G. Müller, *Z. Metallk.* **53,** 321 (1962).

diffusion. In the case of steady-state diffusion through a flat plate, the flux is the same through all planes that are parallel to the surface of the plate, i.e. J is a constant. Equation (2.5.1) shows the diffusion coefficient under these conditions to be proportional to the reciprocal of the concentration gradient:

$$D = -\frac{J}{\partial c/\partial x} \tag{2.5.2}$$

In order to determine the dependence of D on the concentration it is simpler experimentally, to use a hollow cylinder with radial flux. Then the flux under steady-state conditions is a function of the radial distance r, but the mass m of substance diffusing through any cylindrical area of length l in a given time t is independent of r. The flux J through such a cylindrical area is the mass m divided by the area and by the time:

$$J = \frac{m}{2\pi rlt} \tag{2.5.3}$$

Substituting this in (2.5.2) results in

$$D = -\frac{m}{2\pi rlt(\partial c/\partial r)} = -\frac{m}{2\pi lt(\partial c/\partial \ln r)} \tag{2.5.4}$$

Measuring (a) the mass of the diffusing substance passing per unit time through the wall of a given hollow cylinder and (b) its concentration in the wall as a function of r, the diffusivity at different concentrations can be calculated with the aid of equation (2.5.4). According to this equation $\partial c/\partial \ln r$ is constant when D is constant, i.e. plotting c against $\log r$ will then give a straight line. However, when D is dependent on concentration, the slope of this graph will not be a constant.

The diffusivity of carbon in fcc iron (austenite) at a number of temperatures has been determined by R. P. Smith[1] in the manner just described. Carburizing gas was passed through the inside of a cylinder of pure iron. This gas was a mixture of hydrogen and methane, the CH_4 content being such that soot was deposited on the iron. The inner surface of the cylinder was thus maintained at a carbon content close to its value for equilibrium with graphite. The outer surface was kept at a low carbon content by a flow of hydrogen containing water vapour. The quantity of carbon per second (m/t) passing through the cylinder wall under steady-state conditions was determined by chemical analysis of the outlet gas. After the experiments the carbon distribution in the cylinder was determined by machining successive layers, about 1 mm thick, off the wall using a precision lathe and measuring the mean carbon content of each layer by the usual combustion method. Fig. 51 shows a typical graph of the carbon distribution for a cylinder

1. R. P. Smith, *Acta Metall.* **1**, 578 (1953).

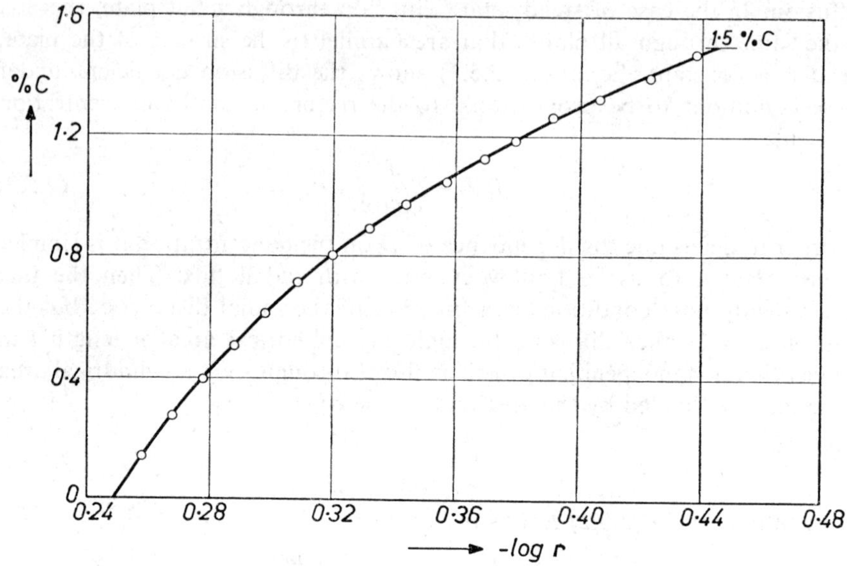

FIG. 51. Steady-state diffusion of carbon at 1000°C through the wall of a hollow iron cylinder. The concentration of carbon in the cylinder in weight per cent is plotted as a function of $-\log r$, where r is the radius in cm. The outer surface was maintained at a carbon content differing little from zero, the inner surface at a carbon content close to the value (about 1·5 wt %) for iron–graphite equilibrium. The data have been taken from Table II of R. P. Smith's article in *Acta Metall.* **1,** 578 (1953).

held at 1000°C during the diffusion experiment. With the aid of equation (2.5.4) and for a number of carbon concentrations, D was calculated from m/t (expressed in grams of carbon per second) and from the slope of the distribution graph after having converted weight per cent carbon to grams per cm^3 at the diffusion temperature.

Since the slope of the distribution curve decreases with increasing carbon content, the diffusivity of carbon in austenite increases with increasing concentration (see equation (2.5.4)). Fig. 52 shows $\log D$ as a function of the carbon content at three temperatures. The figure also gives some results obtained by Wells, Batz and Mehl[(1)] using a non-steady-state method. The determination of the dependence of D on the concentration by means of the latter method is discussed below.

1. C. WELLS, W. BATZ and R. F. MEHL, *Trans. AIME* **188,** 553 (1950).

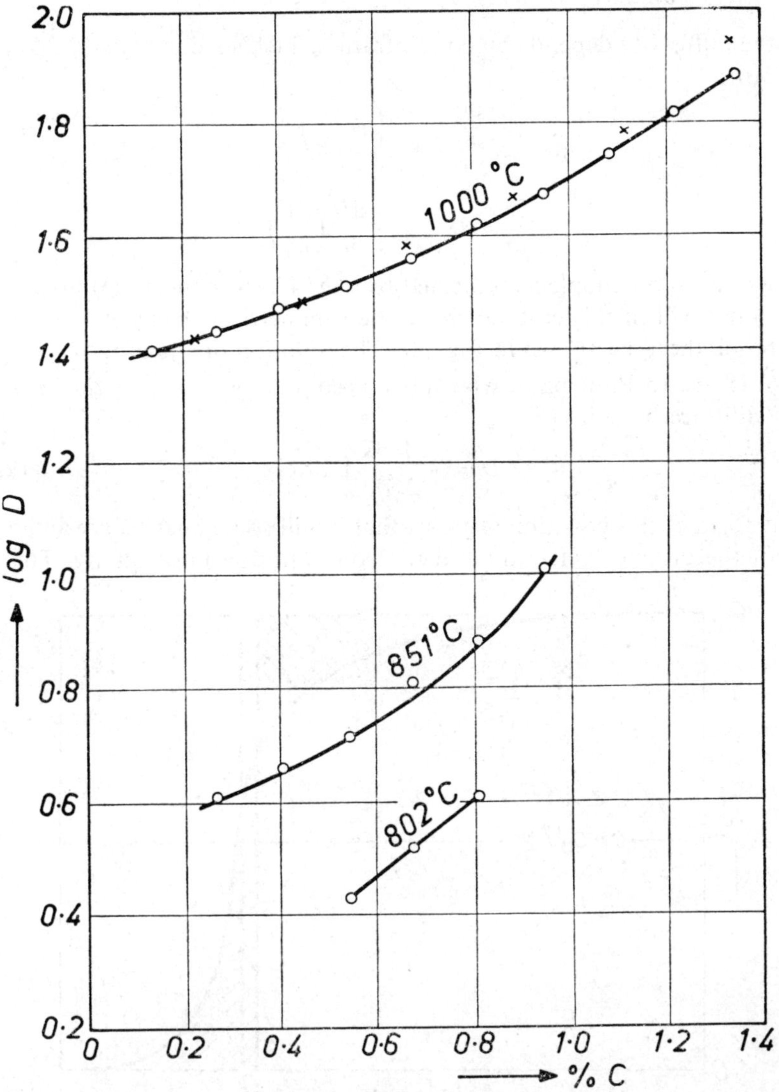

FIG. 52. Dependence of the diffusivity of carbon, in $(\mu m)^2\,s^{-1}$, on its concentration in austenite at three temperatures. Log D is plotted against weight per cent carbon. The circles give the steady-state results taken from Table VI of R. P. Smith's article in *Acta Metall.* **1,** 578 (1953); the crosses represent the non-steady-state results of Wells, Batz and Mehl, taken from Table III of Wells' article in the book *Atom Movements*, American Society for Metals, Cleveland (1951). Wells, Batz and Mehl's results at 851°C and 802°C are not shown in the figure. These are also in good agreement with Smith's results.

Non-steady-state measurements

If the diffusivity depends on concentration, Fick's second law (2.3.5) takes the form:

$$\frac{\partial c}{\partial t} = \frac{\partial}{\partial x}\left(D\,\frac{\partial c}{\partial x}\right) \tag{2.5.5}$$

or

$$\frac{\partial c}{\partial t} = D\,\frac{\partial^2 c}{\partial x^2} + \frac{\mathrm{d}D}{\mathrm{d}c}\left(\frac{\partial c}{\partial x}\right)^2 \tag{2.5.6}$$

Using a diffusion couple as indicated by Fig. 47, the concentration–penetration graph will in this case not have the symmetrical shape of Fig. 48, but will be of the type shown in Fig. 53. The solution of equations (2.5.5) and (2.5.6) is due to Boltzmann who introduced $x/\sqrt{t}$ as a new variable. For a given diffusion time it is:

$$D = -\frac{1}{2t}\frac{\mathrm{d}x}{\mathrm{d}c}\int_0^c x\,\mathrm{d}c \tag{2.5.7}$$

The form of this equation suggests that it will sometimes be advantageous to plot the distance vertically and the concentration horizontally. This has

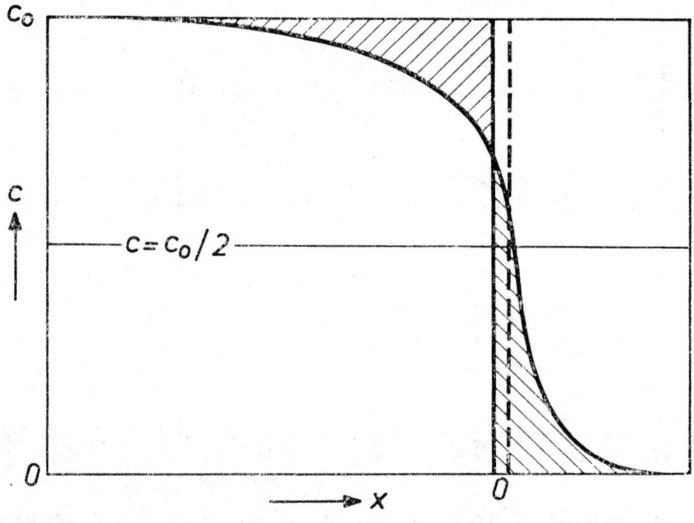

FIG. 53. The concentration in the diffusion couple of Fig. 47 as a function of x after a given time interval t. In constructing this concentration–penetration curve it has been assumed that the diffusivity is not constant at constant temperature and pressure but depends in a given way on the concentration. The broken vertical line corresponds to the original interface; the solid vertical line near to it corresponds to the Matano interface and is defined such that the hatched areas on each side of the line are equal.

been done in Fig. 54 which shows amongst other things that dx/dc tends to infinity as c approaches c_0. According to equation (2.5.7), D should then also tend to infinity unless the value of the integral in the equation is zero for $c = c_0$. In view of the impossibility of D being infinite at any concentration, the value of the integral must be zero for the limits 0 and c_0, which means that the hatched area below the horizontal axis must equal that above it:

$$\int_0^{c_m} x\,\mathrm{d}c = -\int_{c_m}^{c_0} x\,\mathrm{d}c \qquad (2.5.8)$$

In this equation c_m is the concentration in the plane perpendicular to the cylinder axis and at such a place that condition (2.5.8) is satisfied. This plane is called the Matano interface[1]. In general it does not coincide with the original interface. With the concept of the Matano interface it is possible to determine the value of the integral in equation (2.5.7) for any desired concentration. For the concentration corresponding to point A in Fig. 54

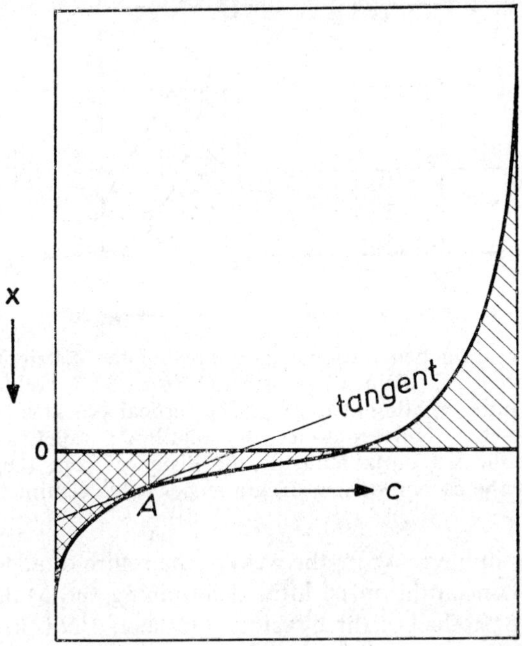

FIG. 54. The curve of Fig. 53 but co-ordinates interchanged.

this value is given by the cross-hatched area. The slope of the tangent at A gives the value of dx/dc. With the aid of equation (2.5.7) the diffusivity is then found as the product of three factors.

1. C. MATANO, *Jap. J. Phys.* **8**, 109 (1933).

As an illustration of the determination of diffusivities by this method, we will discuss one of the experiments carried out by Wells, Batz and Mehl[1] on the diffusion of carbon in austenite. The diffusion couple used in this experiment consisted of an iron cylinder containing 1·38 wt % carbon, welded to a cylinder of almost pure iron. After heating for 24 hours at a temperature of 1000°C consecutive layers, each about 1·3 mm thick and perpendicular to the cylinder axis, were machined off and analysed. The results are plotted on the graph of Fig. 55. The distance corresponding to a point on the graph

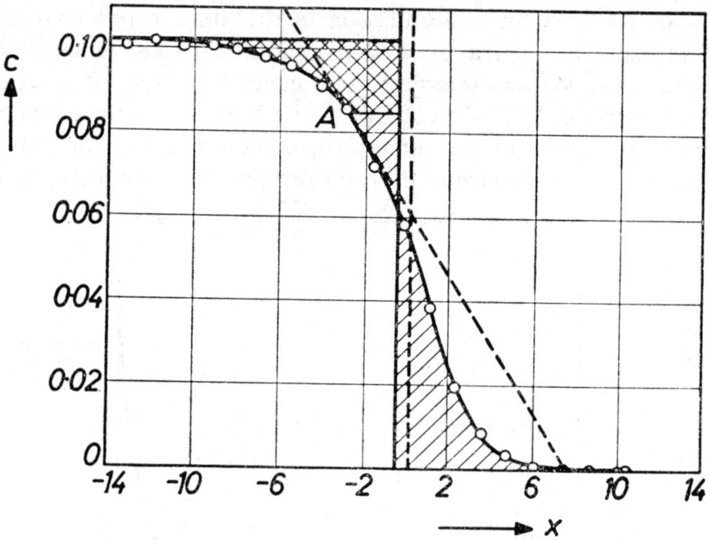

FIG. 55. Concentration–penetration curve showing the distribution of carbon in a couple consisting of a plain carbon steel (1·38 wt % C) welded to relatively pure iron, after 24 hours heating at 1000°C. The vertical axis at $x = 0$ (the broken line) corresponds to the original interface, the solid line parallel to it indicates the Matano interface. Plotted horizontally is the distance from the weld in mm; plotted vertically is the carbon content in grams per cubic centimetre (Wells, Batz and Mehl).

is the number of millimetres from the weld to the centre of a layer. The value of D at a given concentration is, after determining the Matano interface, again found as the product of three factors. Of these, $1/2t$ is known from the duration of the experiment; the two others are given for point A by the slope of the tangent there and by the cross-hatched area. As mentioned above, part of the results obtained by Wells, Batz and Mehl, *loc. cit.*, are given in Fig. 52. Similar experiments could be carried out to study the diffusion of nitrogen in iron.

1. C. Wells, W. Batz and R. F. Mehl, *Trans. AIME* **188,** 553 (1950).

The diffusion in binary *interstitial* alloys, discussed above, is produced by foreign atoms (e.g. C or N) jumping from interstice to interstice, whilst the metal atoms themselves take virtually no part in the diffusion (see Sections 2.7 and 2.9). The distance between the original interface and the Matano interface is in these cases almost completely due to the change in lattice parameter of the alloy with concentration. Diffusion in binary *substitutional* alloys, which in the framework of this book is of lesser importance, is a more complicated process. This diffusion proceeds mainly via unoccupied lattice sites (vacancies), which are present in thermodynamic equilibrium at high temperatures. These vacancies migrate through the lattice, which is possible because they can change places with the atoms A and B. Obviously an atom A or B can only jump in this way if it happens to have a vacancy as neighbour. In general, the probability of an atom A jumping into a neighbouring vacancy is different from that of a B atom, thus causing different diffusivities for A and B.

This explains the occurrence of the so-called Kirkendall effect[1] which was first noticed in the α phase of the copper–zinc system. Rectangular bars of 70–30 brass were electrolytically plated with a thick copper layer. The boundary planes between the two components were marked by means of thin molybdenum wires (the 'markers') in order to be able to detect possible relative displacements of the planes (see Fig. 56). At 785°C a gradual reduction in the distance between the

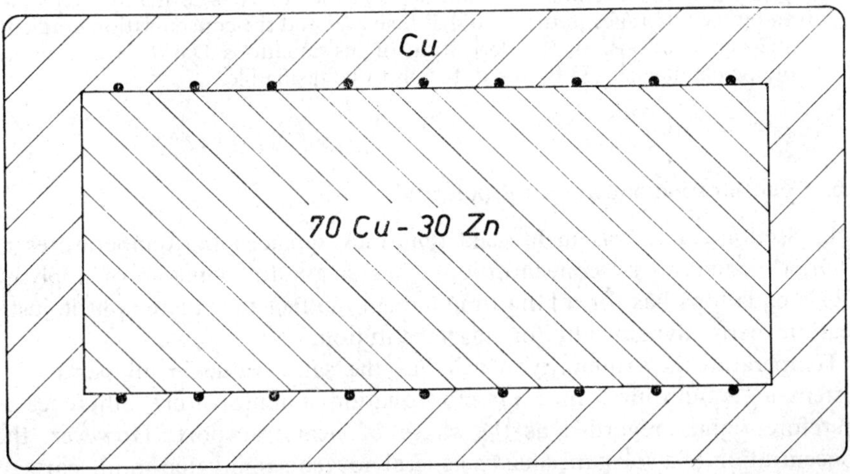

FIG. 56. Diagrammatic sketch indicating a copper–brass couple as used by Smigelskas and Kirkendall in the study of marker movement occurring during diffusion. The black dots indicate the cross-sections of the molybdenum wires.

marked planes took place, i.e. the molybdenum wires moved towards the side rich in zinc. Only 10% of this effect is due to the reduction of the lattice parameter of brass with falling zinc content. The remaining 90% has its cause in the fact that zinc diffuses faster than copper, resulting in more atoms leaving the brass than entering it or, in other words, causing more vacancies to enter the brass than to

1. A. D. SMIGELSKAS and E. O. KIRKENDALL, *Trans. AIME* **171**, 130 (1947).

leave it. In consequence hydrostatic pressures are built up in the diffusion couple, which are relieved by a mass-flow of zinc and copper together. The markers share this movement of the solution *in bulk*, they are as it were sucked into the brass. This wholesale shift does not always occur for the full 100%; in that case the compensation is partly achieved by a precipitation of vacancies to form 'negative crystals', i.e. pores bordered by flat surfaces, in the regions supersaturated with vacancies (the zinc-rich regions in this case).

The Kirkendall effect raises the question as to what exactly we are to understand by diffusion in substitutional alloys. As we saw above, diffusion in a binary diffusion couple can give rise to the shifting or flow of one part of the couple with respect to the other. Even in the case of substitutional alloys the concept of a single diffusion coefficient can still be adhered to and a determination carried out by the Matano method. In that case the diffusion based on the jumps of atoms as well as the bulk motion mentioned above are considered to be parts of the same process. If, on the other hand, it is wished to keep the name of diffusion exclusively for the jumping of atoms then, as was done above, two diffusion coefficients must be distinguished. For an elegant treatment of these problems, we refer the reader to a paper by Darken[1]. It is sufficient to mention here that the relation between the overall diffusivity D and the partial diffusivities D_A and D_B is given by

$$D = \chi D_A + (1 - \chi) D_B \qquad (2.5.9)$$

where χ is the atom fraction of B and $(1 - \chi)$ that of A. The marker velocities are given by the difference in the partial diffusivities and the concentration gradients at the marker locations. In the derivation of his equations Darken assumed the total volume change caused by the diffusion to be negligible.

2.6. Concentration and chemical potential

In Section 2.2 it was mentioned that Fick replaced the temperatures in Fourier's equation by concentrations. The successful outcome of applying Fick's equations has for a long time hidden the fact that there is little justification thermodynamically for this substitution.

Temperature is a quantity which has the same value in all parts of a system at equilibrium. In Fourier's equation a temperature difference is therefore rightly regarded as the cause of heat transport. However, the concentration of a substance has in general not the same value in all parts of an arbitrary system after it has reached a state of equilibrium. In thermodynamics it is known that it is rather the chemical potential μ (partial molar Gibbs free energy) of a dissolved substance, that is the fundamental quantity analogous to temperature. The 'driving force' in an isothermal diffusion process is therefore the negative gradient of the chemical potential of the diffusing substance; the concentration gradient has much less fundamental significance. Diffusion may even cause differences in concentration that did not exist before.

1. L. S. Darken, *Trans. AIME* **175,** 184 (1948).

In the case of diffusion in single-phase binary alloys at constant temperature and pressure, the diffusion laws discussed above can usually be applied, provided the dependence of diffusivity on concentration is taken into account. If a phase consists of more than two components, however, the gradients of concentration and chemical potential can even have opposite signs[1]. A diffusion flow then occurs in the direction of increasing concentration, which is often referred to as uphill diffusion. Interesting examples of this type of diffusion have been studied by Darken[2].

In his experiments Darken used welded specimens consisting of two alloy steels (compare Fig. 47). Two of his specimens were wholly austenitic during the diffusion period of 10 to 14 days at 1050°C. The two parts of the first specimen had roughly equal carbon content (about 0·45 %), but very different silicon contents (3·80 and 0·05 %, respectively). Fick's law would predict a negligible flux of carbon; in actual fact the carbon flux in the diffusion

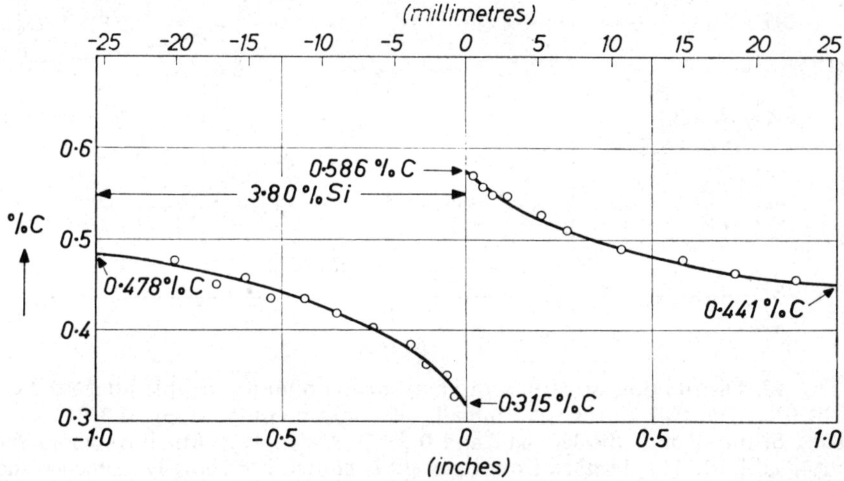

FIG. 57. Distribution of carbon in an austenitic diffusion couple resulting from a 13-day annealing at 1050°C. The carbon and silicon contents were initially 0·478 % C and 3·80 % Si throughout the left half of the couple and 0·441 % C and 0·05 % Si throughout the right-hand part. The distance from the weld is plotted horizontally in inches and millimetres (Darken).

couple was such that the concentration differences after the experiment were much greater than before (Fig. 57). Through the interface carbon diffused from an austenite containing 0·32 % C to an austenite with 0·59 % C. The phenomenon is clearly due to the difference in silicon content of the two

1. W. JOST, *Diffusion und chemische Reaktion in festen Stoffen*, Steinkopff-Verlag, Dresden und Leipzig (1937).
2. L. S. DARKEN, *Trans. AIME* **180**, 430 (1949).

halves and to the much lower diffusivity of silicon compared to that of carbon. The difference in diffusivity of silicon and carbon is so large that it may be assumed, to a first approximation, that only the carbon diffuses.

In the second specimen, uphill diffusion was even more pronounced. In this case it was due to a high silicon content in one half (3·80%) and a high manganese content (6·45%) in the other half, silicon decreasing and manganese increasing the affinity of austenite for carbon. Fig. 58 gives the

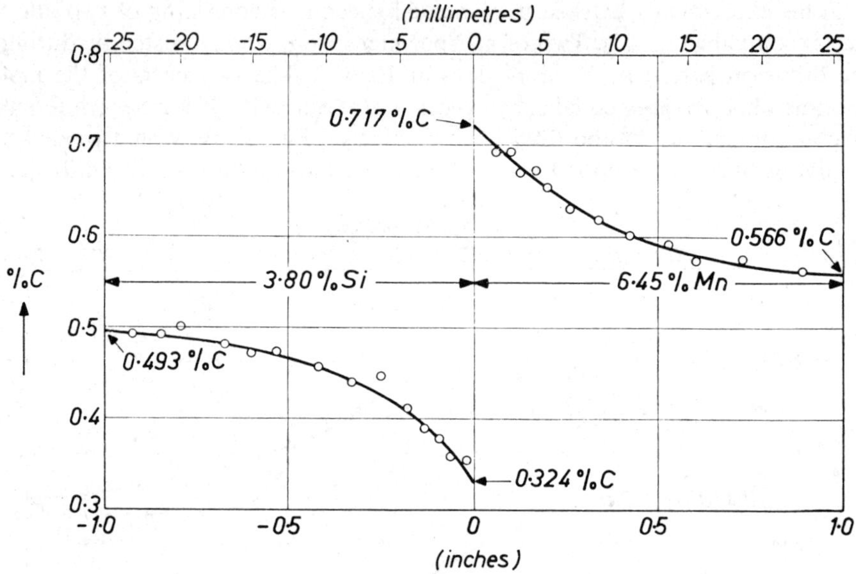

FIG. 58. Distribution of carbon in an austenitic diffusion couple after 10 days at 1050°C. The two halves were initially of uniform composition: 0·493% C + 3·80% Si throughout the left half and 0·566% C + 6·45% Mn throughout the right-hand part. The distance from the weld is plotted horizontally in inches and millimetres (Darken).

results of this experiment and shows that carbon diffused through the interface from an austenite containing 0·32% carbon to an austenite with 0·72% carbon.

The diffusion rate of carbon in austenite is very great at high temperatures. It may therefore be expected that in the region of the joint at 1050°C partial equilibrium will soon be established in the sense that in its immediate vicinity the chemical potential of carbon has the same value on both sides of it. Bearing in mind Figs. 57 and 58 the chemical potential of carbon at 1050°C in austenite containing 3·80% Si + 0·32%C may therefore be expected to have the same value as that in austenite containing 0·05% Si + 0·59% C, or

6·45% Mn + 0·72% C. This has been elegantly confirmed in a thermodynamic investigation by Smith[1], who determined carbon activities in Fe–Si and Fe–Mn alloys by measuring equilibria between these alloys and gas mixtures of hydrogen and methane. Thus, the chemical potential of carbon in Figs. 57 and 58 really is constant through the joint, so that the results can be correctly described by a flux equation of the form

$$J = -M \frac{\partial \mu}{\partial x} \tag{2.6.1}$$

After annealing there is of course no strict discontinuity in the concentration of silicon or manganese at the position of the joint, as there is some slight diffusion of these elements in addition to the large diffusion of carbon. For the same reason the carbon concentration does not change abruptly at the weld, but merely very rapidly in its vicinity. If the specimen were

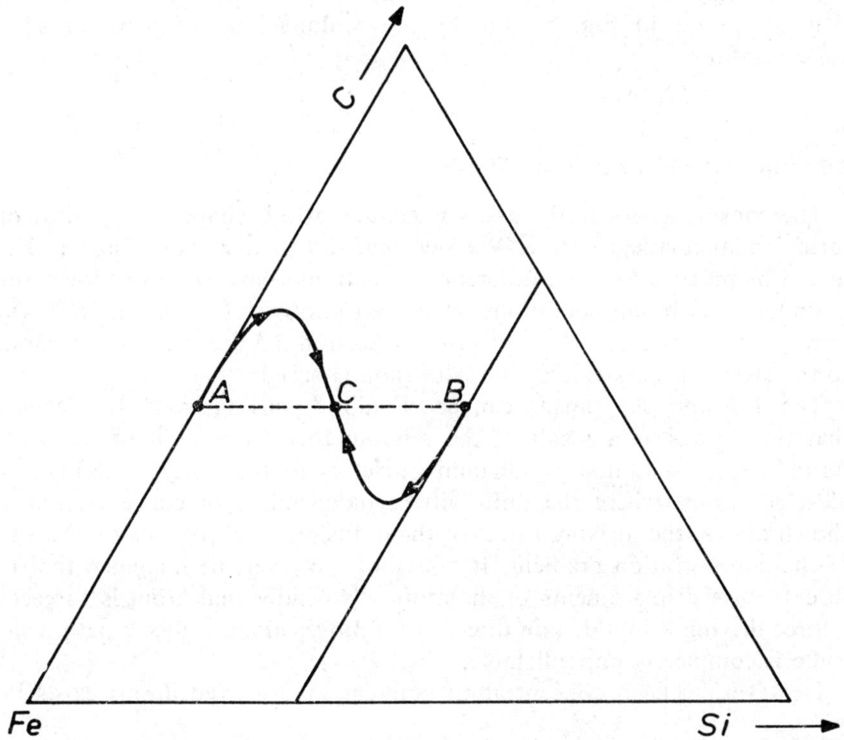

FIG. 59. Diagram showing the change in composition at two points on opposite sides of the weld in the diffusion couple of Fig. 57. Full uniformity of composition (point C) would only be approached after many years of heating at 1050°C (Darken).

1. R. P. SMITH, *J. Am. chem. Soc.* **70**, 2724 (1948).

heated long enough at 1050°C then in the end complete thermodynamic equilibrium would be achieved such that the composition would be the same throughout the whole specimen. It is interesting to consider how the composition at two points at equal distances from, but on opposite sides of the weld would change with time in that case. For the specimen to which Fig. 57 refers this change is shown diagrammatically in Fig. 59 with the aid of a ternary diagram. The points A and B correspond to the initial compositions at the two points. The change in composition with time is shown by the two halves of the continuous line curve. After a sufficiently long time interval, the two halves approach the common point C, corresponding to the final composition of the whole specimen. It will be seen that the path taken by the system to reach the point of lowest free energy departs appreciably from the direct one shown by the broken line. As mentioned before this is caused by the big difference in diffusivities of carbon and silicon. Note in this connection that the points in Fig. 59 initially move along lines of constant silicon concentration.

2.7. The statistical nature of diffusion

The considerations in the first six sections of this chapter are mainly of a formal mathematical nature. We saw that diffusivities can often be determined by means of partial differential equations whose solutions for various boundary and initial conditions have been known for a long time[1]. Only in Section 2.1 and in the small print in Section 2.5 did we refer in passing to the atomic processes on which diffusion is based.

The diffusion phenomena can be visualized starting from the viewpoint that they occur as a result of the random thermal motion of the atoms. As before, we shall first of all limit ourselves to the study of diffusion in ideal solutions, where the diffusivity is independent of concentration. As shown above, the 'driving force' of the diffusion then depends on the value of the concentration gradient. It must not, however, be imagined that real forces cause displacements of the atoms. No individual atom is subject to a force driving it in a certain direction; it moves along a zigzag path, whose route is completely unpredictable.

However, when a concentration gradient of dissolved atoms exists in a

1. Only a few solutions have been discussed here. For a comprehensive collection of solutions the reader should consult books on differential equations, on diffusion, or on the conduction of heat in solids (the last because of the analogy between heat conduction and diffusion processes). We mention here one book in each of the three categories: (a) P. Frank and R. von Mises, *Die Differential- und Integralgleichungen der Mechanik und Physik*, Braunschweig (1935); (b) J. Crank, *Mathematics of Diffusion*, Oxford (1956); (c) H. S. Carslaw and J. C. Jaeger, *Conduction of Heat in Solids*, Oxford (1959).

rod, then the concentration on one side of a cross-section perpendicular to the gradient will be higher than that on the other side, and just because the motion of the individual atoms is completely random there will be more going from the high-concentration side to the low-concentration side than vice versa. Considered macroscopically, there will thus be a resulting flow in the direction of decreasing concentration, notwithstanding the fact that for each separate atom a jump in the direction of higher concentration is just as likely as a jump towards the low-concentration side.

Assuming a simple initial distribution of the dissolved atoms, it is possible by statistical reasoning to predict their distribution over the rod at later moments. The possibility of forecasting the behaviour of a large number of individuals when it is completely impossible to do so for one is quite normal in statistics. For instance, all life insurance is based on the fact that whilst it is impossible to state whether a certain healthy person, of given age, will survive the following ten years, the *percentage* of all healthy people of the same age who will survive ten years can be stated with reasonable accuracy.

The Gaussian error function, already mentioned in Section 2.3, plays an important part in statistical considerations. We shall therefore now discuss this function in some detail. It is easily seen that the function $z = \exp(-y^2)$ has a maximum value ($z = 1$) at $y = 0$. The graphic representation of the function is symmetrical with respect to the z axis because the power of y is even (Fig. 60). The curve falls rapidly on both sides of the z axis, but never

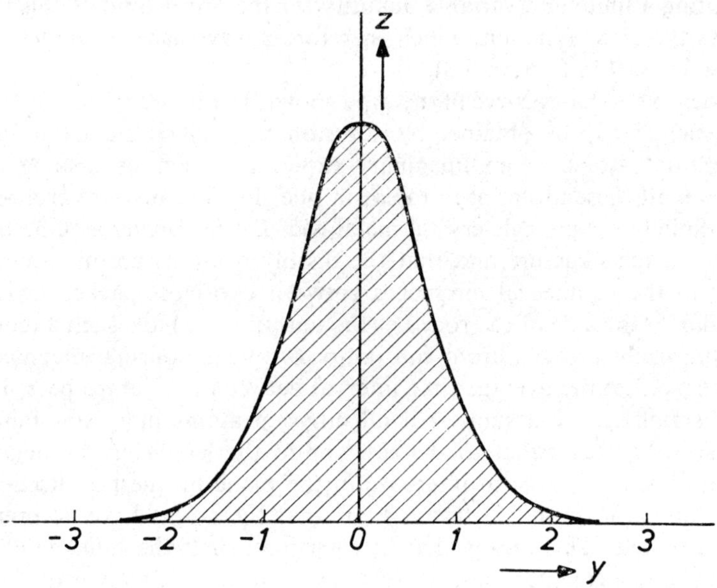

FIG. 60. The function $z = \exp(-y^2)$. The hatched area has the value $\sqrt{\pi}$.

actually reaches the y axis. However, it can be seen from the figure and from Table 8 that even for $y = 3$ the value of the function differs only slightly from zero.

TABLE 8

y	0·0	0·5	1·0	1·5	2·0	2·5	3·0
$\exp(-y^2)$	1·0000	0·7788	0·3679	0·1054	0·0183	0·0019	0·0001

It can be proved that the hatched area below the graph equals $\sqrt{\pi}$:

$$\int_{-\infty}^{+\infty} \exp(-y^2)\,\mathrm{d}y = \sqrt{\pi} \tag{2.7.1}$$

hence for reasons of symmetry:

$$\int_{0}^{\infty} \exp(-y^2)\,\mathrm{d}y = \frac{\sqrt{\pi}}{2} \tag{2.7.2}$$

or

$$\frac{2}{\sqrt{\pi}}\int_{0}^{\infty} \exp(-y^2)\,\mathrm{d}y = 1 \tag{2.7.3}$$

Substituting a finite and variable quantity for the upper limit of this integral produces the error function, which therefore always has a value less than 1 (see also Table 7 in Section 2.3).

In order to see that a curve of the type shown in Fig. 60 represents a distribution such as can be obtained by diffusion, we shall consider the diffusion of interstitial atoms in an imaginary experiment. Let us assume that at time $t = 0$ all these atoms are present in one single, transverse cross-section of an infinitely long, single-crystal metal rod. Let us further assume that the metal has a fcc structure and that the dissolved atoms are nitrogen atoms situated in the octahedral interstices between two close-packed layers perpendicular to the axis of the rod. Let the temperature have such a (constant) value that each nitrogen atom can jump to a neighbouring interstice from time to time. Let the average time interval between two jumps be τ. For the sake of simplicity we assume that all nitrogen atoms jump simultaneously at times τ, 2τ, etc., either to the right or to the left. Thus we neglect the migration from interstice to interstice in the plane in question itself (in the y and z directions), and further assume that the atoms have no preference for right or left. This means that at each jump half the number of atoms will jump one elementary unit of distance λ in the x direction to the right, and the other half the same distance λ to the left, always assuming that the number of atoms involved is so large that statistical deviations from the half can be neglected. For convenience we indicate this large number by $2^8 = 256$, meaning 256 multiplied by a large number.

As indicated in Fig. 61(a), at time $t = 0$ all 256 atoms are in one particular cross-section. After the first collective jump, 128 atoms are at $x = \lambda$ and 128 at $x = -\lambda$. At time $t = 8\tau$, the atoms have spread themselves over the region

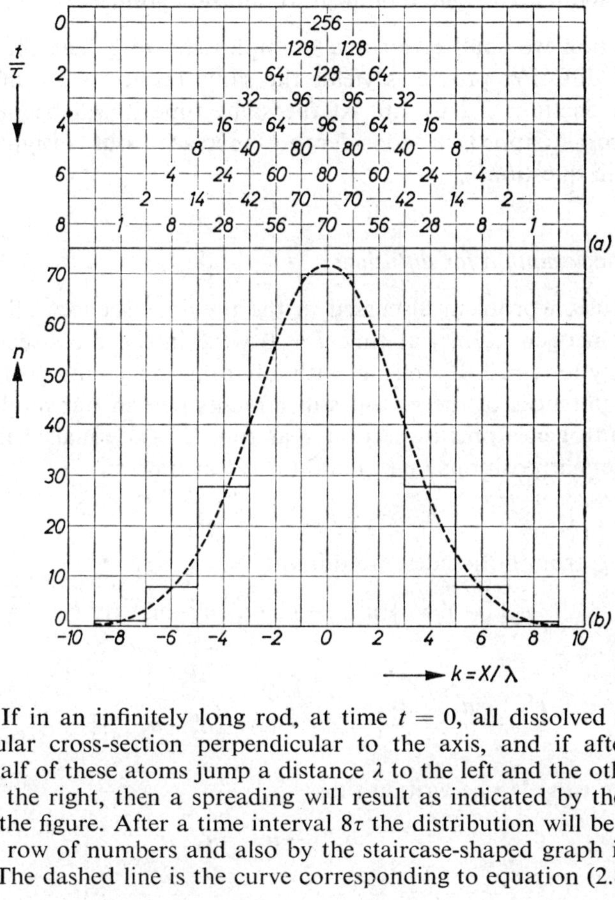

FIG. 61. If in an infinitely long rod, at time $t = 0$, all dissolved atoms are in one particular cross-section perpendicular to the axis, and if after each time interval τ half of these atoms jump a distance λ to the left and the others the same distance to the right, then a spreading will result as indicated by the numbers in part (a) of the figure. After a time interval 8τ the distribution will be as shown by the bottom row of numbers and also by the staircase-shaped graph in part (b) of the figure. The dashed line is the curve corresponding to equation (2.8.7).

from -8λ to $+8\lambda$ as indicated by the bottom row of figures in Fig. 61(a)[(1)]. This last distribution is represented by the histogram in Fig. 61(b). Gradually reducing the distance λ in the figure, i.e. letting λ tend to the real distance

1. The reader should note the relation between the figures in Fig. 61(a) and those of Pascal's triangle, which gives the coefficients of Newton's binomial theorem.

between two neighbouring interstices, and letting the number of jumps increase in the same proportion, the histogram will approach more and more closely to the dashed bell-shaped curve corresponding to the curve in Fig. 60.

2.8. Atomic and macroscopic treatment of diffusion problems

In this section we shall give some examples showing that the statistical treatment of diffusion problems yields the same results as the macroscopic treatment of Sections 2.2 to 2.6. At the same time it will become evident that this atomic approach considerably increases our insight into the phenomena in question.

The differential equation for diffusion

In the diffusion problem discussed in the previous section, all interstitial atoms (e.g. nitrogen atoms) at time $t = 0$ were in the cross-section $x = 0$ of an infinitely long rod. If we start with any other distribution of the atoms among the interstices of the rod, it still remains true in our model that the number of nitrogen atoms in section x at time $(t+\tau)$ equals half the sum of the numbers at positions $(x+\lambda)$ and $(x-\lambda)$ at time t:

$$n_x^{t+\tau} = \tfrac{1}{2}(n_{x+\lambda}^t + n_{x-\lambda}^t) \tag{2.8.1}$$

Subtracting n_x^t from both sides, we obtain:

$$n_x^{t+\tau} - n_x^t = \tfrac{1}{2}(n_{x+\lambda}^t - n_x^t) - \tfrac{1}{2}(n_x^t - n_{x-\lambda}^t) \tag{2.8.2}$$

and dividing by τ:

$$\frac{n_x^{t+\tau} - n_x^t}{\tau} = \frac{1}{2\tau}\left[(n_{x+\lambda}^t - n_x^t) - (n_x^t - n_{x-\lambda}^t)\right]$$

This equation can also be written thus:

$$\frac{n_x^{t+\tau} - n_x^t}{\tau} = \frac{\lambda^2}{2\tau} \frac{\dfrac{n_{x+\lambda}^t - n_x^t}{\lambda} - \dfrac{n_x^t - n_{x-\lambda}^t}{\lambda}}{\lambda} \tag{2.8.3}$$

If we now let τ and λ tend to zero, we are left with

$$\frac{\partial n}{\partial t} = \frac{\lambda^2}{2\tau} \frac{\partial \dfrac{\partial n}{\partial x}}{\partial x} = \frac{\lambda^2}{2\tau} \frac{\partial^2 n}{\partial x^2} \tag{2.8.4}$$

Since n is proportional to the concentration c of the dissolved atoms, (2.8.4) can also be written in the form

$$\frac{\partial c}{\partial t} = \frac{\lambda^2}{2\tau}\frac{\partial^2 c}{\partial x^2} \tag{2.8.5}$$

Strictly speaking, the step from (2.8.3) to the limit (2.8.4) is not justified, since τ and λ always have finite values. However, for practical purposes there is no objection to this extrapolation which, in fact, is an often applied procedure in physics and chemistry. Our idealized atomic model of jumping atoms thus gives the same basic equation (2.3.5) as obtained in Section 2.3 in a formal mathematical manner. An important advance, however, is that the diffusivity D is now expressed in terms of the jump distance λ and the average time τ between two elementary jumps:

$$D = \frac{\lambda^2}{2\tau} \tag{2.8.6}$$

It is instructive to compare this 'microscopic' equation with the 'macroscopic' one (2.3.10) derived earlier.

Comparison of the 'jump method' and the calculation by means of a differential equation

The solution of the differential equation

$$\frac{\partial n}{\partial t} = D\frac{\partial^2 n}{\partial x^2}$$

when at time $t = 0$ all N dissolved atoms are at $x = 0$, is given by

$$n(x, t) = \frac{N}{2\sqrt{\pi D t}}\exp\left(-\frac{x^2}{4Dt}\right) \tag{2.8.7}$$

The correctness of this solution can easily be verified by partial differentiation of (2.8.7), once with respect to t and twice with respect to x. Equation (2.8.7) is also in agreement with the condition $n = 0$ at $t = 0$ for any value of x except for $x = 0$. Finally it is to be noted that the total number of atoms equals N for any value of t. This total is given by

$$n(\text{tot.}) = \int_{-\infty}^{+\infty} \frac{N}{2\sqrt{\pi D t}}\exp\left(-\frac{x^2}{4Dt}\right)\mathrm{d}x$$

or

$$n(\text{tot.}) = \frac{N}{\sqrt{\pi}}\int_{-\infty}^{+\infty}\exp\left[-\left(\frac{x}{2\sqrt{Dt}}\right)^2\right]\mathrm{d}\left(\frac{x}{2\sqrt{Dt}}\right)$$

Hence, using equation (2.7.1):

$$n(\text{tot.}) = N$$

The number of atoms in a slice of thickness dx perpendicular to the rod axis is given by

$$\mathrm{d}n = \frac{N}{2\sqrt{\pi D t}} \exp\left(-\frac{x^2}{4Dt}\right) \mathrm{d}x \tag{2.8.8}$$

We shall use this equation to calculate the distribution of the dissolved atoms after eight jump times in the imaginary experiment of Fig. 61. This is not strictly permissible, since equation (2.8.8) only applies when there is continuity, whilst in the experiment in question the atoms are distributed discontinuously in groups over discs of thickness 2λ. However, notwithstanding this, we substitute the following values in equation (2.8.8): $N = 256$, $D = \lambda^2/2\tau$, $x = k\lambda$, $t = 8\tau$ and $\mathrm{d}x = 2\lambda$. We then obtain for the number of atoms in each slice:

$$\Delta n = \frac{128}{\sqrt{\pi}} \exp\left(-\frac{k^2}{16}\right) \tag{2.8.9}$$

According to Fig. 61 the quantity k in this equation should take the successive values -8, -6, -4, -2, 0, 2, 4, 6, 8. This gives the calculated distribution:

$$1-8-27-56-72-56-27-8-1$$

whereas the jump method (Fig. 61) gave

$$1-8-28-56-70-56-28-8-1$$

Thus even for this very discontinuous case, equation (2.8.8) gives a satisfactory result. The broken line curve in Fig. 61 corresponds to equations (2.8.7) and (2.8.8).

A more complicated case of diffusion

The jump method can also be applied to more complicated cases of diffusion, as for instance that discussed in Section 2.3 where diffusion took place from a long cylinder with an initial concentration c_0 of the dissolved substance, to a long cylinder with zero initial concentration. This problem can be idealized in much the same way as the previous case. In the initial state, each slice of thickness λ to the left of the interface contains an equal number of dissolved atoms. We shall again denote this number by 256. After each time interval τ, we let half the number of foreign atoms jump a distance λ to the right, and the other half an equal distance to the left. At first, as shown by Fig. 62(a), this will only cause the concentration in the immediate vicinity of the interface to change. Gradually, however, the change progresses, and after eight jump times we reach the situation represented by the bottom row of figures in Fig. 62(a) and the histogram of Fig. 62(b).

t/τ																		
0	256	256	256	256	256	256	256	256	256	0	0	0	0	0	0	0	0	0
	256	256	256	256	256	256	256	256	128	128	0	0	0	0	0	0	0	0
2	256	256	256	256	256	256	256	192	192	64	64	0	0	0	0	0	0	0
	256	256	256	256	256	256	224	224	128	128	32	32	0	0	0	0	0	0
4	256	256	256	256	256	240	240	176	176	80	80	16	16	0	0	0	0	0
	256	256	256	256	248	248	208	208	128	128	48	48	8	8	0	0	0	0
6	256	256	256	252	252	228	228	168	168	88	88	28	28	4	4	0	0	0
	256	256	254	254	240	240	198	198	128	128	58	58	16	16	2	2	0	0
8	256	255	255	247	247	219	219	163	163	93	93	37	37	9	9	1	1	0

(a)

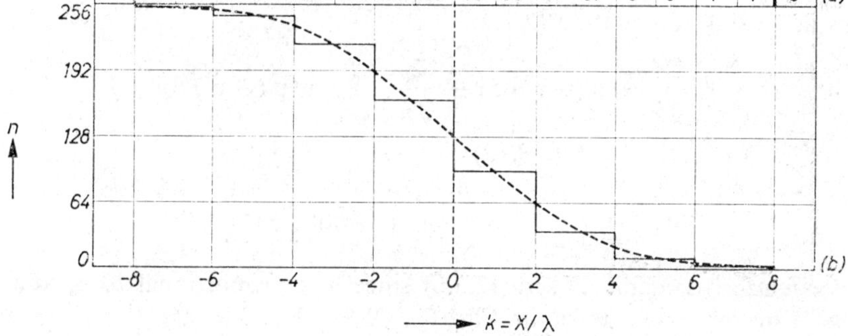

FIG. 62. If, in an infinitely long rod, the distribution of the dissolved atoms at time $t = 0$ is as indicated by the top row of numbers in part (a) of the figure, then jumping as in the previous figure will after time intervals τ lead to distributions as indicated by the subsequent rows of numbers. After a time interval 8τ the distribution will be as shown by the bottom row of numbers and by the staircase-shaped graph in part (b) of the figure. The broken line is the curve corresponding to equation (2.8.13).

Each distribution obtained after an arbitrary number of jump times can be regarded as a superposition of the distributions caused by the separate groups, each of 256 atoms. Using the continuum approach, the distribution after a certain time can be found by integration. First, however, it is necessary to replace equation (2.8.7) by

$$n(x, t) = \frac{N}{2\sqrt{\pi D t}} \exp\left(-\frac{(x-\xi)^2}{4Dt}\right) \tag{2.8.10}$$

if the group of 256 atoms under consideration is, at time $t = 0$, not at the position $x = 0$ but at $x = \xi$. This does not need any further explanation since it will be clear that at any given time the number of atoms n at the point x is determined only by the distance $(x-\xi)$ between x and the source of the particles.

In the problem under discussion here and represented by Figs. 47 and 62 the sources extend from $x = -\infty$ to $x = 0$; the number of atoms n at x, originating from all sources is therefore at time t:

$$n(x, t) = \frac{N}{2\sqrt{\pi D t}} \int_{\xi=-\infty}^{\xi=0} \exp\left(-\frac{(x-\xi)^2}{4Dt}\right) \mathrm{d}\xi \tag{2.8.11}$$

If we introduce the new variable $y = (x-\xi)/2\sqrt{Dt}$, then $\mathrm{d}y = -\mathrm{d}\xi/2\sqrt{Dt}$ and equation (2.8.11) changes into

$$n(x, t) = \frac{N}{\sqrt{\pi}} \int_{x/2\sqrt{Dt}}^{+\infty} \exp(-y^2)\, \mathrm{d}y \tag{2.8.12}$$

Now according to equation (2.7.3)

$$\frac{2}{\sqrt{\pi}} \int_0^{x/2\sqrt{Dt}} \exp(-y^2)\, \mathrm{d}y + \frac{2}{\sqrt{\pi}} \int_{x/2\sqrt{Dt}}^{\infty} \exp(-y^2)\, \mathrm{d}y = 1$$

so that in connection with (2.8.12):

$$n(x, t) = \frac{N}{2}\left[1 - \frac{2}{\sqrt{\pi}} \int_0^{x/2\sqrt{Dt}} \exp(-y^2)\, \mathrm{d}y\right] \tag{2.8.13}$$

This equation is identical with (2.3.6) since c is proportional to n, and c_0 to N. For the case shown in Fig. 62, $N = 256$, $D = \lambda^2/2\tau$, $t = 8\tau$ and $x = k\lambda$, so that

$$x/2\sqrt{Dt} = k/4$$

In this special case equation (2.8.13) therefore takes the form

$$n = 128\left(1 - \operatorname{erf}\frac{k}{4}\right) \tag{2.8.14}$$

With the use of Table 7 and bearing in mind that $\operatorname{erf}\omega = -\operatorname{erf}(-\omega)$, this equation gives the broken line curve in Fig. 62, which agrees very well with the histogram obtained by the discontinuous jump method.

2.9. Interstitial diffusion in fcc and bcc metals

Diffusion in fcc metals

In the cases discussed above the dissolved atoms only jumped in the direction of the x axis, either to the right or to the left. Jumps in other directions were not considered. In order to remove this limitation, let us consider again the interstitial diffusion in a face-centred cubic metal and assume that one of the cube directions of the lattice (the x direction) coincides with the axis of the above-mentioned infinitely long single-crystal cylinder. The dissolved atoms are again assumed to be in the octahedral interstices (the larger ones) of the metal lattice. One such interstice is situated at the centre of the unit cell, while equivalent ones have their centres at the midpoints of the cube edges (Fig. 63). Each interstice is thus symmetrically

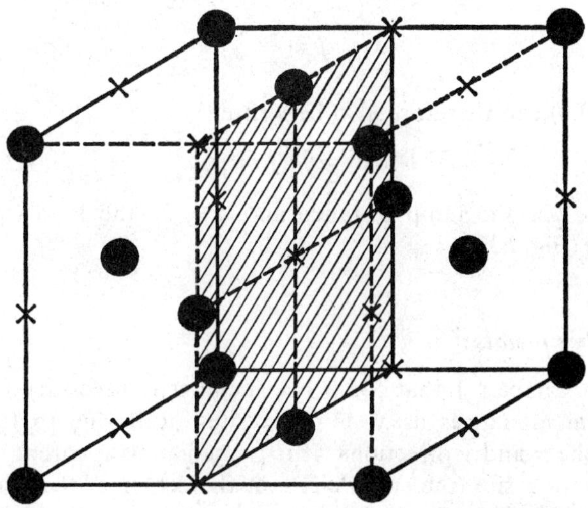

FIG. 63. The centres of the larger interstices in a face-centred cubic stacking of atoms are located at the centre and the midpoints of the edges of the unit cell. In the figure these centres are indicated by crosses. Note that they also form a fcc lattice. For the purpose of the discussions in this section the concentration gradient is supposed to be perpendicular to the hatched plane.

surrounded by twelve similar ones. A jump to four of these leaves the value of x unchanged. When an interstitial atom makes a unit jump, the probability that it moves to the right or to the left is one third each; the probability that the jump leaves x unchanged is also one third.

Instead of equation (2.8.1) we now have:

$$n_x^{t+\tau} = \tfrac{1}{3}n_{x+\lambda}^t + \tfrac{1}{3}n_{x-\lambda}^t + \tfrac{1}{3}n_x^t \tag{2.9.1}$$

or

$$n_x^{t+\tau} - n_x^t = \tfrac{1}{3}(n_{x+\lambda}^t - n_x^t) - \tfrac{1}{3}(n_x^t - n_{x-\lambda}^t) \tag{2.9.2}$$

If we manipulate this equation in the same way as (2.8.2), we find after taking the limit

$$\frac{\partial n}{\partial t} = \frac{\lambda^2}{3\tau}\frac{\partial^2 n}{\partial x^2} \tag{2.9.3}$$

and hence

$$D = \frac{\lambda^2}{3\tau} \tag{2.9.4}$$

Here λ refers to the displacement in the x direction, not to the real jump length, which is $\sqrt{2}$ times greater. The average number of jumps that an atom makes per second, i.e. the frequency of jumping f, is given by

$$f = \frac{1}{\tau} \tag{2.9.5}$$

Equation (2.9.4) can therefore also be written:

$$D(\text{fcc}) = \tfrac{1}{6}l^2 f = \tfrac{1}{12}a^2 f \tag{2.9.6}$$

where $l = \lambda\sqrt{2}$ is the jump distance and $a = 2\lambda$ the lattice parameter of the fcc metal (Fig. 63).

Diffusion in bcc metals

We saw in Chapter 1 that for a bcc metal it is necessary to distinguish the octahedral interstices as x, y and z sites. Referring to Fig. 9 we call FC and FE the x and y directions. It is immediately apparent that an interstitial atom at a z site (the small circle in the centre of the figure) has four possibilities of jumping: it can jump over a distance $\lambda = a/2$ in any one of the directions $+x$, $-x$, $+y$, or $-y$. It cannot jump in the z direction, because on the z axis it is caught between two atoms of the matrix lattice. A similar argument applies to interstitial atoms in the x and y sites, as may be seen from the table below.

Atom location	Jump direction	Jump probability
x sites	$+x$ or $-x$ $+y$ or $-y$ $+z$ or $-z$	$w = 0$ $w = \frac{1}{4}$ $w = \frac{1}{4}$
y sites	$+x$ or $-x$ $+y$ or $-y$ $+z$ or $-z$	$w = \frac{1}{4}$ $w = 0$ $w = \frac{1}{4}$
z sites	$+x$ or $-x$ $+y$ or $-y$ $+z$ or $-z$	$w = \frac{1}{4}$ $w = \frac{1}{4}$ $w = 0$

This survey shows that only two thirds of the total number of interstitial atoms have the possibility of moving in the $+x$ or $-x$ direction. Since the jump probability for both directions is $\frac{1}{4}$, only $\frac{2}{3} \times \frac{1}{4} = \frac{1}{6}$ of all interstitial atoms will move in the $+x$ direction when jumping and equally one sixth in the $-x$ direction. Equation (2.8.1) or (2.9.1) should therefore be replaced by

$$n_x^{t+\tau} = \tfrac{1}{6}n_{x+\lambda}^t + \tfrac{1}{6}n_{x-\lambda}^t + \tfrac{4}{6}n_x^t \tag{2.9.7}$$

It will be easily seen that the diffusivity in this case is given by

$$D(\text{bcc}) = \tfrac{1}{6}l^2 f = \tfrac{1}{24}a^2 f \tag{2.9.8}$$

where $l = \lambda$ is the jump distance, $a = 2\lambda$ the lattice parameter and f the jump frequency.

It was shown above that for fcc as well as for bcc structures the diffusivity is given by $D = l^2 f/6$, while that for one-dimensional diffusion is $D = l^2 f/2$ (Section 2.8). The reason why D is smaller by a factor of three in the former case is simply the outcome of going from one-dimensional to three-dimensional diffusion. We could have saved ourselves the trouble of a separate derivation for bcc crystals by realizing that the equation $D = l^2 f/6$ must be valid for *all* cubic crystals since, taken over a long period of time, the number of jumps of an atom in the positive and negative x directions must be one third of the total number of jumps. It is immaterial in which order the jumps take place.

2.10. The influence of temperature

As would be expected, the jump frequency depends strongly on temperature. At high temperatures f will be high, but it will approach zero with falling temperature. Many experiments have shown that the dependence of diffusivity on temperature, over a large temperature range, can be represented by an equation of the form:

$$D = D_0 \exp(-q/kT) \tag{2.10.1}$$

where k is Boltzmann's constant and T the absolute temperature (see Fig. 64). Usually q in equation (2.10.1) is replaced by $\Delta H = N_0 q$ and k by $R = N_0 k$, where N_0 is Avogadro's number and R the gas constant:

$$D = D_0 \exp(-\Delta H/RT) \tag{2.10.2}$$

For reasons which will become clear below, D_0 is called the frequency factor and ΔH the activation enthalpy. In many cases D_0 and ΔH can be regarded as constants. For the discussion of a more complicated case the reader is referred to Section 2.14.

Equations of the form (2.10.1) are characteristic for processes in which an energy barrier of magnitude q must be overcome by thermal fluctuations. At this point, an appropriate question might be what is the nature of this

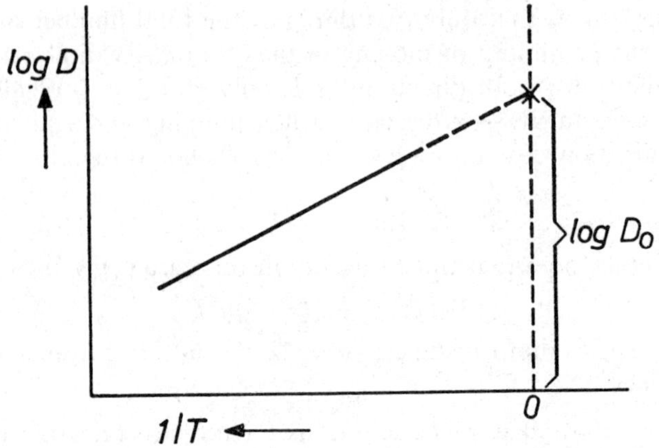

FIG. 64. In agreement with equation (2.10.1) a linear relationship is usually found between the logarithm of the diffusivity and the reciprocal of the absolute temperature. The activation enthalpy of the diffusion can be derived from the slope of the line and D_0 from its intersection with the vertical axis $1/T = 0$.

energy barrier in interstitial diffusion? Since the interstitial sites are equilibrium positions they correspond to potential energy minima which may be regarded as minima in a potential energy surface. They lie at the bottom of troughs, surrounded on all sides by energy barriers. In passing from one trough to another, a diffusing particle is likely to take a route over the lowest saddle in the energy surface. Thus the maximum energy existing at a point midway between any two adjacent equilibrium sites (Fig. 65) lies at the bottom of a saddle point in the potential energy surface.

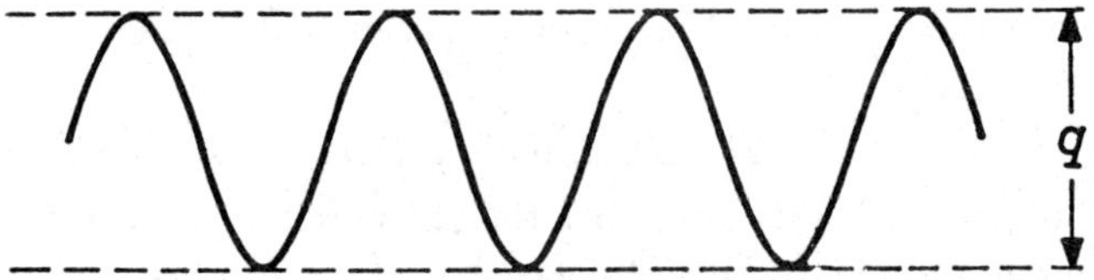

FIG. 65. Potential energy of a crystal as a function of the position of a diffusing interstitial atom. The minima correspond to the equilibrium positions, the maxima to the energy barriers to be overcome by the atom in jumping from one equilibrium position to another.

Between successive jumps from one equilibrium position to another, a diffusing atom executes many oscillations in an interstice. It can jump only when an energy fluctuation of sufficient magnitude q occurs. This energy fluctuation is, in general, not required only to increase the amplitude of oscillation of the diffusing atom. Even though only one atom jumps, the problem is basically a many-body problem because the jumping atom is surrounded by other atoms interacting with it. Two extreme cases can be visualized: in the first, the activation energy is mainly required in the initial site; in the second case it is required mainly in the space between two sites. These two possibilities will now be discussed in more detail.

(a) The activation energy is mainly required to cause the nucleus of the diffusing atom to execute oscillations violent enough to bring it to the point of maximum energy. This might hold for the diffusion of a hydrogen atom in a metal, even in a close-packed metal, but only if it jumps as a proton and is thus not handicapped by the constriction through which it has to pass on its way to a new site.

(b) The activation energy is mainly required to move apart a number of metal atoms between two interstices in order to allow the interstitial atom to squeeze through the narrow intervening space. This could apply to the diffusion of carbon, nitrogen or oxygen in the close-packed fcc metals.

In general, the activation energy q needed for diffusion will be large compared to the average energy of the atoms. The probability that a given atom or a specified small group of atoms possesses a relatively high energy q or an amount of energy in excess of q follows from the Maxwell–Boltzmann distribution law and is given by

$$p = \exp(-q/kT) = \exp(-\Delta H/RT) \qquad (2.10.3)$$

The presence of this factor in equations (2.10.1) and (2.10.2) is thus explained. In Section 2.12 it will be shown that, in general, not only the activation enthalpy should be taken into account but also the *activation entropy* of the diffusion.

Tables 9 and 10 give experimental values of D_0 and ΔH for several cases of interstitial diffusion. The dependence of these quantities on the concentration is only known for the diffusion of carbon in fcc iron. In all other cases the investigators give only one value for D_0 and one for ΔH and often even state explicitly that these values are independent of the concentration.

As shown in Table 9, the activation enthalpies for the diffusion of hydrogen in metals, nearly all lie between 23 and 46 kJ/gram atom (5,500 and 11,000 cal/gram atom), respectively. An exception is the diffusion of hydrogen in bcc iron where ΔH, as determined by various methods and investigators (see Section 2.13), is only about 12·5 kJ (3,000 cal). The values of D_0(H) are nearly all in the range from 0·1 to 1 $mm^2\,s^{-1}$. The only frequency factor far outside this range is D_0(H, Al) which, according to the table, has the value 21 $mm^2\,s^{-1}$. A fresh determination of this value would seem desirable.

Table 10 shows the activation enthalpies for the diffusion of carbon, nitrogen and oxygen in metals to be appreciably greater than those of hydrogen.

TABLE 9

Interstitial diffusion in metals

Diffusion of hydrogen, deuterium and tritium

Interst.	Metal	Crystal struct.	D_0 $mm^2\ s^{-1}$	ΔH kJ/gram atom	Investigators
H	Al	fcc	21	45·6	Eichenauer and Pebler(1)
H	Cu		1·1	38·5	Eichenauer and Pebler(1)
H	Cu		1·1	40·8	Eichenauer, Löser and Witte(2)
D	Cu		0·62	37·8	Eichenauer, Löser and Witte(2)
H	Ag		0·28	31·4	Eichenauer, Künzig and Pebler(3)
H	Au		0·056	23·6	Eichenauer and Liebscher(4)
H	Fe(γ)		1·5	50·2	Sykes, Burton and Gegg(5)
H	Fe(γ)		1·1	41·6	Geller and Sun(6)
H	Ni		0·52	39·9	Ebisuzaki, Kass and O'Keeffe(7)
H	Ni		0·45	36·0	Hill and Johnson(8)
H	Ni		1·1	42·7	Edwards(9)
H	Ni		0·67	39·6	Eichenauer, Löser and Witte(2)
D	Ni		0·48	38·5	Eichenauer, Löser and Witte(2)
H	Pd		0·84	23·9	Jost and Widmann(10)
D	Pd		0·67	23·9	Jost and Widmann(10)
H	Pd		0·43	23·4	Katz and Gulbransen(11)
H	Pd		0·61	25·1	Simons and Flanagan(12)
H	Th		0·29	40·8	Peterson and Westlake(13)
H	Ti(α)	hcp	1·8	51·9	Wasilewski and Kehl(14)
H	Ti(α)		3·0	61·5	Papazoglou and Hepworth(40)
H	Zr(α)		0·070	29·5	Mallett and Albrecht(15)
T	Zr(α)		0·15	37·9	Cupp and Flubacher(16)
H	Ti(β)	bcc	0·20	27·8	Wasilewski and Kehl(14)
H	Zr(β)		0·53	34·8	Gelezunas, Conn and Price(17)
H	Fe(α)		0·076	9·6	Sykes, Burton and Gegg(5)
H	Fe(α)		0·22	12·1	Geller and Sun(6)
H	Fe(α)		0·088	12·8	Stross and Tompkins(18)
H	Fe(α)		0·093	11·3	Eichenauer, Künzig and Pebler(3)
H	Fe(α)		0·14	13·4	Johnson and Hill(19)
H	Fe(α)		0·14	13·7	Wagner and Sizmann(20)
H	Nb		2·1	39·2	Albrecht, Goode and Mallett(21)
H	Nb		0·30	17·4	Wert(47)
H	Ta		0·055	15·1	Wert(47)

TABLE 10

Interstitial diffusion in metals

Interst.	Metal	Crystal struct.	D_0 mm² s⁻¹	ΔH kJ/gram atom	Investigators
			Diffusion of carbon, nitrogen and oxygen		
C	Fe(γ)	fcc	67	156·8	Smith[22]
C	Fe(γ)		45	152·8 (0·1 at % C)	Wells, Batz and Mehl[23]
C	Fe(γ)		5·0	113·0 (6·8 at % C)	Wells, Batz and Mehl[23]
N	Fe(γ)		91	168·4	Grieveson and Turkdogan[24]
O	Fe(γ)		580	168·8	Swisher and Turkdogan[25]
O	Cu		1·7	66·9	Pastorek and Rapp[42]
O	Ag		0·37	46·0	Eichenauer and Müller[26]
N	Th		0·21	94·0	Gerds and Mallett[43]
O	Th		$1{\cdot}3 \times 10^4$	205·0	Peterson[27]
N	Ti(α)	hcp	1·2	189·3	Wasilewski and Kehl[28]
O	Zr(α)		20	171·6	Mallett, Albrecht and Wilson[29]
O	Zr(α)		22	196·6	Béranger and Lacombe[30]
O	Hf(α)		66	212·6	Pemsler[31]
N	Ti(β)	bcc	3·5	141·4	Wasilewski and Kehl[28]
N	Zr(β)		1·5	128·4	Mallett, Belle and Cleland[32]
O	Ti(β)		8·3	130·6	Claisse and Koenig[33]
O	Zr(β)		4·5	118·0	Mallett, Albrecht and Wilson[29]
C	Fe(α)		0·39	80·3 (cf. Section 2.14)	Lord and Beshers[34]
N	Fe(α)		0·49	76·8 (cf. Section 4.9)	Lord and Beshers[34]
O	Fe(α)		3·7	97·5	Swisher and Turkdogan[25]
C	Cr		0·9	110·9	Zemskiy and Spasskiy[46]
N	Cr		0·03	101·7	De Morton[35]
N	Cr		$3{\cdot}5 \times 10^{-4}$	85·8	Klein and Clauer[36]
N	Cr		1·6	115·0	Klein[37]
C	Mo		3·4	171·6	Rudman[38]
N	Mo		0·3	117·2	Evans and Eyre[44]
N	Mo		2·3	138·1	Frauenfelder[41]
N	Mo		0·43	108·8	Jehn and Fromm[45]
N	W		0·24	118·8	Jehn and Fromm[45]
C	V		0·45	114·2	Powers and Doyle[39]
C	Nb		0·40	138·1	Powers and Doyle[39]
C	Ta		0·61	161·1	Powers and Doyle[39]
N	V		0·92	142·5	Powers and Doyle[39]
N	Nb		0·86	146·0	Powers and Doyle[39]
N	Ta		0·56	158·4	Powers and Doyle[39]
O	V		1·3	121·3	Powers and Doyle[39]
O	Nb		2·1	112·6	Powers and Doyle[39]
O	Ta		0·44	106·5	Powers and Doyle[39]

1. W. EICHENAUER and A. PEBLER, *Z. Metallk.* **48,** 373 (1957).
2. W. EICHENAUER, W. LÖSER and H. WITTE, *Z. Metallk.* **56,** 287 (1965).
3. W. EICHENAUER, H. KÜNZIG and A. PEBLER, *Z. Metallk.* **49,** 220 (1958).
4. W. EICHENAUER and D. LIEBSCHER, *Z. Naturf.* **17a,** 355 (1962).
5. C. SYKES, H. H. BURTON and C. C. GEGG, *J. Iron Steel Inst.* **156,** 155 (1947).
6. W. GELLER and T. H. SUN, *Arch. Eisenhütt Wes.* **21,** 423 (1950).
7. Y. EBISUZAKI, W. J. KASS and M. O'KEEFFE, *J. chem. Phys.* **46,** 1378 (1967).
8. M. L. HILL and E. W. JOHNSON, *Acta metall.* **3,** 566 (1955).
9. A. G. EDWARDS, *Br. J. appl. Phys.* **8,** 406 (1957).
10. W. JOST and A. WIDMANN, *Z. phys. Chem.* **(B) 45,** 285 (1940).
11. O. M. KATZ and E. A. GULBRANSEN, *Rev. scient. Instrum.* **31,** 615 (1960).
12. J. W. SIMONS and T. B. FLANAGAN, *J. phys. Chem., Ithaca* **69,** 3581 (1965).
13. D. T. PETERSON and D. G. WESTLAKE, *J. phys. Chem., Ithaca* **64,** 649 (1960).
14. R. J. WASILEWSKI and G. L. KEHL, *Metallurgia* **50,** 225 (1954).
15. M. W. MALLETT and W. M. ALBRECHT, *J. electrochem. Soc.* **104,** 142 (1957).
16. C. R. CUPP and P. FLUBACHER, *J. nucl. Mater.* **6,** 213 (1962).
17. V. L. GELEZUNAS, P. K. CONN and R. H. PRICE, *J. electrochem. Soc.* **110,** 799 (1963).
18. T. M. STROSS and F. C. TOMPKINS, *J. chem. Soc.* 230 (1956).
19. E. W. JOHNSON and M. L. HILL, *Trans. AIME* **218,** 1104 (1960).
20. R. WAGNER and R. SIZMANN, *Z. angew. Phys.* **18,** 193 (1964).
21. W. M. ALBRECHT, W. D. GOODE and M. W. MALLETT, *J. electrochem. Soc.* **106,** 981 (1959).
22. R. P. SMITH, *Trans. AIME* **230,** 476 (1964).
23. C. WELLS, W. BATZ and R. F. MEHL, *Trans. AIME* **188,** 553 (1950).
24. P. GRIEVESON and E. T. TURKDOGAN, *Trans. AIME* **230,** 407 (1964).
25. J. H. SWISHER and E. T. TURKDOGAN, *Trans. AIME* **239,** 426 (1967); see also W. FRANK, H. J. ENGELL and A. SEEGER, *Trans. AIME* **242,** 749 (1968).
26. W. EICHENAUER and G. MÜLLER, *Z. Metallk.* **53,** 321, 700 (1962).
27. D. T. PETERSON, *Trans. AIME* **221,** 924 (1961).
28. R. J. WASILEWSKI and G. L. KEHL, *J. Inst. Metals* **83,** 94 (1954).
29. M. W. MALLETT, W. M. ALBRECHT and P. R. WILSON, *J. electrochem. Soc.* **106,** 181 (1959).
30. G. BÉRANGER and P. LACOMBE, *J. nucl. Mater.* **16,** 190 (1965).
31. J. P. PEMSLER, *J. electrochem. Soc.* **111,** 1185 (1964).
32. M. W. MALLETT, J. BELLE and B. B. CLELAND, *J. electrochem. Soc.* **101,** 1 (1954).
33. F. CLAISSE and H. P. KOENIG, *Acta Metall.* **4,** 650 (1956).
34. A. E. LORD and D. N. BESHERS, *Acta Metall.* **14,** 1659 (1966).
35. M. E. DE MORTON, *J. appl. Phys.* **33,** 2768 (1962).
36. M. J. KLEIN and A. H. CLAUER, *Trans. AIME* **233,** 1771 (1965).
37. M. J. KLEIN, *J. appl. Phys.* **38,** 167 (1967).
38. P. S. RUDMAN, *Trans. AIME* **239,** 1949 (1967).
39. R. W. POWERS and M. V. DOYLE, *Acta Metall.* **4,** 233 (1956); *J. appl. Phys.* **28,** 255 (1957); *J. Metals, N.Y.* **9,** 1287 (1957); *Acta Metall.* **6,** 643 (1958); *J. appl. Phys.* **30,** 514 (1959); *Trans. AIME* **215,** 655 (1959).
40. T. P. PAPAZOGLOU and M. T. HEPWORTH, *Trans. AIME* **242,** 682 (1968).
41. R. FRAUENFELDER, *J. chem. Phys.* **48,** 3966 (1968).
42. R. L. PASTOREK and R. A. RAPP, *Trans. AIME* **245,** 1711 (1969).
43. A. F. GERDS and M. W. MALLETT, *J. electrochem. Soc.* **101,** 175 (1954).
44. J. H. EVANS and B. L. EYRE, *Acta Metall.* **17,** 1109 (1969).
45. H. JEHN and E. FROMM, *J. less-common Metals* **21,** 333 (1970).
46. S. V. ZEMSKIY and M. N. SPASSKIY, *Physics Metals Metallogr., Wash.* **21,** 129 (1966).
47. C. A. WERT, *Physics Chem. Solids* **31,** 1771 (1970).

They are nearly all in the range from 80 to 200 kJ/gram atom (ca. 20,000 to 50,000 cal/gram atom). A striking exception is ΔH(O, Ag) which is only 46 kJ/gram atom (11,000 cal/gram atom). The values of D_0 cover a much larger range than in the case of hydrogen, namely from about 0·1 to about 100 $mm^2 s^{-1}$. Well outside this already large range are D_0(O, Th) according to Peterson, *loc. cit.*, and D_0(N, Cr) according to Klein and Clauer, *loc. cit.*, which are reported to be $1{\cdot}3 \times 10^4$ $mm^2 s^{-1}$ and $3{\cdot}5 \times 10^{-4}$ $mm^2 s^{-1}$, respectively. It should be noted, however, that De Morton, *loc. cit.*, found D_0(N, Cr) to be approximately 100 times greater, namely 3×10^{-2}. A recent redetermination by Klein, *loc. cit.*, gave a value that is much more in accordance with reasonable expectations, namely $D_0(N, Cr) = 1{\cdot}6$ $mm^2 s^{-1}$.

Table 10 also shows that the activation enthalpies for interstitial diffusion in the close-packed fcc and hcp metals are in general greater than those in the less closely packed bcc metals[1]. This can best be compared for metals which exist both in the bcc and in the fcc or hcp modification. The values given for these metals in the literature are shown separately in Table 11.

TABLE 11

Activation enthalpies, ΔH, for the diffusion of hydrogen, carbon, nitrogen and oxygen in the various modifications of iron, titanium and zirconium

	ΔH(H) kJ/g atom	ΔH(C) kJ/g atom	ΔH(N) kJ/g atom	ΔH(O) kJ/g atom
bcc Fe	12·6	80·3	76·8	97·5
fcc Fe	41·8	156·9	168·4	168·8
bcc Ti	27·8		141·4	130·6
hcp Ti	61·5		189·3	
bcc Zr	34·8		128·4	118·0
hcp Zr	29·5			184·1

Also interesting in this connection are recent experiments by Piper[2] on the diffusion of hydrogen in palladium–copper alloys (0–58 atom % copper). At 25°C the diffusivity of hydrogen is relatively insensitive to alloy composition as long as the crystal structure remains unchanged. However, the phase change from face-centred cubic to body-centred cubic causes the diffusivity

1. The atomic radii of Ti and Zr in their bcc modifications are smaller than in the hcp forms. As a result, the close-packed structures of these metals have lower densities than the more open bcc arrangements.

2. J. PIPER, *J. appl. Phys.* **37**, 715 (1966).

at room temperature to increase by two orders of magnitude, to become 5×10^{-3} mm² s⁻¹. This large increase is due to a decrease in activation enthalpy from about 24·3 kJ/gram atom (5,800 cal/gram atom) (Table 9) to about 10·0 kJ/gram atom (2,400 cal/gram atom).

Only the activation enthalpies of hydrogen in the two modifications of zirconium depart from the general trend. In connection with this it should be noted that Cupp and Flubacher (see Table 9) found the activation enthalpy for the diffusion of tritium in hcp zirconium to be 37·9 kJ/gram atom (9,070 cal/gram atom). They are of the opinion that the difference between this value and the value 29·5 kJ/gram atom (7,060 cal/gram atom) found by Mallett and Albrecht, *loc. cit.*, for the diffusion of normal hydrogen in hcp zirconium is real and not due to inaccuracies in the experiments. Since the relatively great difference between these two activation enthalpies certainly cannot depend entirely on the difference in zero-point energy between hydrogen and tritium, Cupp and Flubacher regard it as being due to a tunnelling effect, the lighter hydrogen tunnelling more easily through the potential barrier than tritium. Eichenauer, *et al.*, on the other hand, found (Table 9) a *higher* activation enthalpy for the diffusion of hydrogen than for that of deuterium in both copper and nickel. Table 9 also shows that Jost and Widmann, *loc. cit.*, found no difference in activation enthalpy between hydrogen and deuterium in palladium. Thus as regards the diffusion of the various hydrogen isotopes in metals, the situation is still far from clear. This is not only so for the values of ΔH but also, as will be discussed in the next section, for the values of D_0.

2.11. The value of D_0 for the diffusion of hydrogen

In this section we shall show how values of D_0 can be *calculated* for the diffusion of hydrogen in metals, using very simple assumptions.

When estimating the value of D_0 for substitutional as well as for interstitial diffusion, absolute rate theory[1−7] is often used. An essential stage in this method is the calculation of the concentration of 'activated complexes', i.e. regions containing a diffusing atom midway between two

1. S. Glasstone, K. J. Laidler and H. Eyring, *The Theory of Rate Processes*, McGraw-Hill, New York (1941).
2. C. A. Wert and C. Zener, *Phys. Rev.* **76,** 1169 (1949).
3. C. A. Wert, *Phys. Rev.* **79,** 601 (1950).
4. F. Seitz, *Phase Transformations in Solids* (edited by R. Smoluchowski), John Wiley, New York (1951), Chapter 4.
5. C. Zener, *Imperfections in Nearly Perfect Crystals* (edited by W. Shockley), John Wiley, New York (1952), Chapter 11.
6. G. H. Vineyard, *Physics Chem. Solids* **3,** 121 (1957).
7. O. Manley, *Physics Chem. Solids* **13,** 244 (1960).

equilibrium sites. The method employs statistical mechanics, and is only permissible if the atom stays long enough at the saddle point to reach equilibrium with the surrounding lattice. The extremely fast diffusion of hydrogen compared to that of other interstitial elements (C, N, O) makes it likely that it jumps as a proton and passes the saddle point so quickly that full screening by electrons and full additional readjustment of the surrounding lattice during the jump does not take place. For this reason we shall calculate $D_0(H)$ by a procedure, which, similar to that based on the absolute rate theory, involves an approximation of doubtful accuracy, but which leads to virtually the same result in a much simpler way.

We start from the model in which protons execute oscillations in interstices of the metal lattice, and occasionally jump to an adjacent interstice. Based on the discussion in Section 1.6 it is assumed that each proton behaves as a three-dimensional harmonic oscillator as long as it is in an interstice. Its oscillations are then characterized by a single frequency ν. It can jump to an adjacent site when the displacement from its equilibrium position exceeds a certain value.

For each individual proton, equation (2.10.3) gives the fraction of oscillations with sufficiently large amplitude to allow the move to an adjacent site. We can therefore write for the average number of jumps per second:

$$f = n\nu \exp(-\Delta H/RT) \qquad (2.11.1)$$

where n is the number of adjacent interstices. As we know, this is 12 for a fcc metal (Fig. 63) and 4 for a bcc metal (Fig. 9). The diffusivity is given by

$$D = \alpha a^2 f \qquad (2.11.2)$$

where, according to (2.9.6) and (2.9.8) $\alpha = 1/12$ for a fcc metal and $\alpha = 1/24$ for a bcc metal. Combination of equations (2.11.1) and (2.11.2) results in

$$D = n\alpha a^2 \nu \exp(-\Delta H/RT) \qquad (2.11.3)$$

Hence

$$D_0 = n\alpha a^2 \nu \qquad (2.11.4$$

As an example, let us consider the diffusion of hydrogen in palladium. As mentioned in Section 1.8, the hydrogen atoms in hydrogen-rich palladium (β phase of the system Pd–H) are situated in octahedral interstices of the lattice. It seems a reasonable assumption that this also applies to less concentrated solutions (α phase). From neutron-scattering experiments Bergsma and Goedkoop(1) derived the value $\nu = 1.35 \times 10^{13}\ \text{s}^{-1}$ for the protons in hydrogen-rich palladium. Assuming this same value to apply to dilute solutions, then with $n = 12$, $\alpha = 1/12$, $\nu = 1{\cdot}35 \times 10^{13}\ \text{s}^{-1}$ and $a = 0{\cdot}388$ nm, we find for D_0:

$$D_0(\text{H, Pd}) = 2{\cdot}0\ \text{mm}^2\ \text{s}^{-1}$$

1. J. Bergsma and J. A. Goedkoop, *Physica, 's Grav.* **26**, 744 (1960).

Experimentally the following values for $D_0(\mathrm{H, Pd})$ have been obtained: 0·43, 0·61 and 0·84 $\mathrm{mm^2\ s^{-1}}$ (see Table 9). As far as the order of magnitude is concerned the calculated value is therefore not unreasonable, particularly in view of the approximate nature of the calculation. The more complicated calculation gives with the aid of the absolute rate theory[1] almost the same result, namely $D_0(\mathrm{H, Pd}) = 1{\cdot}8\ \mathrm{mm^2\ s^{-1}}$.

On the basis of the classical theory the vibration frequencies of protons, deuterons and tritons in the interstices of a given metal would be expected to be in the ratio $1 : 1/\sqrt{2} : 1/\sqrt{3}$. It then follows from equation (2.11.4) that the values of D_0 should be in the same ratio. According to Table 9, $D_0(\mathrm{H})/D_0(\mathrm{D})$ for nickel is indeed about $\sqrt{2}$, but for palladium it is 1·3 and for copper 1·8. For zirconium $D_0(\mathrm{H})/D_0(\mathrm{T})$ does not have the expected value of $\sqrt{3}$, but according to the table has a value even less than 1 (about 0·5). Clearly, new and better experiments are needed.

2.12. The entropy of activation for interstitial diffusion

Both the simplified calculation in the previous section and the more elaborate one by Simons and Flanagan, *loc. cit.*, based on absolute rate theory give a value of D_0 for the diffusion of hydrogen in palladium which is higher than that found experimentally. This also applies to the diffusion of hydrogen in nickel[2] and in a number of other metals. On the other hand, values of D_0 calculated for the diffusion of carbon, nitrogen or oxygen in metals are often lower than those found experimentally[3].

These discrepancies can be explained qualitatively by realizing that the move of an interstitial atom from an equilibrium site to a saddle-point position changes not only the enthalpy but also the *entropy* of the system. Statistical thermodynamics indicates that the rate of the process is determined, not by the enthalpy of activation ΔH, but rather by the Gibbs free energy of activation $\Delta G = \Delta H - T\Delta S$. The exponential in equation (2.10.2) should therefore be replaced by

$$\exp(-\Delta G/RT) = \exp(-\Delta H/RT)\exp(\Delta S/R) \qquad (2.12.1)$$

In this equation ΔG is the work done in isothermally and reversibly moving one gram atom of interstitials from equilibrium sites to adjacent saddle points.

1. J. W. Simons and T. B. Flanagan, *J. phys. Chem., Ithaca* **69,** 3581 (1965).
2. M. L. Hill and E. W. Johnson, *Acta Metall.* **3,** 566 (1955).
3. C. A. Wert and C. Zener, *Phys. Rev.* **76,** 1169 (1949).

The above implies that equations (2.10.2) and (2.11.4) should be replaced by

$$D = D_0 \exp(-\Delta H/RT) \qquad (2.12.2)$$

$$D_0 = n\alpha a^2 \nu \exp(\Delta S/R) \qquad (2.12.3)$$

In (2.12.3) ΔS is the sum of a configurational and a vibrational contribution. The former corresponds to the increase in lattice distortion caused by moving one gram atom of solute atoms from equilibrium sites to adjacent saddle points and is always positive(1). With the diffusion of carbon, nitrogen or oxygen the increase in lattice distortion is relatively large and thus the total entropy of activation is positive. Equation (2.11.4) therefore produces values of D_0 that are too small.

On the other hand, with hydrogen diffusion the increase in lattice distortion is likely to be negligible if the jumping particles are protons. The major component of ΔS is then the change in vibrational entropy caused by moving interstitials from equilibrium sites to saddle points. It may be assumed that a proton in the transition state retains two of its three degrees of vibrational freedom; if it jumps in the x direction it still has its vibrational freedom in the y and z directions. Due to the closer proximity of metal atoms the frequency of the two remaining vibrational motions will be substantially higher than that of the oscillations at an equilibrium site. This corresponds to a negative value of ΔS and hence to a value of $\exp(\Delta S/R)$ less than unity. Unlike the case with carbon, nitrogen and oxygen, calculation of D_0 for hydrogen by means of equation (2.11.4) will thus give values that are too large.

2.13. Diffusion of hydrogen in iron

Many investigators have studied the diffusion of hydrogen in bcc iron. Some determined the diffusivity by means of the permeation method using the results of measurements of permeability and solubility (Section 2.2); others used the evolution method, i.e. they measured the rate at which hydrogen escapes from solid iron cylinders or spheres (Section 2.4). Table 9 gives the results obtained by six groups of researchers in this field. The values found by Sykes *et al.*, Geller *et al.* and Wagner *et al.* are derived from permeation measurements at 250°C and at higher temperatures; those obtained by Stross *et al.*, Eichenauer *et al.* and Johnson *et al.* derive from evolution measurements carried out at temperatures higher than 150°C or 200°C. The results show a satisfactory agreement.

1. C. Zener in *Imperfections in Nearly Perfect Crystals* (ed. W. Shockley), John Wiley, New York (1952), p. 289.

When the diffusivity of hydrogen in iron at 20°C is calculated from equation (2.10.2) and the six pairs of values for D_0 and ΔH (Table 9), values between 5×10^{-4} and $1{\cdot}5 \times 10^{-3}$ mm² s⁻¹ are obtained. Table 12 gives the diffusivities at various temperatures calculated with the aid of Geller and Sun's equation *loc. cit.*:

$$D(\text{H}, \alpha\ \text{Fe}) = 0{\cdot}22 \exp(-12{,}100/RT) \quad \text{mm}^2\ \text{s}^{-1} \qquad (2.13.1)$$

For comparison the diffusivities of carbon and nitrogen in iron are also entered in the table. As was earlier mentioned in Chapter 1, the much larger values of D(H) as compared to D(C) and D(N) lead to the assumption that hydrogen jumps as proton.

It must be emphasized that the values of D(H) at 20°C and 100°C in Table 12 are extrapolations using equation (2.13.1). There is a difference of opinion on whether this extrapolation is permissible. Much lower values are found when the diffusivity of hydrogen in iron at 20°C and 100°C is derived

TABLE 12

Diffusivities of hydrogen, carbon and nitrogen in bcc iron

Temp. °C	D(H) mm² s⁻¹	D(C) mm² s⁻¹	D(N) mm² s⁻¹
20	$1{\cdot}5 \times 10^{-3}$	$2{\cdot}0 \times 10^{-15}$	$8{\cdot}8 \times 10^{-15}$
100	$4{\cdot}4 \times 10^{-3}$	$3{\cdot}3 \times 10^{-12}$	$8{\cdot}3 \times 10^{-12}$
200	$1{\cdot}0 \times 10^{-2}$	$1{\cdot}0 \times 10^{-9}$	$1{\cdot}7 \times 10^{-9}$
300	$1{\cdot}7 \times 10^{-2}$	$4{\cdot}3 \times 10^{-8}$	$5{\cdot}3 \times 10^{-8}$
400	$2{\cdot}5 \times 10^{-2}$	$5{\cdot}9 \times 10^{-7}$	$6{\cdot}0 \times 10^{-7}$
500	$3{\cdot}3 \times 10^{-2}$	$4{\cdot}1 \times 10^{-6}$	$3{\cdot}6 \times 10^{-6}$
700	$4{\cdot}9 \times 10^{-2}$	$6{\cdot}1 \times 10^{-5}$	$4{\cdot}4 \times 10^{-5}$
900	$6{\cdot}3 \times 10^{-2}$	$3{\cdot}6 \times 10^{-4}$	$2{\cdot}3 \times 10^{-4}$

from evolution measurements. As an illustration we mention that from this type of measurement Johnson and Hill[1] assumed the diffusivity above 200°C to be represented by

$$D(\text{H}, \alpha\ \text{Fe}) = 0{\cdot}14 \exp(-13{,}400/RT) \quad \text{mm}^2\ \text{s}^{-1} \qquad (2.13.2)$$

and at temperatures below 200°C by

$$D(\text{H}, \alpha\ \text{Fe}) = 12 \exp(-32{,}700/RT) \quad \text{mm}^2\ \text{s}^{-1} \qquad (2.13.3)$$

Fig. 66 represents the results of their measurements graphically. For a temperature of 20°C equation (2.13.3) gives a diffusivity of only $1{\cdot}8 \times 10^{-5}$ mm² s⁻¹. Similar results were obtained by various other investigators, not

1. E. W. JOHNSON and M. L. HILL, *Trans. AIME* **218,** 1104 (1960).

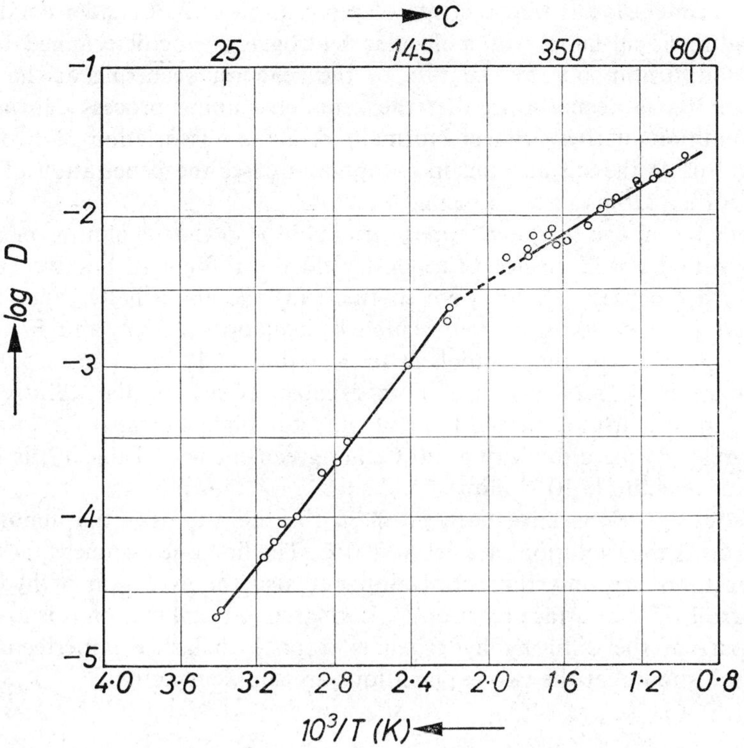

FIG. 66. The logarithm of the diffusivity of hydrogen in iron as a function of the reciprocal of the absolute temperature, according to Johnson and Hill. The text explains why the solid line which is derived from evolution rates below 200°C does not correspond to diffusion rates.

only for pure iron but also for low-alloy steels of different compositions. From the many investigations on the rate of escape of hydrogen from low-alloy steels we only mention those by Hobson(1) and by Coe and Moreton(2).

Still lower values are found when the diffusivity is derived from permeation measurements. Unidirectional supply of molecular hydrogen causes no measurable quantities of the gas to penetrate an iron wall at 20°C, even with a pressure on one side approaching 100 MN m^{-2} and virtually zero on the other side. However, from this it cannot be concluded that the diffusivity of hydrogen in iron at 20°C is negligible since with a unidirectional supply of *atomic* hydrogen, permeation does in general occur readily. The

1. J. D. HOBSON, *J. Iron Steel Inst.* **189,** 315 (1958).
2. F. R. COE and J. MORETON, *J. Iron Steel Inst.* **204,** 366 (1966).

obvious conclusion is that the rate of permeation at 20°C under conditions of unidirectional supply of molecular hydrogen is not determined by the rate of diffusion, but by the rate of the reaction occurring at the entry surface. It has been shown that the rate-determining process depends on the condition of this surface. Normally it is the dissociation of hydrogen into atoms at the surface and in exceptional cases the penetration of these chemisorbed atoms into the metal.

From the above it would appear probable also that evolution measurements below 200°C do not in general yield the diffusivity but the rate of some other process. If this were so then: (a) the abnormally rapid fall in diffusivity below 200°C as represented by equation (2.13.3) and Fig. 66 is only apparent; (b) the symbol D in equation (2.13.3) is not justifiable; (c) equation (2.13.2) will give more reliable values for the diffusivity of hydrogen in α iron than (2.13.3) not only for high, but also for low temperatures; (d) the diffusivity at 20°C, in agreement with Table 12, is of the order of magnitude 10^{-3} mm^2 s^{-1}.

Two explanations have been given in the literature for the abnormally fast drop of the evolution rate below 200°C. The first one assumes that at low temperatures not only the permeation but also the evolution of hydrogen is controlled by a surface reaction[(1)]. In the second explanation it is assumed that part of the diffusing hydrogen is trapped in lattice imperfections[(2)]. We shall show that the two explanations are supplementary.

The role of surface reactions during evolution and permeation of hydrogen

When iron contains dissolved hydrogen, its escape from the metal into an evacuated space or into a space filled with an inert gas requires four successive steps: (a) diffusion of hydrogen atoms or protons in the interior of the metal; (b) transition of these particles from the dissolved state into the state of adsorption at the surface; (c) formation of hydrogen molecules at the exit surface; (d) release of the adsorbed molecules. If escape of the dissolved hydrogen takes place from a specimen stored in water, then, according to Owe Berg[(3)], it is possible that the formation of H_2 in the adsorption layer takes place by means of the reaction

$$H_2O + H \rightarrow OH + H_2$$

after which the adsorbed water molecule is re-formed because the adsorbed hydroxyl radical (not to be confused with the hydroxyl ion) reacts with another hydrogen atom.

1. J. D. Fast, *Philips tech. Rev.* **6,** 365 (1941) and **7,** 74 (1942).
2. L. S. Darken and R. P. Smith, *Corrosion* **5,** 1 (1949).
3. T. G. Owe Berg, *Corrosion* **14,** 562t (1958).

If it is accepted that it is always possible to derive the diffusivity of hydrogen in iron from its rate of escape from the metal, then it follows that environmental and surface conditions never affect this rate noticeably. Many experiments, however, have shown that the escape of hydrogen from steel may be controlled by the rate of a surface reaction rather than by the rate of diffusion of hydrogen from the interior of the steel to the surface. Some experimental results obtained by Hudson *et al.*[(1−3)] are of interest here: (a) hydrogen escapes faster from steel when it is under water than when it is in an atmosphere of dry nitrogen (Fig. 67); (b) the rate of escape of hydrogen from sheets of certain steels is independent of the thickness of the sheets; (c) the evolution rate often depends on the roughness of the steel

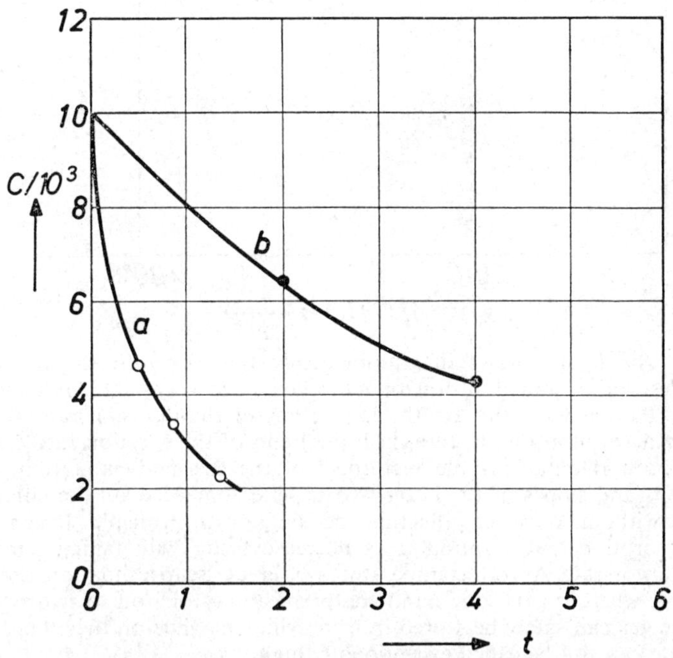

FIG. 67. Rate of escape of hydrogen from cold-drawn capped SAE 1020 steel precharged with the gas. Graph *a* shows the evolution rate at 38°C from samples immersed in water, curve *b* that from samples at the same temperature in an atmosphere of dry nitrogen gas. *C* is the concentration of the hydrogen in mm^3 per 100 grams Fe and *t* is the time in hours (Hudson *et al.*).

1. R. M. HUDSON, W. G. NORRIS and G. L. STRAGAND, *Ind. Engng Chem. int. Edn* **51,** 319 (1959).
2. R. M. HUDSON, K. J. RIEDY and G. L. STRAGAND, *Corrosion* **16,** 115t (1960) and **18,** 79t (1962).
3. R. M. HUDSON and G. L. STRAGAND, *Corrosion* **16,** 253t (1960).

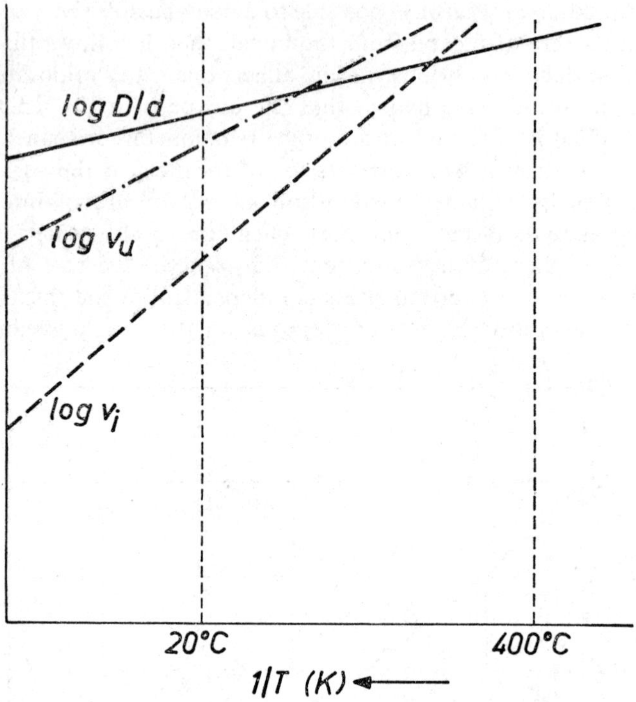

FIG. 68. The figure shows diagrammatically that for iron the processes of sorption, desorption and diffusion of hydrogen have very different temperature coefficients. Plotted vertically are the logarithms of the diffusion rate D/d (solid line), of the desorption rate v_u (dot-dash line) and of the sorption rate v_i (broken line); plotted horizontally is the reciprocal of the thermodynamic temperature. The positions and slopes of the latter two depend somewhat on the condition of the surface without, however, affecting the diagram in principle. It can be seen that the desorption and sorption rates decrease faster with falling temperature than the diffusion rate. At room temperature this gives rise to a situation, unexpected at first sight, where on the one hand the permeability of iron to hydrogen is so low that the gas can safely be stored in iron cylinders while on the other hand the diffusion rate can still be called exceptionally high.

surface. These facts indicate that surface reactions can be rate-determining in the escape of hydrogen from steel. For pure iron Dadian and Talbot-Besnard[1] have systematically studied the influence of surface roughness on the rate of escape of hydrogen. They did this by scoring the surface with well defined scratches at regular intervals.

1. M. DADIAN and S. TALBOT-BESNARD, *C. r. hebd. Séanc. Acad. Sci., Paris* **260,** 1940 (1965) and *Mém. scient. Revue Métall.* **63,** 375 (1966).

From the fact that at temperatures above 150° or 200°C diffusion is rate-determining in the evolution of hydrogen from iron and steel, it follows that the rate of desorption from the surface layer increases more rapidly with the temperature than the rate of diffusion in the interior does. Since at high temperatures diffusion is also rate-determining for the permeation of hydrogen through iron walls, the same argument shows that the rate of sorption of hydrogen in the surface layer, which, at 20°C, is much lower than the rate of desorption, increases even more. Fig. 68 gives a diagrammatic representation of these relations.

The role of traps in the evolution and permeation of hydrogen

Darken and Smith, *loc. cit.*, did not consider slow surface reactions to be the cause of the apparent low diffusivity of hydrogen in steel below 200°C, but attributed it to the presence of 'traps' in the interior of the metal. By traps are meant sites where a hydrogen atom has a lower energy and thus a longer average time of stay than in a normal interstice.

McNabb and Foster[1] developed Darken and Smith's idea further in a formal mathematical theory. Their analysis is based on a model in which the hydrogen atoms are assumed to move in a random manner through the crystal lattice, but capable of being captured and delayed at certain fixed sites (traps), uniformly distributed throughout the metal. They do not commit themselves as to the nature or origin of the traps but make suggestions concerning the laws governing the interchange of hydrogen between diffusing and trapped populations. Experiments in support of their theory have been published by Foster *et al.*[2].

Various concepts have been put forward to account for the traps. Most investigators who have concerned themselves with this problem also visualize the hydrogen in the traps in the form of atoms or protons. In these considerations the traps are regarded as abnormal interstices present at the grain boundaries and around dislocation lines, foreign atoms and inclusions. It has, however, not been possible to prove that there is an interaction between dissolved hydrogen and the various kinds of lattice imperfections in iron or steel at 0°C or higher temperatures. We present arguments below in favour of another view in which the trapped hydrogen is visualized not as atoms or protons in potential troughs of greater depth than those corresponding to normal interstices but as molecules H_2, i.e. as hydrogen that is no longer in solution in the metal.

From many experiments it is known that below 200°C, hydrogen in iron shows a tendency to precipitate in the form of H_2 in all lattice imperfections

1. A. McNabb and P. K. Foster, *Trans. AIME* **227**, 618 (1963).
2. P. K. Foster, A. McNabb and C. M. Payne, *Trans. AIME* **233**, 1022 (1965).

where there is room for hydrogen molecules. This precipitation can take place at the grain boundaries, on the surfaces of inclusions, in internal cavities and in crack nuclei formed by the coalescence of a number of dislocations. In extreme cases this leads to cracks, to blisters on the surface and to a reduction of the ductility or even brittle fracture of the metal. The precipitation of H_2 under discussion is also indicated by solubility values below 200°C that are higher than could be expected from extrapolation of the values found at higher temperatures[1]. In our opinion this exceptionally high 'solubility' at low temperatures only appears because a large part of the hydrogen present in the metal is not actually in solution.

The linearity of the log D versus $1/T$ graph below 200°C (Fig. 66) suggests that the diffusion rate in this temperature range is controlled by a single process with an activation enthalpy of 33 kJ/gram atom (7·8 kcal/gram atom) according to equation (2.13.3). We visualize this process as a re-entering into solution of the precipitated hydrogen. An activation enthalpy of 33 kJ/gram atom is a likely value since (a) the heat of solution of hydrogen in bcc iron is about 27 kJ/gram atom (6·5 kcal/gram atom)[2,3] and (b) the activation enthalpy for the dissolution of H_2 from micro-voids is unlikely to deviate much from the heat of solution.

Diffusible and non-diffusible hydrogen

In the technical literature on hydrogen in steel a distinction is often made between diffusible and non-diffusible or residual hydrogen. This is based on the fact that, during vacuum heating, part of the hydrogen present in the steel escapes below 500°C whereas another part is only released above 800°C. From this it could be concluded that the residual hydrogen is present in the steel as a chemical compound. It has been suggested that the compound is H_2O resulting from the reduction of FeO by dissolved hydrogen. Above 800°C the bonded hydrogen would, according to this suggestion, change into the lattice-dissolved or diffusible form by the reverse reaction

$$H_2O + Fe \rightarrow FeO + 2[H]$$

Hill and Johnson, *loc. cit.*, however, were able to prove that the residual hydrogen is certainly not present in the steel in the form H_2O. They charged iron containing 0·006% carbon and 0·06% oxygen with hydrogen by heating it for a week at 550°C in 6·7 MN m^{-2} H_2. After this treatment the iron contained an appreciable amount of residual hydrogen in addition to an approximately equal amount of 'diffusible' hydrogen. In order to determine

1. See e.g. M. L. HILL and E. W. JOHNSON, *Trans. AIME* **221**, 622 (1961).
2. M. H. ARMBRUSTER, *J. Am. Chem. Soc.* **65**, 1043 (1943).
3. W. GELLER and T. H. SUN, *Arch. Eisenhütt. Wes.* **21**, 423 (1950).

whether H_2O is the source of the residual hydrogen the oxygen content of the metal was increased to 0·12% before charging it with hydrogen. It must be assumed that during this process CO was formed with a considerable reduction of the carbon content. After charging the oxygen-rich metal with hydrogen in the manner described above, it contained very little residual hydrogen.

The experimental results of Hill and Johnson make it likely that non-diffusible or residual hydrogen is present as CH_4 in micro-voids in the steel. Various investigators[1-3] have conclusively proved that CH_4 can be formed in carbon-containing iron and steel by means of the reaction

$$Fe_3C+4[H] \rightarrow CH_4+3Fe$$

The effect of stress on the diffusivity of hydrogen in steel

It is well known that hydrogen can be the cause of delayed brittle fracture of some steels with high tensile strength (e.g. AISI 4340). This hydrogen-induced failure begins to occur at relatively low applied stresses. In an attempt to explain this phenomenon Troiano and co-workers[4,5] suggest that the diffusivity of hydrogen in steel is strongly affected by the presence of stresses. In their opinion, the hydrogen stays in the dissolved state but segregation takes place because the hydrogen atoms are attracted by highly stressed regions within the metal. However, the authors offer no experimental evidence of the effect of stress on the diffusion of hydrogen in steel. Such an effect can hardly be expected if it is assumed that hydrogen jumps as proton (cf. Table 12).

Beck *et al.*[6] were able to show that the diffusivity of hydrogen in steel at 25°C is not changed by stress. As far as we know, this is the first published investigation of the effect of stress on the diffusivity of hydrogen in steel. It supports our explanation[7] for the occurrence of delayed brittle fracture. According to this explanation the cause should not be sought in stress-induced diffusion of dissolved hydrogen to preferred areas, but in its precipitation as molecular hydrogen in crack nuclei formed by the accumulation and coalescence of dislocations. The high hydrogen pressures in the

1. I. Class, *Stahl Eisen* **80,** 1117 (1960).
2. L. C. Weiner, *Corrosion* **17,** 109 (1961).
3. H. H. Podgurski, *Trans. AIME* **221,** 389 (1961).
4. J. G. Morlet, H. H. Johnson and A. R. Troiano, *J. Iron Steel Inst.* **189,** 37 (1958) and *Trans. AIME* **212,** 528 (1958).
5. E. A. Steigerwald, F. W. Schaller and A. R. Troiano, *Trans. AIME* **215,** 1048 (1959) and **218,** 832 (1960).
6. W. Beck, J. O'M. Bockris, J. McBreen and L. Nanis, *Proc. R. Soc.* **A 290,** 220 (1966).
7. J. D. Fast and D. J. van Ooijen, *Philips tech. Rev.* **24,** 252 (1962/3).

crack nuclei resulting from this may cause them to grow into macroscopic cracks.

2.14. Diffusion of carbon in iron

As indicated by Table 12, the value of the diffusivity of hydrogen in bcc iron at 900°C is only about forty times that at 20°C. For the diffusivity of carbon in bcc iron this ratio is not 40 but around 10^{11}. To measure such a large range of diffusivities it is necessary to employ several experimental methods. It has even been possible to extend the diffusion measurements of carbon in iron down to −40°C. At temperatures around 0°C and less, Wert(1) used the elastic after-effect. At higher temperatures, up to 125°C, he determined diffusivities by the internal friction method. Both these methods will be discussed in following chapters. At still higher temperatures, up to about 200°C, Wert calculated diffusivities from precipitation experiments. Before Wert did his experiments, Stanley(2) had determined diffusivities of carbon in iron at temperatures above 500°C by the 'classical' method, i.e. by chemical analysis following bulk diffusion from an iron cylinder containing carbon to another cylinder of almost pure iron (cf. Section 2.3).

Wert showed that the data available in 1950 could with good approximation be represented by a straight line when plotting log D versus $1/T$ (Fig. 69). This led him to the conclusion that the activation enthalpy (ΔH) has a constant value, i.e. 84·1 kJ gram atom^{-1} (20,100 cal/gram atom), over the whole range of temperatures from −35° to 800°C. Later studies, however, proved the relation to be somewhat more complicated than originally visualized.

These complications do *not* occur in the region of low temperatures. Many measurements(3−7) of anelastic and magnetic effects in the temperature range −40° to 350°C confirm that the diffusivities of carbon in bcc iron lie on a straight line when plotted as log D versus $1/T$. From all available data (including those of Wert) Lord and Beshers, *loc. cit.*, using the principle of least squares, derived the following equation for this straight line:

$$D = 0{\cdot}394 \exp(-80{,}150/RT) \quad \text{mm}^2\ \text{s}^{-1} \tag{2.14.1}$$

1. C. A. Wert, *Phys. Rev.* **79,** 601 (1950).
2. J. K. Stanley, *Trans. AIME* **185,** 752 (1949).
3. W. R. Thomas and G. M. Leak, *Phil. Mag.* **45,** 986 (1954).
4. R. R. Hasiguti and G. Kamoshita, *J. phys. Soc. Japan* **9,** 646 (1954).
5. L. Guillet and B. Hocheid, *Revue Métall., Paris* **53,** 122 (1956).
6. R. E. Maringer, *J. appl. Phys.* **31,** 229S (1960) and **35,** 2375 (1964).
7. A. E. Lord and D. N. Beshers, *Acta Metall.* **14,** 1659 (1966).

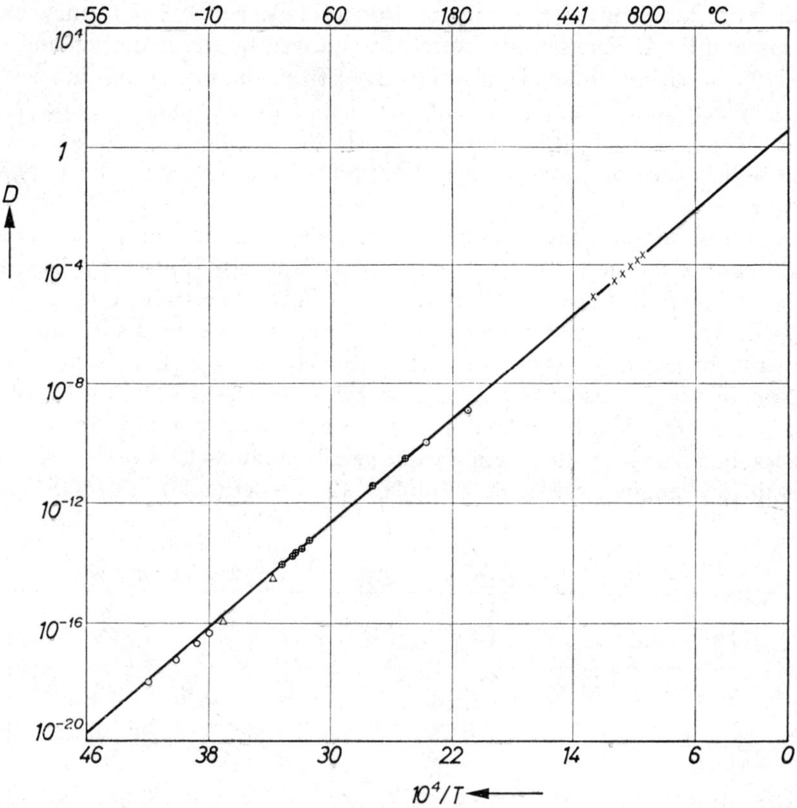

FIG. 69. Diffusivity of carbon in bcc iron as a function of temperature (log D versus $1/T$) according to data available in 1950. These data were obtained by Wert and Stanley using various experimental methods in the temperature range $-35°$ to 800°C. According to Wert the data fit a straight line covering about fourteen decades of the diffusivity.

However, measurements by Smith[1] and Homan[2] combined with earlier results by Stanley, *loc. cit.*, prove convincingly that at high temperatures the log D versus $1/T$ plot departs from linearity.

Smith determined diffusivities of carbon in bcc iron in the region 500° to 865°C by means of a decarburization method. For his measurements he used iron–carbon alloys in which two phases coexisted at each of the decarburization temperatures. In the temperature range 900° to 727°C the two

1. R. P. SMITH, *Trans. AIME* **224,** 105 (1962).
2. C. G. HOMAN, *Acta Metall.* **12,** 1071 (1964).

phases were bcc iron (α Fe) and fcc iron (γ Fe); below 727°C they were bcc iron and Fe_3C. The samples were decarburized by wet hydrogen and the diffusivity of carbon in bcc iron was derived from the rate of motion of the boundary between the single-phase α region and the two-phase region $\alpha+\gamma$ (above 727°C) or $\alpha+Fe_3C$ (below 727°C). In the calculations use was made of the well known positions of the phase boundaries in the Fe–Fe_3C phase diagram.

The diffusivities at high temperatures, measured by Smith, are much larger than equation (2.14.1) suggests. This was also found by Homan, *loc. cit.*, who studied the diffusion of carbon in the temperature range 616° to 844°C, using tracer techniques. Fig. 70 represents the results of the measurements by Stanley, Smith and Homan. The broken line is an extrapolation of the results at low temperatures as given by equation (2.14.1). From the results, Smith concludes that the activation enthalpy for carbon diffusion in α iron increases from a value smaller than 84 kJ mole^{-1} (20 kcal/mole) at low temperatures to 92 kJ mole^{-1} (22·0 kcal/mole) near 500°C and

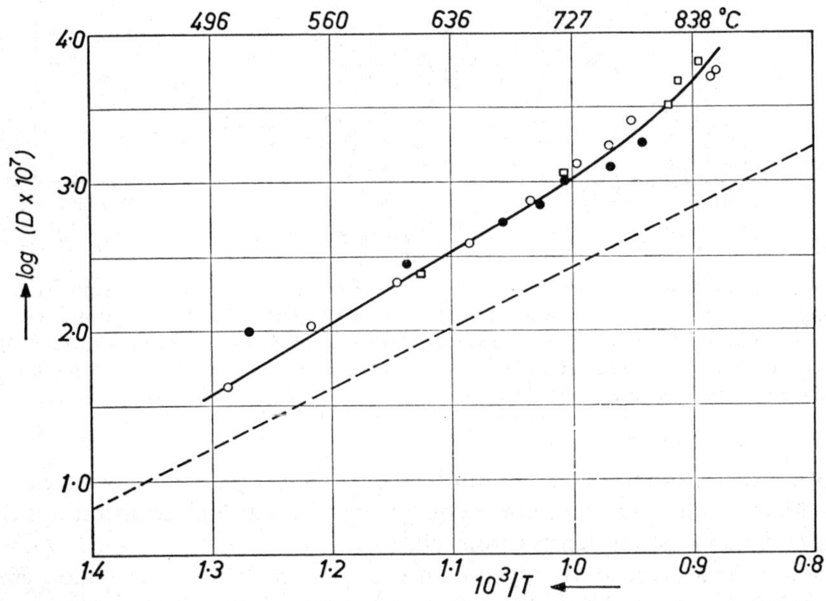

FIG. 70. According to more recent investigations the picture of diffusion of carbon in bcc iron is less simple than suggested by Fig. 69. At high temperatures the log D versus $1/T$ graph deviates from linearity. The figure gives the results of measurements by Stanley (shaded circles), Smith (open circles) and Homan (squares). The broken line is an extrapolation of the results at low temperatures as represented by equation (2.14.1).

to 105 kJ mole^{-1} (25·0 kcal/mole) at 850°C. A more satisfactory conclusion would be that at high temperatures other diffusion mechanisms become active in addition to the 'normal' mechanism.

In his theoretical considerations Homan, *loc. cit.*, assumes two diffusion mechanisms to be active at high temperatures. One is the normal interstitial diffusion of unpaired carbon atoms, the other the diffusion of carbon–vacancy pairs. He explains the accelerated diffusion with the assumption that the activation enthalpy for the motion of the pairs is only about half that for unpaired carbon atoms. At low temperatures the diffusion of pairs is insignificant since then hardly any unoccupied lattice sites are present in the metal under conditions of thermodynamic equilibrium. Experiments by Damask and co-workers(1–6) on the migration of carbon in iron–carbon alloys containing many vacancies produced by irradiation with neutrons, show that formation of carbon–vacancy pairs does indeed occur. However, the binding energy of a carbon atom to a vacancy, according to these experiments, is only 40 kJ mole^{-1} (9·5 kcal/mole). This implies that Homan's explanation of the diffusion anomaly is not tenable. In the temperature range of this anomaly, carbon–vacancy complexes do not exist since they dissociate above 250°C.

McLellan *et al.*(7) therefore offer another explanation. They assume that at high temperatures a small part of the carbon occupies tetrahedral sites which according to Beshers(8) are energetically only slightly less favourable than octahedral sites (cf. Chapter 1). Denoting the octahedral and tetrahedral sites by o and t, respectively, the possible jumps that an interstitial atom can make in the bcc lattice are as follows: (1) the route of an atom from one o site to another o site will be o–t–o; (2) the route of an atom from one t site to another t site will be either t–o–t or (3) t–t (see Figs. 71 and 9). In a solution where both o and t sites are occupied, all three diffusion routes will contribute to the overall diffusivity. Therefore D will be a weighted linear sum of three partial diffusivities: D(o–t–o), D(t–t) and D(t–o–t). According to McLellan *et al.*, D(t–o–t) must equal D(o–t–o) since the corresponding routes are additions of an o–t and a t–o jump.

In their quantitative analysis of the problem, McLellan *et al.* come to the conclusion that even at high temperatures 0·1 %, at the most, of the dissolved

1. H. Wagenblast and A. C. Damask, *Physics Chem. Solids* **23,** 221 (1962) and *Acta Metall.* **10,** 333 (1962).
2. F. E. Fujita and A. C. Damask, *Acta Metall.* **12,** 331 (1964).
3. R. A. Arndt and A. C. Damask, *Acta Metall.* **12,** 341 (1964).
4. H. Wagenblast, F. E. Fujita and A. C. Damask, *Acta Metall.* **12,** 347 (1964).
5. R. A. Johnson and A. C. Damask, *Acta Metall.* **12,** 443 (1964).
6. R. A. Johnson, G. J. Dienes and A. C. Damask, *Acta Metall.* **12,** 1215 (1964).
7. R. B. McLellan, M. L. Rudee and T. Ishibachi, *Trans. AIME* **233,** 1938 (1965).
8. D. N. Beshers, *J. appl. Phys.* **36,** 290 (1965).

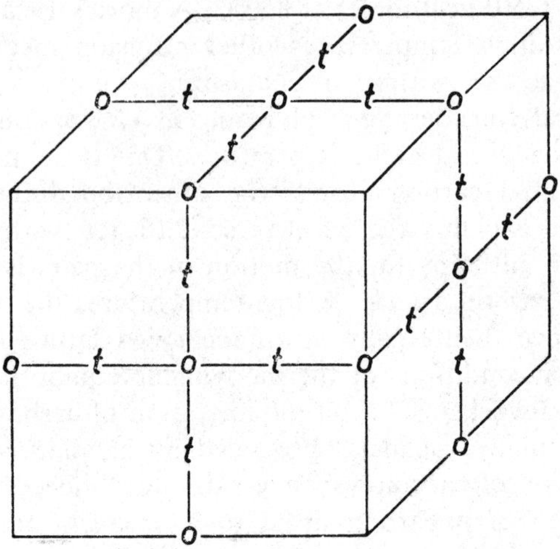

FIG. 71. Unit cell of the bcc lattice showing the interstitial sites for three cell faces. There are no interstitial sites in the interior of the cube. The tetrahedral and octahedral sites are marked t and o, respectively. The lattice atoms are not shown.

carbon atoms is situated at t sites. They make an important contribution to the total diffusion, notwithstanding this low concentration, since the activation enthalpy for t–t jumps is relatively small. At low temperatures only *D*(o–t–o) is measured because then all carbon or nitrogen atoms occupy o sites.

If, at low temperatures, a significant fraction of the interstitial atoms were at t sites, measurements of internal friction (see next chapter) would show two Snoek damping peaks, one corresponding to o–t–o jumps and one to t–t jumps. The latter would be much smaller than the former even with equal occupation of o and t sites since the strain dipoles caused by atoms at t sites are much smaller than those formed by atoms at o sites[1].

After the above discussion of the diffusion of carbon in bcc iron we give some details of the diffusion of carbon in fcc iron insofar as these have not already been mentioned in Section 2.5. In that section we noted that the high solubility of carbon in γ iron makes it possible to determine the dependence of D on the carbon concentration[2,3]. The study by Wells *et al.* showed that plotting log D against $1/T$ results in a straight line, provided the carbon concentration is constant. This has been observed for six different

1. L. J. DIJKSTRA, *Philips Res. Rep.* **2,** 357 (1947).
2. C. WELLS, W. BATZ and R. F. MEHL, *Trans. AIME* **188,** 553 (1950).
3. R. P. SMITH, *Acta Metall.* **1,** 578 (1953).

concentrations (1, 2, 3, 4, 5 and 6% C). For each of these the following well-known relation thus holds:

$$D = D_0 \exp(-\Delta H/RT) \qquad (2.14.2)$$

where D_0 and ΔH are independent of temperature. The values for D_0 and ΔH were determined in the usual way for each of the concentrations (see Fig. 64 and its description). Fig. 72 shows the dependence of ΔH, and Fig. 73 that of D_0 on the carbon content in atom per cent.

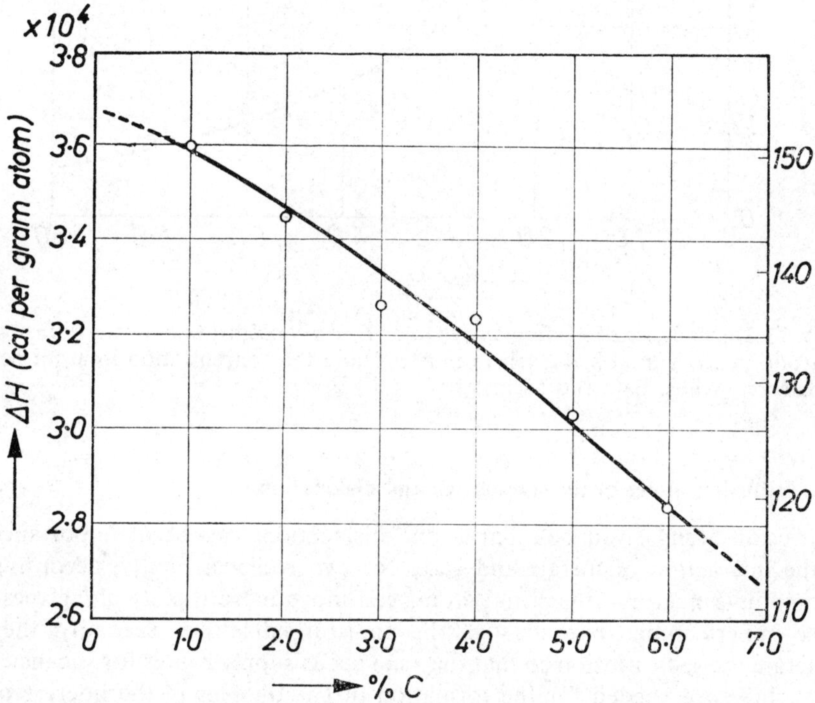

FIG. 72. Dependence of the activation enthalpy ΔH on concentration for the diffusion of carbon in fcc iron. ΔH is given in cal per gram atom and kJ/g atom; the concentration is given in atom per cent carbon (Wells, Batz and Mehl).

It appears that the complications noted above for the diffusion of carbon in bcc iron are not shown by nitrogen. A single Arrhenius equation (2.14.2) covers all available data for nitrogen diffusion in α and δ iron from $-50°$ to 1470°C. A compilation of these data has been given by Lord and Beshers[1] (see Section 4.9).

1. A. E. LORD and D. N. BESHERS, *Acta Metall.* **14**, 1659 (1966).

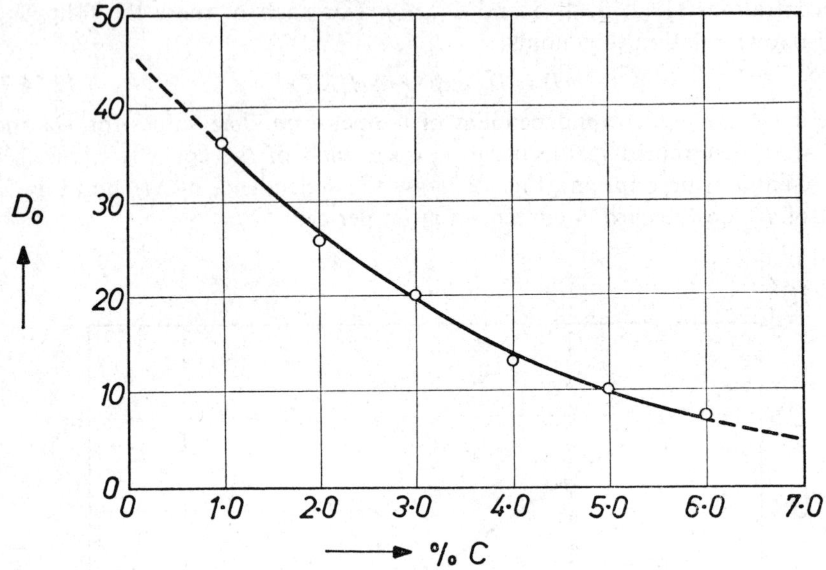

FIG. 73. Dependence of the frequency factor D_0 on concentration for the diffusion of carbon in fcc iron. D_0 is given in $mm^2\,s^{-1}$ and the concentration in atom per cent carbon (Wells, Batz and Mehl).

2.15. Diffusion along grain boundaries and dislocations

Diffusion along grain boundaries and dislocations can be of importance for the interaction of metals and gases for two reasons. Firstly, according to some investigators, interstitial atoms can move more rapidly along these lattice imperfections than through the undisturbed lattice. Secondly, they accelerate the self-diffusion so that they can act as supply routes for vacancies where these are needed for the formation of gas bubbles in the interior of the metal.

High diffusivity paths

Since the grains in a polycrystalline metal in general show great differences in orientation, the arrangement of the atoms at the grain boundaries is much less orderly than in the interior of the grains. Only twin boundaries have no disorganized transition zone. There is no difference of opinion on the structure of boundaries between grains with only a small difference in orientation. They are narrow regions containing many dislocations, the concentration of which increases with the difference in orientation between the adjoining crystals. For a detailed treatment of the structure of small-angle boundaries we refer the reader to books on dislocations[1]. The simplest type of small-angle boundary, the tilt boundary, will be discussed below.

Different opinions do exist on the atomic structure of boundaries between grains with a large difference in orientation. Many accept a conception introduced by Mott[2], which visualizes a large-angle boundary as being constituted of islands of good fit (coherent areas) separated by areas of poor fit. Nevertheless, there is no doubt at all that the atoms, both in small-angle and in large-angle boundaries are in many places less closely packed than in an undisturbed crystal. The activation enthalpy for diffusion will therefore often be lower at the grain boundaries than inside the grains.

A lower activation enthalpy for diffusion in the boundaries does not necessarily imply that the diffusion in polycrystalline material takes place mainly along the boundaries; it will in many cases be overcompensated by the fact that only a very small fraction of the total volume is available for grain-boundary diffusion. Transport will take place mainly along the boundaries, only if

$$\frac{D(\mathrm{b})}{D(\mathrm{l})} \gg \frac{\text{average grain diameter}}{\text{average grain boundary width}} \qquad (2.15.1)$$

where $D(\mathrm{b})$ and $D(\mathrm{l})$ are the grain-boundary and lattice diffusivities. If the high-diffusivity path along a grain boundary is taken to be 0·5 nm wide and the grain size 50 μm, $D(\mathrm{b})/D(\mathrm{l})$ must be much greater than 10^5 to satisfy this condition. The diffusing substance will not penetrate along the grain boundaries much further than in the grains themselves when the ratio of the two diffusion coefficients is smaller. This is due to the faster penetration along the boundaries becoming insignificant as a result of sideways diffusion into the grains adjoining the boundaries.

The lower activation enthalpy for grain-boundary diffusion implies that the diffusivity in the boundaries will decrease more slowly with falling temperature than that in the lattice. In some cases therefore diffusion along the grain boundaries will predominate at low temperatures whilst at high temperatures the diffusion will mainly take place in the interior of the grains. An illustration of this is self-diffusion in silver with an initial grain size of 35 μm. The diffusivities measured in the temperature range 500° to 600°C are significantly higher in this material than in single crystals of silver[3]. This is shown by Fig. 74; above 700°C both samples have the same

1. See e.g. W. T. Read, *Dislocations in Crystals*, McGraw-Hill, New York (1953).
2. N. F. Mott, *Proc. phys. Soc.* **60**, 391 (1948).
3. R. E. Hoffman and D. Turnbull, *J. appl. Phys.* **22**, 634 and 984 (1951); D. Turnbull *Atom Movements*, Am. Soc. Metals, Cleveland (1951), p. 129.

value of D but at temperatures well below 700°C the values for D measured in the polycrystalline samples reflect primarily the effect of diffusion along grain boundaries. The slopes of the two lines in Fig. 74 indicate that the activation enthalpy for grain-boundary diffusion of silver in silver is about one half that for lattice diffusion.

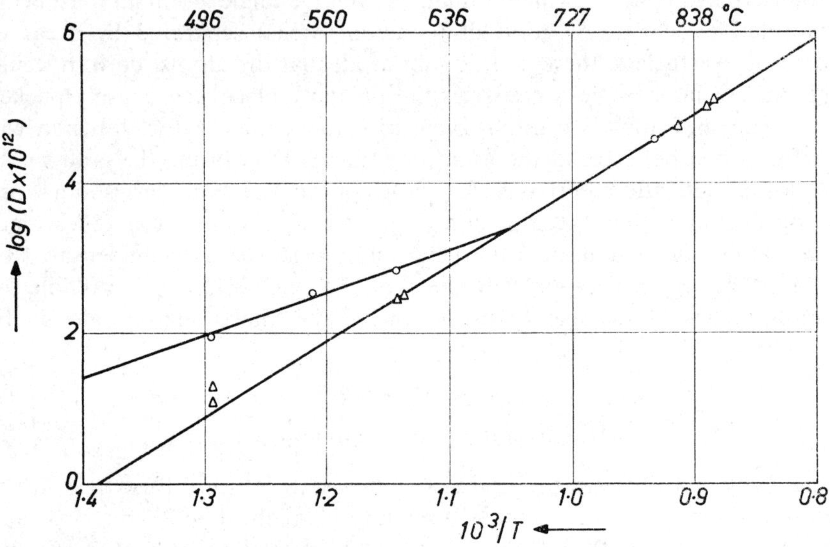

FIG. 74. Self-diffusion coefficient of silver (ordinate) in single crystals (triangles) and in polycrystalline samples with an average grain diameter of 35 μm (circles) plotted against reciprocal of the absolute temperature (abscissa)—after Hoffman and Turnbull.

The above phenomenon was studied by allowing a radioactive silver isotope to diffuse into ordinary, non-radioactive silver. Overall diffusivities (Fig. 74) were determined by measuring the radioactivity of successive thin layers machined off the metal. The preferential diffusion along the grain boundaries has also been shown in a more direct manner by means of photography, not only for silver but also for other metals. As an illustration we shall now discuss self-diffusion in bcc iron.

Leymonie and Lacombe[1] covered a flat face of a sheet of iron with a 0·2 μm-thick layer of radioactive iron. After heating it for a certain time at constant temperature, successive layers were removed electrolytically. The fresh free surface was then brought into contact with a photographic film. In many instances the

1. C. LEYMONIE and P. LACOMBE, *Revue Métall., Paris* **54,** 653 (1957). See also their contribution to the book *La Diffusion dans les Métaux* (editors J. D. Fast, H. G. van Bueren and J. Philibert), Philips Technical Library, Eindhoven (1957).

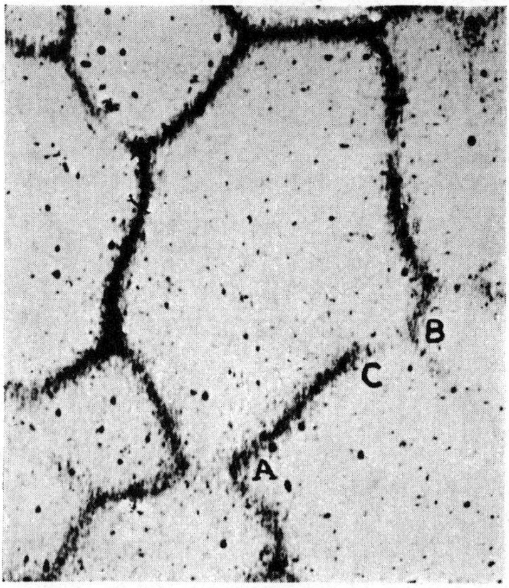

FIG. 75. Radiograph of armco iron showing grain boundaries (magnification 50 ×). The history of the sample is as follows. After secondary recrystallization the metal was electroplated with radioactive iron about 0·2 μm thick and kept at a temperature of 725°C for three days in an atmosphere of pure hydrogen. After this a layer of 20 μm was removed from the metal. The radiograph was obtained by bringing the newly exposed fresh surface in close contact with a photographic film and maintaining the contact for 24 hours. The radiograph shows that at 725°C the diffusion of the radioactive iron takes place mainly along the grain boundaries (Leymonie and Lacombe).

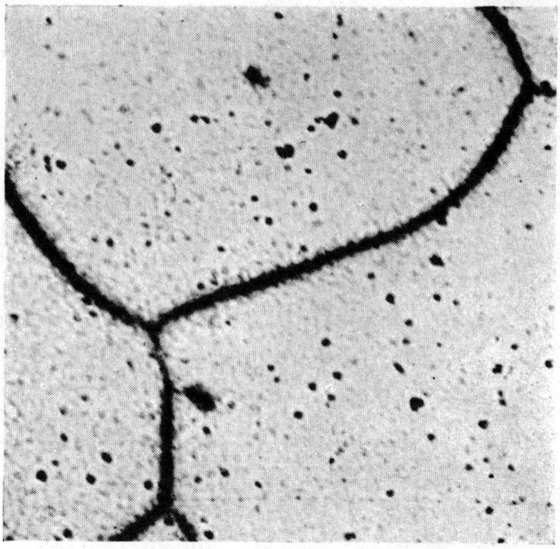

FIG. 76. Radiograph of high-purity iron showing intergranular penetration of radioactive iron (magnification 75 ×). For the history of the sample see Fig. 75 (Leymonie and Lacombe).

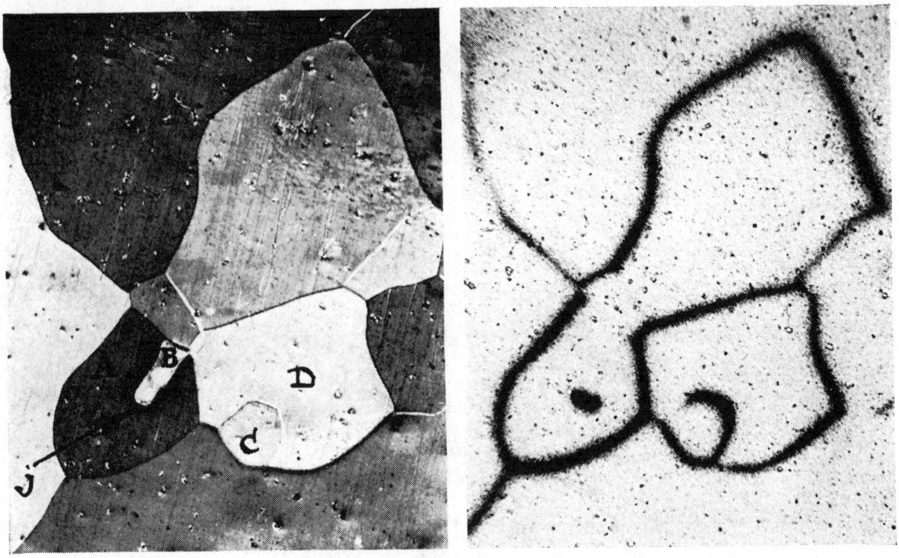

a *b*

FIG. 77. A comparison of the micrograph (a) and the radiograph (b) of the same iron sample clearly shows the influence of the orientation of the grain boundaries with respect to the adjacent grains on the intergranular penetration of radioactive iron at 650°C. Examples of this are the boundary between grains C and D and the boundary between grains A and B. The latter two are in twin position with respect to each other; the radioactive iron has only penetrated along the non-coherent part of this boundary (Leymonie and Lacombe, magnification 75 ×).

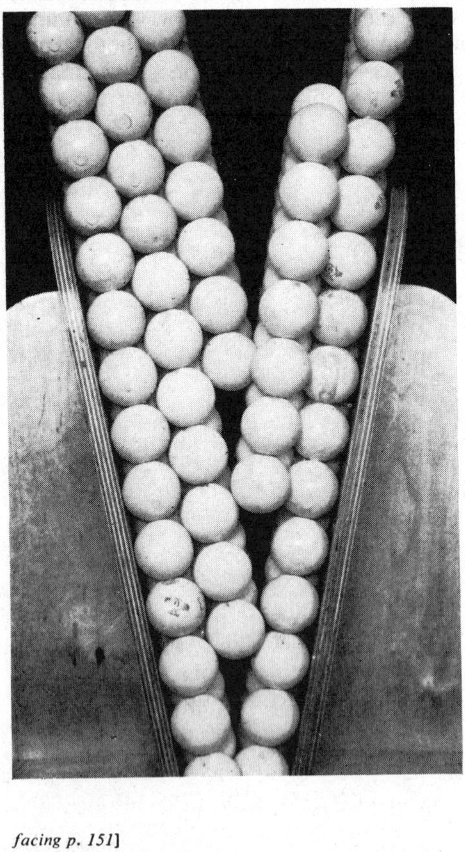

FIG. 78. Okkerse, Tiedema and Burgers used lead samples consisting of only two crystals separated by a symmetrical small-angle tilt boundary in their study of self-diffusion in lead. The figure shows a model of such a boundary built up from table-tennis balls. The diffusivity in this type of boundary was shown to be strongly anisotropic. When the direction of the diffusion is perpendicular to the plane of the figure, i.e. parallel to the dislocation lines, fast penetration takes place through the dislocation 'pipes'. No preferential grain boundary diffusion occurs in the direction perpendicular to the dislocation lines.

film blackened along lines exactly matching the grain boundaries in the metal. Figs. 75 and 76 give examples of this effect for the self-diffusion at 725°C in armco iron and in very pure iron (see Plate III, opposite page 150). After heating at a temperature above 850°C the grain boundaries can no longer be made visible on the auto-radiographs, indicating that at the higher temperatures grain-boundary diffusion no longer predominates.

The two photographs (Figs. 75 and 76) not only show that at 725°C self-diffusion in bcc iron takes place preferentially along the grain-boundaries but also that it is not of the same magnitude for all boundaries. This is evident from the uneven blackening at the boundaries and is even more clearly shown in Fig. 77 (Plate IV, facing this page) where photograph (b) is a radiograph showing the intergranular penetration of radioactive iron in armco iron after heating for 30 days at 650°C and removal of a 60 μm layer. For comparison, (a) in the figure is the corresponding normal photomicrograph. Some grain boundaries, clearly visible in the micrograph are absent in the radiograph. It can also be seen that the amount of diffusion is determined not only by the difference in orientation but also by the position of the boundary with respect to the two lattices[1]: in the radiograph only a part of the boundary between grains C and D is visible. The twin position of the grains A and B is also interesting: diffusion has been pronounced along the incoherent part of the boundary but negligible along the coherent part. This is to be expected since in the latter case there is no transition region, i.e. no high-diffusivity path.

A distinct relation between grain-boundary diffusion and diffusion along dislocations has been found by using samples consisting of only two crystals, separated by a symmetrical tilt boundary. A model of such a boundary is shown in Fig. 78 (Plate IV, facing this page). It contains a regular array of parallel edge dislocations perpendicular to the plane of the figure. When the diffusion takes place in the direction of the dislocation lines then, as experiments with lead[2] and silver[3] have shown, preferential grain-boundary self-diffusion occurs. However, when the diffusion takes place in a direction at right angles to the dislocation lines, preferential grain-boundary diffusion is absent. The experiments indicate that edge dislocations constitute paths of easy diffusion (pipe diffusion). For a theoretical treatment of diffusion along grain boundaries and dislocations we refer the reader to a number of publications[4-9].

1. In this connection we remind the reader that the position of a grain boundary depends on five parameters: three to fix the orientation of one lattice with respect to the other and two to define the position of the interface with respect to one of the lattices.
2. B. Okkerse, T. J. Tiedema and W. G. Burgers, *Acta Metall.* **3,** 300 (1955).
3. R. E. Hoffman, *Acta Metall.* **4,** 97 (1956).
4. J. C. Fisher, *J. appl. Phys.* **22,** 74 (1951).
5. R. T. P. Whipple, *Phil. Mag.* **45,** 1225 (1954).
6. H. S. Levine and C. J. MacCallum, *J. appl. Phys.* **31,** 595 (1960).
7. T. Suzuoka, *Trans. Japan Inst. Metals* **2,** 25 (1961).
8. L. C. Luther, *J. chem. Phys.* **43,** 2213 (1965) and **45,** 1080 (1966).
9. J. P. Stark, *J. appl. Phys.* **36,** 3938 (1965).

Formation of gas bubbles

In the interior of a solid metal that is free of pores, a gas bubble can only form in a place where the metal atoms can escape by diffusion or, in other words, when vacancies can diffuse towards the location in question. This has been clearly shown in experiments by Barnes and co-workers[1–4], who also demonstrated which lattice imperfections can act as sources or conductors for vacancies.

In these experiments a plane face of a copper, aluminium, beryllium or uranium sample is, at relatively low temperature, bombarded with α particles from a cyclotron. In the interior of the metal these particles become helium atoms by taking up electrons. Since in these experiments the particles are mono-energetic, all helium atoms finish up in a narrow band beneath the surface and parallel to it. The thickness of such a helium-rich layer is about 10 μm. Barnes regards this layer as a strongly supersaturated interstitial solution of helium in the metal. At room temperature no diffusion or precipitation occurs, even at a concentration of 1 atom % helium.

On heating the bombarded metal to a relatively high temperature, the helium atoms become mobile by capturing vacancies. One vacancy must be provided for an interstitial helium atom to become substitutional and a further one is necessary to enable it to diffuse through the metal. When two atoms meet, it will be more economical in energy when they stay together; this small group of two helium atoms plus two vacancies may become the nucleus of a bubble when further atoms and vacancies reach it and become attached before it breaks up. The extremely low solubility of helium in metals (cf. Chapter 1) implies that only a very small number of helium atoms and very few vacancies per gas atom are required to form a stable nucleus.

When the bubbles have attained a size that justifies the use of macroscopic concepts like gas pressure p and surface tension σ, the stability condition for a bubble is

$$p = 2\sigma/r$$

where r is its radius. Applying this equation to a bubble with a radius of 0·2 μm and assuming the surface tension of the solid metal to be 1 N m^{-1}, the bubble will be stable when the helium pressure inside it is about 10 MN m^{-2}. As long as the bubbles are small the surface tension can therefore balance a large part of the internal gas pressure.

1. R. S. Barnes, G. B. Redding and A. H. Cottrell, *Phil. Mag.* **3,** 97 (1958).
2. R. S. Barnes, *Colloque sur la Diffusion à l'Etat solide* (editor: G. Chaudron), North Holland Publ. Co., Amsterdam (1959), p. 57.
3. R. S. Barnes, *Phil. Mag.* **5,** 635 (1960).
4. R. S. Barnes and D. J. Mazey, *Proc. R. Soc.* **A 275,** 47 (1963).

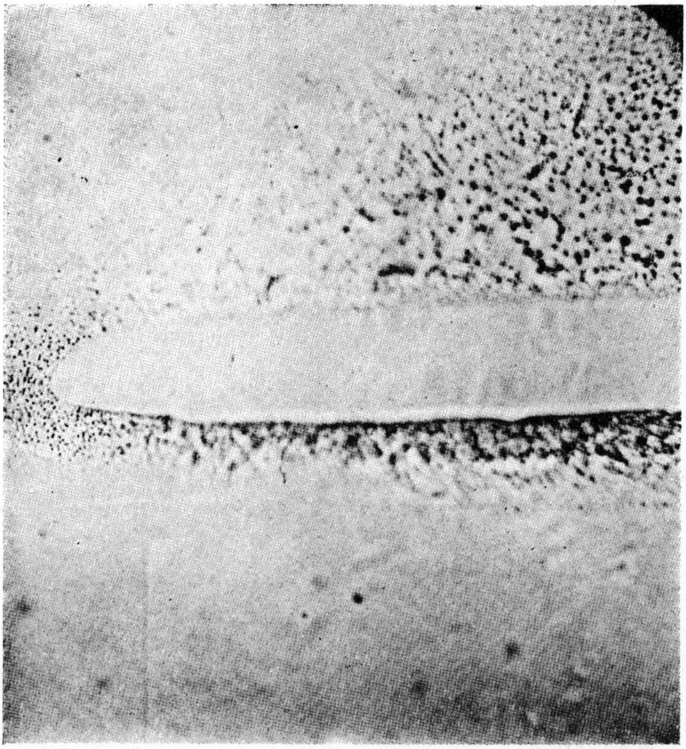

FIG. 79. In experiments carried out by Barnes *et al.* well annealed copper was bombarded with 6·4 pJ (40 MeV) α particles until a metal layer of about 10 µm (lying about 1 mm below the bombarded face of the specimen) contained approximately one atom per cent helium. The specimen showed no bubbles after this treatment. Only after heating above 600°C did bubbles appear. The figure shows a section, cut perpendicularly through the helium-rich layer after the specimen had been heated at 800°C for one hour. Near the free surface (lying on the left of the figure) vacancies have apparently migrated inwards, allowing the helium to precipitate from solution with the formation of bubbles. The heat treatment was inadequate to fill the whole layer with bubbles.

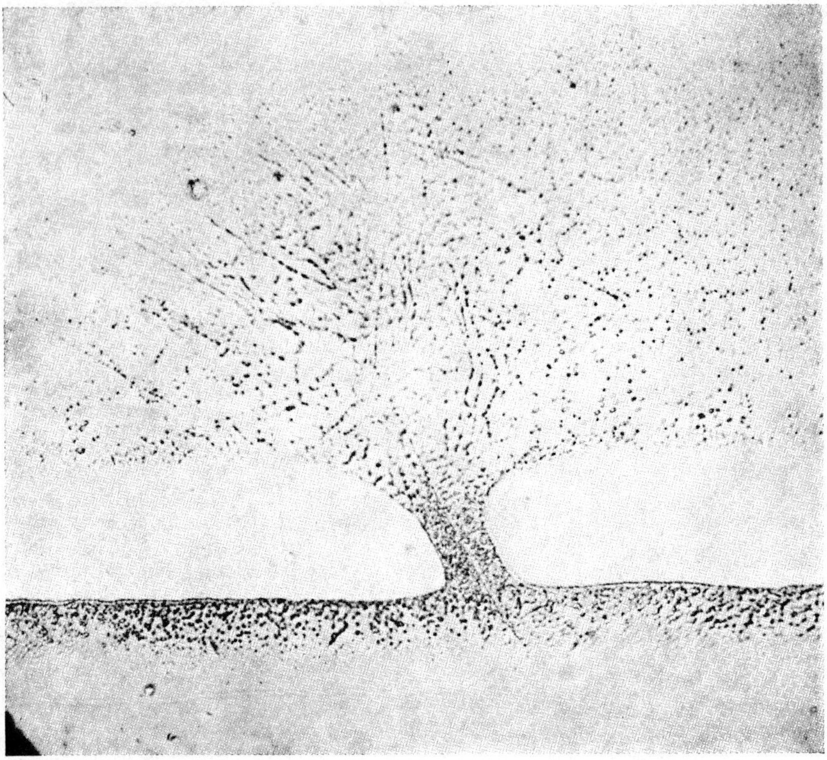

FIG. 80. The figure relates to the same piece of copper as partly seen in Fig. 79. It shows an area where a grain boundary crosses the helium-rich layer. The width of the bubble-filled zone at both sides of the boundary suggests that this boundary supplied vacancies at the same rate as the free surface during the heating at 800°C.

Bubble formation in a bombarded single crystal does not start in the middle of the helium-rich region where the concentration is highest, but on its periphery. This is the best proof that precipitation of helium from solution is only possible when there is a supply of vacancies. In polycrystalline samples, the free surface and the grain boundaries are the predominant sources of vacancies. Fig. 79 shows the effect of a free surface, Fig. 80 that of a grain boundary (see Plates V and VI, between pages 152 and 153). In both cases bubble formation has only taken place near the vacancy source. Much longer heating times would be necessary to precipitate all the helium present.

Voids in the metal can also act as vacancy sources. Since there always is a counterflow of metal atoms, equal in number to the flow of vacancies, a void will eventually be completely filled and cease to be a source. Not all grain boundaries behave in the same manner, some act as sources of vacancies, while others merely act as supply routes, conducting vacancies to the areas where they are needed. In Barnes' experiments coherent twin boundaries were never seen to generate or conduct vacancies. This is not surprising since these are perfectly fitting boundaries and it also agrees with results of radiographic studies mentioned earlier (cf. Fig. 77, Plate IV).

Experiments by Murray[(1)] appear to indicate that dislocations also play a significant role in the formation of helium bubbles in aluminium. He deduces from his experimental results that helium precipitates in the form of cylindrical cavities along the dislocations. The helium was generated in aluminium containing 0·1 atom % lithium by irradiating this alloy with thermal neutrons. Murray's experiments are less convincing than those of Barnes *et al.* because under neutron irradiation one tritium atom is formed for each helium atom according to the reaction

$$^{6}\mathrm{Li} + \mathrm{n} \rightarrow {}^{3}\mathrm{H} + {}^{4}\mathrm{He}$$

From these experiments it is not possible to decide whether, and if so how, bubble formation has been affected by the tritium.

Available space does not permit a detailed discussion of all the interesting experiments by Barnes *et al.*, nor of related studies by many other investigators. We only mention that Barnes and Mazey[(2)] were able to show that, in copper at 800°C, small helium bubbles can migrate as a whole, probably under the influence of a temperature gradient. The main mechanism of this migration consists in the diffusion of metal atoms along the surface of the bubble from the leading to the trailing part. It goes without saying that these investigations are of great importance in connection with the swelling of the fuel elements in nuclear reactors, caused by

1. G. T. Murray, *J. appl. Phys.* **32,** 1045 (1961).
2. R. S. Barnes and D. J. Mazey, *Proc. R. Soc.* **A 275,** 47 (1963) and *Mém. Scient. Revue Métall.* **63,** 81 (1966).

the introduction of the inert gases krypton, xenon and helium during irradiation[1]. A unified theory of the diffusion-controlled mobilities of pores in solids has recently been presented by Nichols[2].

After heating nickel specimens at 1,000° or 1,200°C in a vacuum furnace, Mykura[3] observed that normal grain-boundary grooves[4] had developed in pure nickel, but ridges had developed on nickel specimens containing 0·05% carbon and an unknown percentage of oxygen. He attributed this phenomenon to the formation of internal gas bubbles (CO) and the associated diffusion of vacancies along the grain boundaries into the specimens, each vacancy adding one atom to a ridge. The ridges had the usual grain boundary grooves superimposed on them. Fig. 81 shows the change of shape of a ridge with time.

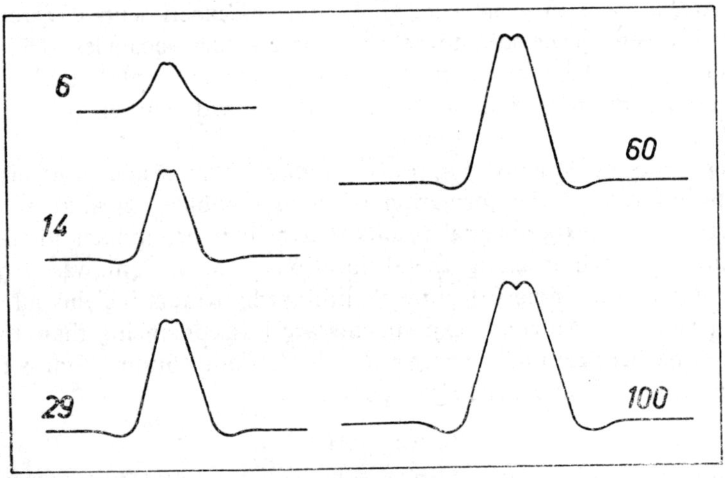

FIG. 81. During the heating of impure nickel specimens at 1000°C in a vacuum furnace, ridges develop at the grain boundaries. The figure shows the change of ridge shape with time for a 0·2 mm thick specimen. The effect was first observed by Mykura and attributed by him to the diffusion of vacancies needed for the formation of internal gas bubbles, along the grain boundaries and into the specimen. The numbers in the figure give the time of heating in hours.

1. See also: R. S. BARNES, *J. nucl. Mater.* **11,** 135 (1964) and R. S. NELSON, *J. nucl. Mater.* **19,** 149 (1966).
2. F. A. NICHOLS, *J. nucl. Mater.* **30,** 143 (1969).
3. H. MYKURA, *Phil. Mag.* **4,** 907 (1959).
4. W. W. MULLINS, *J. appl. Phys.* **28,** 333 (1957) and **30,** 77 (1959). See also his article in the book *Metal Surfaces* (editors: W. D. Robertson and N. A. Gjostein), Am. Soc. Metals, Cleveland (1963).

As expected, the ridge height varied widely for different boundaries. Mykura measured the orientations of several pairs of grains and in all cases, large orientation differences were associated with large ridges and small orientation differences with small ridges. The ridge shown in Fig. 81 had a height of a few microns after heating at 1,000°C for a hundred hours.

Diffusion of gas atoms along extended lattice imperfections

Many workers are of the opinion that not only the rate of self-diffusion along grain boundaries and dislocations is abnormally high but also the rate of diffusion of interstitial atoms, particularly those of oxygen, nitrogen and carbon. We shall discuss some of the experiments which appear to support this view below and, at the end of this section, arguments against accelerated diffusion of interstitials along extended lattice imperfections will be summarized.

Let us first consider the diffusion of oxygen in silver. The grain growth in silver after cold working and recrystallization is retarded by oxygen, even if present in minute amounts(1). Klein and Huggins(2) were able to show that this behaviour is not due to the formation of a fine dispersion of an impurity oxide, but to oxygen in solid solution. Grain-growth restraint even occurred in the interior of samples that had been heated in oxygen at 800°C for only thirty minutes before cold rolling from 5 to 2·5 mm. Klein and Huggins concluded from their experiments that the rate of migration of oxygen in silver at 800°C corresponds very roughly to an apparent bulk diffusivity between 10^{-1} and 10^{-2} $mm^2\ s^{-1}$. According to the data in Table 10 the real volume diffusivity at 800°C is only about 10^{-3} $mm^2\ s^{-1}$. The unexpected high mobility of oxygen was assumed to be due to transport of trace amounts along extended structural imperfections such as grain boundaries or dislocations. No direct proof for this interpretation is offered by Klein and Huggins.

When determining the diffusivity of oxygen in zirconium Davis, *et al.*(3) obtained two straight lines when plotting log D against $1/T$. The intersection of the two lines corresponds to a temperature of approximately 700°C. Similar results were obtained by Debuigne and Lehr(4) and also by Béranger and Lacombe(5). Fig. 82 gives the results obtained by the latter pair, showing the straight lines intersecting at a

1. J. C. Chaston, *J. Inst. Metals* **71,** 23 (1945).
2. M. J. Klein and R. A. Huggins, *Trans. AIME* **224,** 903 (1962).
3. M. Davis, K. R. Montgomery and J. Standring, *J. Inst. Metals* **89,** 172 (1960–61).
4. J. Debuigne and P. Lehr, *Mém. Scient. Revue Métall.* **60,** 911 (1963).
5. G. Béranger and P. Lacombe, *J. nucl. Mater.* **16,** 190 (1965).

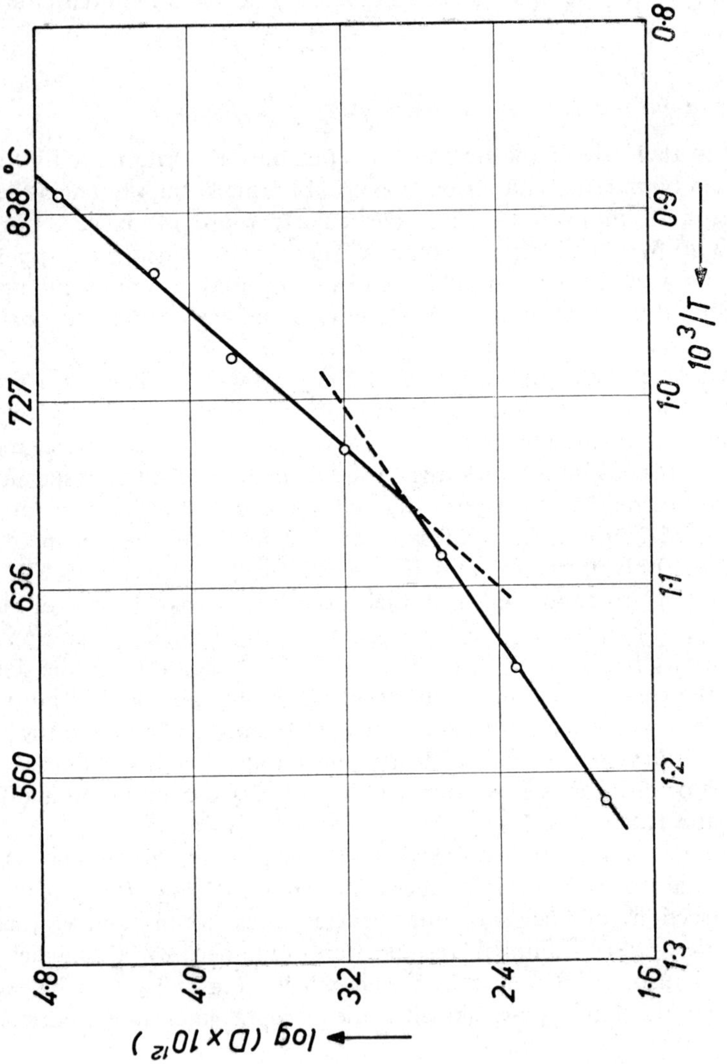

FIG. 82. Diffusion of oxygen in zirconium (log D versus $1/T$) after Béranger and Lacombe.

temperature between 650° and 700°C and corresponding to the equations

$$D = 22 \exp(-197{,}000/RT) \qquad (2.15.2)$$

for the temperature range 650° to 850°C and

$$D = 7{\cdot}9 \times 10^{-3} \exp(-128{,}000/RT) \qquad (2.15.3)$$

for the temperature range 550° to 650°C[1]. All three groups of researchers assume that grain boundary diffusion predominates at low temperatures (below 700° or 650°C) and lattice diffusion at high temperatures but direct proof of this interpretation could not be offered. Some support for it is found in an investigation by Sainfort, *et al.*[2] who, with the aid of a high-sensitivity X-ray diffractometer could prove that oxygen that has penetrated zirconium at a temperature below 770°C is not uniformly distributed in the grains.

From observations with an electron microscope, E. Smith[3] inferred that the diffusion of *carbon in α iron* occurs at a much higher rate along the dislocations than through the undisturbed lattice. He studied the precipitation of carbon from supersaturated solutions in bcc iron by means of the thin film technique. According to his results, the dislocations encourage the nucleation at 100°C of an initial carbide whose structure and composition were not determined. The same was observed for the nucleation of cementite at 200°C. More important in the context of this section, is his statement that the growth of the initial precipitates on the dislocations was faster than the growth of those in the matrix, the dislocations appearing to act as channels for the diffusion of carbon. Also, according to Smith, the growth of the cementite particles is accelerated by dislocations acting as supply routes for carbon.

Contrary to the above, convincing experimental evidence exists that the diffusion of carbon in α iron is slower along a dislocation than through the undisturbed lattice. As mentioned in Chapter 1 the carbon atoms find sites in the dislocations where they have a lower energy than elsewhere. The dislocations can therefore be expected to act as sinks for carbon atoms and to stimulate the nucleation of carbide particles. On the other hand, it is likely that the jump of a carbon atom from a favourable position in the dislocation to any other site will require a higher activation enthalpy than a jump in a perfect crystal. These expectations are convincingly confirmed

1. Equations (2.15.2) and (2.15.3) do not completely correspond to the lines in Fig. 82 which are drawn through the points which according to Béranger and Lacombe are the most reliable.

2. G. Sainfort, R. Jacquesson and P. Laurent, *Colloque sur la Diffusion à L'Etat solide* (editor: G. Chaudron), North Holland Publ. Co., Amsterdam (1959), p. 79.

3. E. Smith, *Direct Observation of Imperfections in Crystals* (editors: J. B. Newkirk and J. H. Wernick), John Wiley, New York (1961), p. 203.

by measurements of the internal friction of α iron containing dislocations and carbon or nitrogen atoms. These and similar measurements on interstitial solutions in other bcc metals are discussed in the next chapters.

It may be assumed that the preceding does not apply only to bcc metals but that in general interstitials are *more stable* but *less mobile* in dislocations than in the undisturbed crystal.

3

INTERNAL FRICTION

3.1. Introduction

In recent decades the measurement of internal friction has been of valuable assistance for the acquisition of information on the behaviour of gases in metals. In the next chapter it will be shown that internal friction caused by displacements of gas atoms and other interstitially dissolved atoms in bcc metals can supply us with data on diffusivities, concentrations, terminal solubilities, surface reactions, precipitation phenomena, mutual interactions and interactions between interstitials and other lattice imperfections. This chapter deals in a preliminary, qualitative manner with internal friction of the relaxation type caused by foreign interstitials in bcc metals and fcc alloys. In the framework of this book internal friction of the hysteresis type is of little importance, but in order to preclude confusion the difference between relaxation and hysteresis will be explained.

In principle the measurement of internal friction is simple: the metal sample is caused to vibrate and the dissipation of energy (the transformation of vibration energy into heat) occurring as a result of internal causes is measured either directly or indirectly. In the atomic picture, any internal friction is ascribed to relative displacements of lattice imperfections under the influence of the vibration(1). Among the imperfections capable of directly or indirectly causing internal friction of the relaxation type are, apart from interstitial atoms, conduction electrons, phonons (acoustical quanta), vacancies, substitutional atoms, grain boundaries and dislocations. The latter can give rise to both relaxation and hysteresis phenomena. In ferromagnetic metals internal friction of the hysteresis type can be caused not only by the movement of dislocations but also by movement of domain walls.

3.2. Relaxation

In Section 1.4 it was mentioned that the existence of three types of octahedral interstices in a bcc metal leads to a redistribution of dissolved carbon,

1. C. M. Zener, *Elasticity and Anelasticity of Metals*, University Press, Chicago (1948).

nitrogen or oxygen atoms when a uniaxial tensile or compressive stress is applied to the metal. When a rod of bcc iron containing small quantities of nitrogen or carbon in solid solution is elastically stretched it will be noted that a gradually decreasing tensile stress is required to keep the strain constant. In other words, a *relaxation* of the metal—or rather of the tensile stress—occurs. If, on the other hand, a constant stress is applied then a large part of the strain takes place immediately, but it is followed by a gradual small increase. The latter is called elastic *after-effect*. Relaxation and after-effect are symptoms of one and the same phenomenon and, in uncomplicated cases, the one can be derived from the other. Both effects are therefore often referred to as relaxation.

Owing to the anelastic effects under discussion, the strain lags behind the stress when the latter is a periodic function of time (Fig. 83). The phase angle between the two is usually called the loss angle. As long as it is small, this angle is a direct measure of the fraction of the vibration energy that is dissipated during each vibration cycle. As in electrical circuit theory the loss angle δ is also indicated by Q^{-1} (Q = quality of the tuned circuit). In free oscillations the energy loss results in a reduction of the amplitude, i.e. in a damping of the oscillation. Besides the expression 'internal friction', the term 'damping' is also used to express the quantity Q^{-1}.

Since any relaxation is related to the time necessary for the internal equilibrium to adjust itself, the damping due to interstitials is dependent on the quotient of two frequencies, the jump frequency of the interstitial atoms and the frequency of oscillation of the sample containing the interstitials. There is practically no internal friction when the jump frequency is either much smaller or much larger than the frequency of oscillation. In the first case the internal equilibrium cannot follow the changing conditions at all; in the second case the internal equilibrium can adjust itself continuously, resulting at the same time in the modulus of elasticity or modulus of rigidity being smaller than that in the first case (Fig. 84). The damping is large when the two frequencies are of the same order of magnitude (cf. Section 3.4).

Whilst the internal friction of a relaxing material is strongly dependent on the frequency, it is virtually independent of amplitude, at least for small amplitudes.

3.3. Hysteresis

Dissipation of energy does not only arise from relaxation but also from hysteresis. In the simplest cases hysteresis can be recognised by the fact that there is loss of energy even when the frequency of vibration approaches zero, resulting in a closed loop in the stress–strain (σ, ϵ) diagram. After completing a cycle ϵ does not in general return to zero for $\sigma = 0$ and there is a remanent

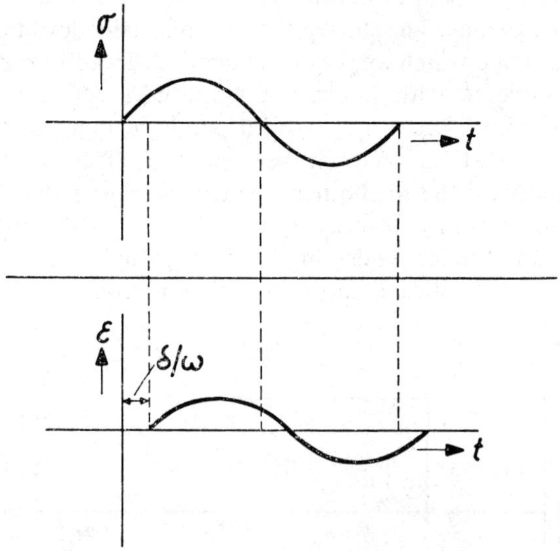

FIG. 83. When a metal, capable of relaxation, is subjected to an elastic stress σ which is a sine function of time, then the resulting strain ϵ always passes through zero somewhat later than σ (δ is the loss angle and ω the angular frequency of vibration).

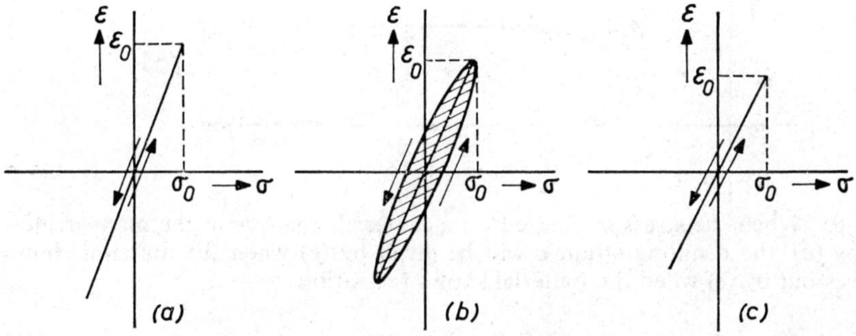

FIG. 84. This figure also refers to a material capable of relaxation and subjected to a stress $\sigma = \sigma_0 \sin \omega t$. The lagging of the strain behind stress is shown here in a σ, ϵ diagram (*b*). The hatched area represents the energy dissipated per cycle per unit of volume. At very low frequencies the material will behave as in (*a*), i.e. as a non-relaxing material. However, in that case, the amplitude ϵ_0 for the same value of σ_0 is greater than in (*b*). The same material will also show proportionality between ϵ and σ at relatively high frequencies (*c*). Its stiffness is then greater than in (*b*) and, obviously, than in (*a*).

deformation. Often the shape and the area of the loop in the σ, ϵ diagram are independent of frequency. The shape of the loop usually deviates considerably from the elliptic shape which applies to relaxation. In addition the loop does not uniformly increase with increasing amplitude, which means that for hysteresis the internal friction in general depends strongly on the amplitude.

There is thus a distinct contrast between the effects of hysteresis and relaxation. Simplifying the phenomena, it can be stated that in the case of relaxation the dissipation of energy is strongly dependent on frequency but independent of amplitude, whilst in the case of hysteresis the dissipation depends strongly on amplitude and only little on frequency.

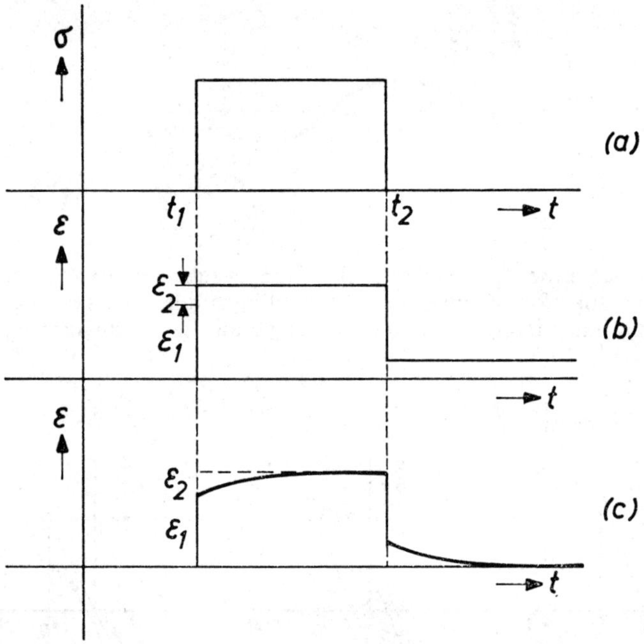

FIG. 85. When the stress σ, exerted on a material, changes in the manner indicated by (*a*), the resulting strain ϵ will be given by (*b*) when the material shows hysteresis but by (*c*) when the material shows relaxation.

Fig. 85 diagrammatically illustrates the difference between hysteresis and relaxation for a constant stress that is applied from time $t = t_1$ to time $t = t_2$ and then removed (*a*). A body exhibiting hysteresis will immediately respond to the stress by a strain $\epsilon = \epsilon_1 + \epsilon_2$, where ϵ_2 is the remanent part of the strain, remaining after the stress is removed (*b*). A material exhibiting relaxation requires time for ϵ_2 to adjust itself and behaves as indicated by (*c*).

The study of the various causes of relaxation is helped by the fact that they generally occur in different frequency regions. Similarly, the fact that the different causes of hysteresis become noticeable at different amplitudes is helpful in the study of hysteresis effects.

3.4. Damping due to interstitials in bcc metals

The magnitude of the internal friction caused by foreign interstitials in bcc metals obviously depends on the concentration of the interstitials and

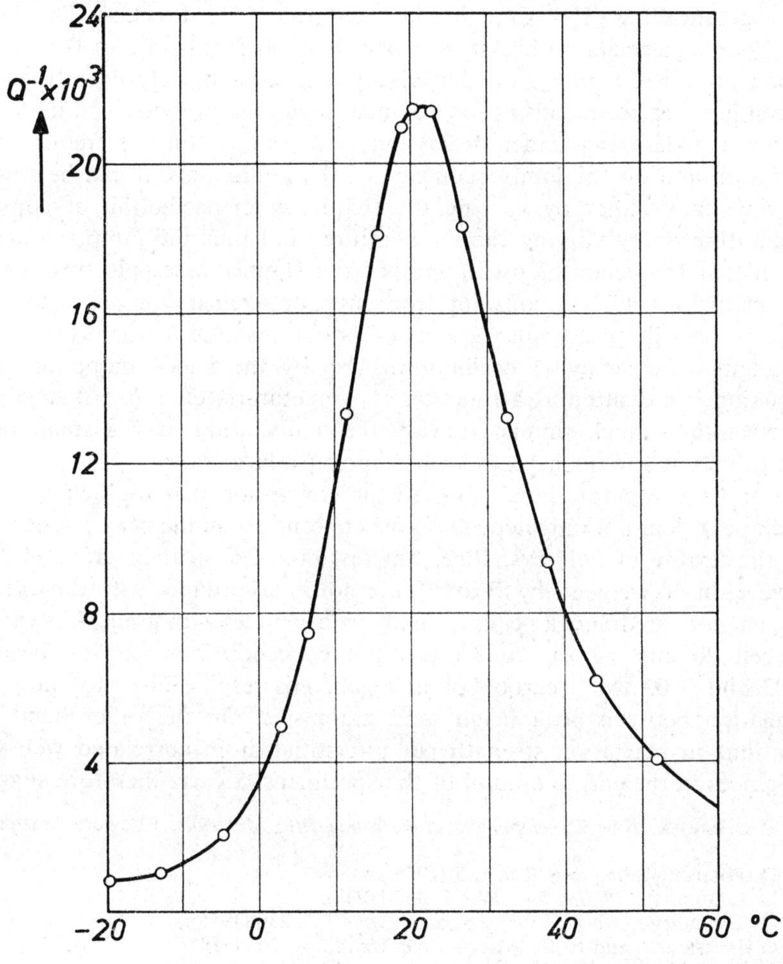

FIG. 86. Damping of the free torsional oscillations of a polycrystalline wire of bcc iron containing 0·027 wt % dissolved nitrogen. Frequency of oscillation 0·77 Hz (Fast).

on the magnitude of the tetragonal distortion caused by one interstitial atom. It occurred to Snoek[1] that for carbon-bearing bcc iron this distortion could be derived from the well-known tetragonality of martensite. In the theory expounded by Snoek and also by Polder[2], the internal friction in a single crystal of carbon-bearing iron is calculated for the case in which it is subjected to a periodically changing uniaxial stress. According to this theory the height of the damping peak is, at low carbon concentrations, proportional to the concentration and inversely proportional to the absolute temperature. It is strongly dependent on the direction of the uniaxial stress and is greatest in a $\langle 100 \rangle$ direction but zero in a $\langle 111 \rangle$ direction. The theory is in good agreement with experiment[3]. It applies not only to the case of carbon or nitrogen in α iron, but also to solutions of carbon, nitrogen or oxygen in other bcc metals as for instance vanadium, niobium and tantalum.

Since the damping under discussion is dependent on the frequency of oscillation and on the jump frequency of the interstitials, it can be studied experimentally either by varying the frequency of oscillation of constant temperature or by varying the temperature (and thus the jump frequency) at constant frequency of oscillation. In the simpler cases plotting internal friction against $1/T$ at constant frequency or against log ω at constant temperature will give similar graphs (T is the absolute temperature and ω the angular frequency of oscillation). Usually the Snoek damping under discussion is measured as a function of temperature (cf. Fig. 86) since it is experimentally much simpler to vary the temperature over a small range than to vary the frequency of oscillation over a large range.

Several investigators have come to the conclusion that the height of the Snoek peak does not only depend on the concentration, the peak temperature and the texture of polycrystalline samples, but also on their grain size[4–8]. However, measurements by Swartz[9] are not in accordance with this conclusion. He measured Snoek peaks of many iron samples with grain sizes varying between 20 and 74 μm and interstitial concentrations varying between 0·003 and 0·02 wt % carbon or nitrogen. His results failed to show any correlation between peak height and grain size. On the other hand, the variations in relaxation strength per interstitial atom correlated well with differences in the elastic moduli of the specimens. Swartz therefore suggests

1. J. L. Snoek, *New Developments in Ferromagnetic Materials*, Elsevier, Amsterdam (1947).
2. D. Polder, *Philips Res. Rep.* **1**, 5 (1945).
3. L. J. Dijkstra, *Philips Res. Rep.* **2**, 357 (1947).
4. G. Lagerberg and A. Josefsson, *Acta Metall.* **3**, 236 (1955).
5. G. Lagerberg and E. G. Wolff, *Acta Metall.* **6**, 136 (1958).
6. H. J. Seemann and W. Dickenscheid, *Acta Metall.* **6**, 62 (1958).
7. P. Stark, B. L. Averbach and M. Cohen, *Acta Metall.* **6**, 149 (1958).
8. J. D. Fast, *Métaux Corros. Inds.* **36**, 431 (1961).
9. J. C. Swartz, *Acta Metall.* **17**, 1511 (1969).

that the 'grain-size dependence' found by several people is non-existent and that all variations in relaxation strength per interstitial atom must be ascribed to variations in polycrystalline texture.

3.5. Damping due to interstitials in fcc metals

In the crystal of a face-centred cubic element any isolated foreign interstitial atom causes an elastic distortion with cubic symmetry. As long as their concentration is low, a redistribution of such atoms will cause no deformation of the crystal and, conversely, a deformation will cause no redistribution. Low concentrations of isolated interstitial atoms in fcc metals cannot therefore be the cause of internal friction. On the other hand, high concentrations can be the cause of damping since, in that case, the interstitials are partly present as nearest-neighbour pairs (i–i pairs).

In the absence of external forces the pair-axes will be distributed at random over the available crystallographic directions. Relaxation by reorientation of the pairs will take place when a tensile stress is applied in a given direction. An example of this kind of relation has been described by Tsien(1); his experiments are a continuation of the work carried out by him together with Kê and Mišek(2). Tsien measured the internal friction of nickel containing different amounts of carbon in solid solution and observed damping peaks with heights proportional to the square of the carbon content. The obvious explanation of this is that the internal friction here is caused by pairs of carbon atoms. Let us imagine that a C–C pair occupies interstices A and B of the unit cell in Fig. 87 (Ni atoms not shown); a vertical tensile stress will then generally cause one of the two carbon atoms to jump to one of the four positions indicated by crosses. This will result in a configuration in which the angle between the C–C axis and the stress direction is a minimum. The relaxation effects caused by C–C pairs in nickel have also been studied by Diamond and Wert(3).

Relaxation and internal friction in fcc metals can also be caused by atom pairs, each consisting of an interstitial and substitutional atom (i–s pairs). As an example we take the research by Mah and Wert(4) on the internal friction of the rare-earth metal ytterbium of about 99% purity. According to their experiments introduction of nitrogen or oxygen leads to a damping peak in the temperature region 100° to 200°C when the frequency of vibration is about 1 Hz. The height of the peak increases with increasing concentration

1. C. T. Tsien, *Scientia sin.* **10**, 930 (1961).
2. T. S. Kê, C. T. Tsien and K. Mišek, *Scientia sin.* **4**, 519 (1955).
3. S. Diamond and C. Wert, *Trans. AIME* **239**, 705 (1967).
4. G. Mah and C. A. Wert, *Trans. AIME* **230**, 16 (1964).

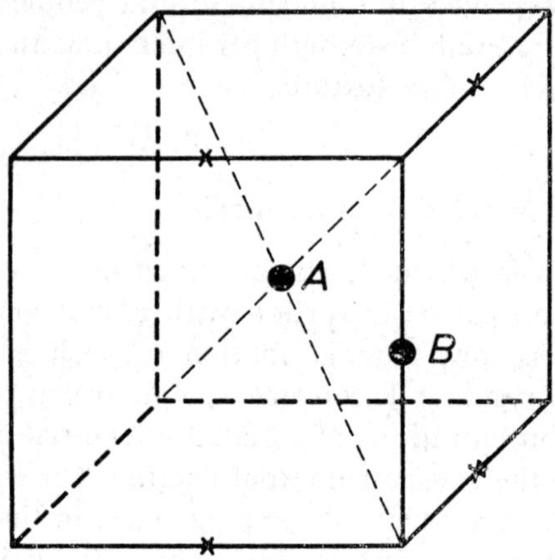

FIG. 87. When the octahedral interstices A and B of a unit cell of a fcc metal are occupied by two interstitial atoms (e.g. C, N or O), then a vertical tensile stress will tend to push one of the two atoms to one of the four interstices indicated by a cross. The metal atoms are not shown.

of dissolved oxygen or nitrogen. To understand this let us imagine that the ytterbium (fcc) contains dissolved oxygen and in addition a substitutional impurity X which has a greater affinity for oxygen than Yb and which also has a larger atomic radius than Yb. If the affinity difference is sufficiently large, each oxygen atom will for preference occupy an octahedral interstice with an X atom as one of its six nearest neighbours. Under compression in a vertical ⟨100⟩ direction it will be more favourable, as far as energy is concerned, for the oxygen atom to be at the side of an X atom than either just above or below it. The reverse holds when the atomic radius of the impurity X is smaller than that of Yb. Under a periodically changing load in the direction mentioned, the atom pairs X–O will, over a certain temperature range, continually change their orientation and thus cause internal friction. It is not to be expected that the activation enthalpy of this reorientation agrees exactly with that of the diffusion of oxygen in pure ytterbium.

Extreme cases of the type of damping under discussion have been studied by Russian and Chinese investigators(1–3). They measured the internal friction of fcc alloys (austenitic steels) containing high percentages of foreign substitutional atoms in addition to interstitially dissolved atoms. In these alloys (e.g. in a steel containing 25% Cr, 20% Ni and 0·3% C) they found

1. K. M. ROZIN and B. N. FINKEL'SHTEIN, *Dokl. Akad. Nauk SSSR* **91**, 811 (1953).
2. T. S. KÊ and C. T. TSIEN, *Scientia sin.* **5**, 625 (1956).
3. T. L. WU and C. M. WANG, *Scientia sin.* **7**, 1029 (1958).

an internal friction that could be explained only as arising from a redistribution of carbon atoms. Since the diffusivity, and therefore the jump frequency, of the carbon atoms is much lower in fcc than in bcc iron (see previous chapter, especially Tables 10 and 11) it can be expected that the damping peak in the fcc alloys under discussion occurs at a higher temperature than the Snoek peak in bcc iron, the damping being greatest at the temperature at which the jump frequency is comparable with that of the macroscopic vibration. At a vibration frequency of 1 Hz the peak is indeed at about 300°C in the case of the earlier-mentioned Fe–Cr–Ni–C alloy but at about 40°C in the case of a bcc Fe–C alloy. The value of the diffusivity of carbon in the fcc alloy at the temperature of 300°C where the peak occurs, as calculated from the results of the damping measurements (see next chapter), differs

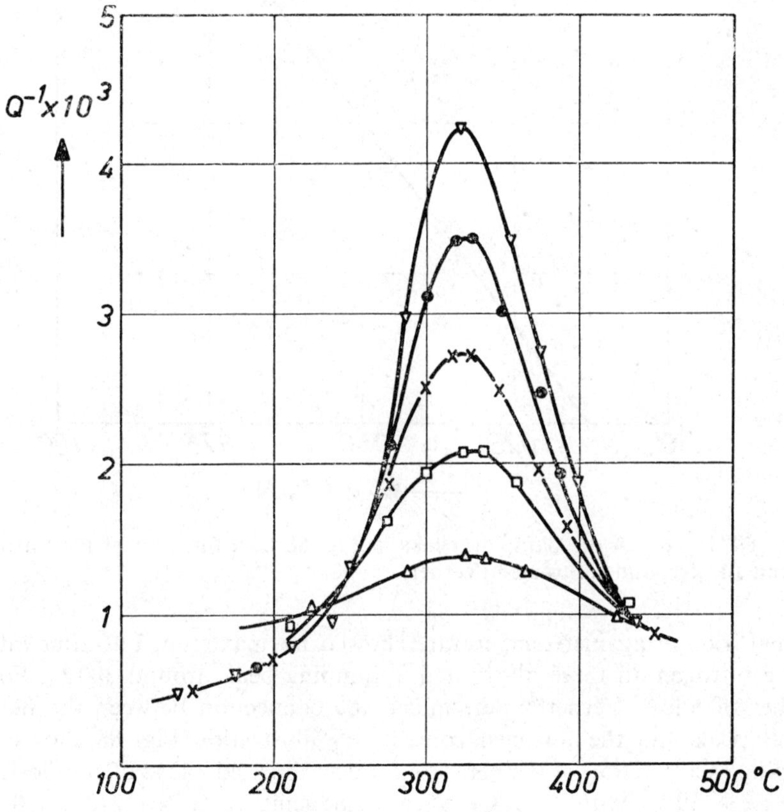

FIG. 88. Damping of the free torsional oscillations of five wires of fcc alloys containing 0·22 wt % N (lowermost curve), 0·36 wt % N, 0·47 wt % N, 0·60 wt % N and 0·77 wt % N (uppermost curve). The metallic composition of the alloys is 74% Fe, 19% Mn, 7% Cr. Frequency of oscillation 2 Hz (Verner).

only little from the value at this temperature obtained by extrapolation of the diffusivities of carbon in fcc iron measured above 900°C.

As is to be expected, nitrogen dissolved in fcc alloys causes relaxation effects similar to those caused by carbon. Verner *et al.*[(1)] carried out measurements on three decarburized fcc alloys of compositions 70% Fe + 30% Ni, 72% Fe + 28% Mn and 71% Fe + 20% Ni + 9% Mn. The graph of

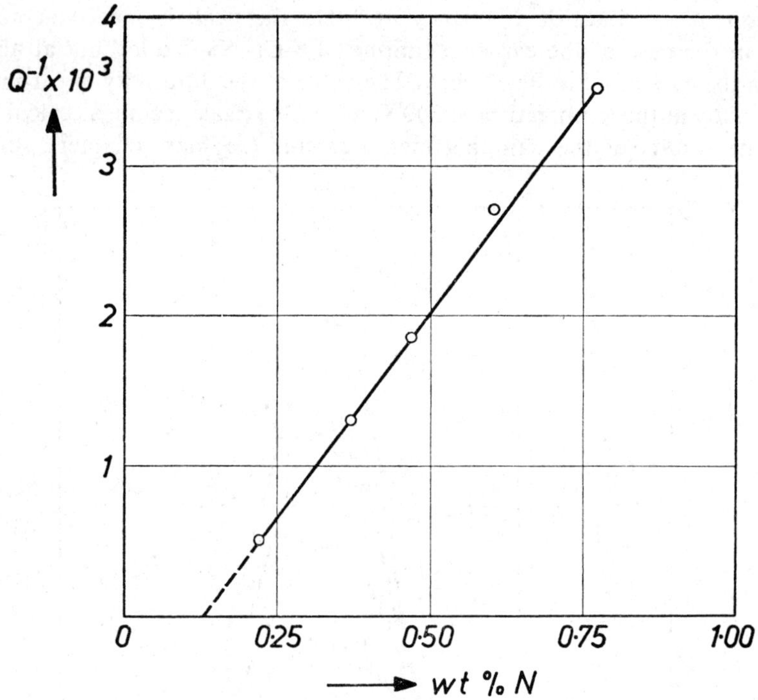

FIG. 89. Height of the damping peaks in Fig. 88 as a function of the nitrogen content. Background subtracted (Verner).

internal friction against temperature showed no maximum but after introducing nitrogen all three alloys had a damping peak around 300°C. For a number of alloys Verner[(2)] determined the connection between the height of this peak and the nitrogen content. As illustration Fig. 88 shows the damping as a function of temperature for decarburized alloys of composition 74% Fe + 19% Mn + 7% Cr after introducing respectively 0·22, 0·36,

1. V. D. VERNER, B. N. FINKEL'SHTEIN and A. V. SHALIMOVA, *Soviet Phys. Solid St.* **3**, 2441 (1962).
2. V. D. VERNER, *Phys. Metals Metallogr.* **14**, 70 (1962).

0·47, 0·60 or 0·77 wt % N and quenching from 1,150°C. All five alloys were completely austenitic. Fig. 89 shows that the height of the peak is a linear function of the nitrogen content. This means that the damping is caused by jumps of nitrogen atoms (reorientation of i–s pairs) and that here *pairs* of nitrogen atoms play only a minor part, if any.

A comparison of Figs. 86 and 88 shows that the damping under discussion, when expressed per nitrogen atom, is about 150 times greater in bcc iron than in the fcc alloys. This great difference also applies to carbon and is due to the fact that the octahedral interstices in these alloys deviate much less from cubic symmetry than they do in bcc iron.

In the next chapter we shall discuss examples of relaxation due to i–s and i–i defects in bcc metals. This type of relaxation in bcc metals takes place in addition to the well-known Snoek relaxation caused by isolated i defects (Section 3.4).

3.6. The cold-work peak

Iron–nitrogen and iron–carbon alloys

When a bcc metal, containing an interstitial impurity, is plastically deformed at room temperature it will exhibit not only the normal Snoek peak, but in addition an internal friction maximum at a higher temperature. This 'cold-work peak' was mentioned by Snoek[(1)] in his original paper on the internal friction of iron containing carbon or nitrogen. It has been studied extensively by other people[(2–8)]. Much higher cold-work peaks can be produced in iron containing nitrogen than in carbon-bearing iron. Figure 90 gives the cold-work peak of a sample of nitrogen-bearing iron as measured by Kê, *loc. cit.*, Fig. 91 that of a carbon-bearing sample as measured by Kamber *et al.*, *loc. cit.*

The nitrogen cold-work peak is found to be fully formed within the time required for its measurement, even when the measurement is done very quickly. The height of the carbon cold-work peak, however, is initially nearly zero. When a carburized specimen is heated at 100°C the carbon Snoek peak completely ages away in a relatively short time, and several hours of heating at 240°C are needed to obtain the maximum height of the cold-work peak. This strongly suggests that the carbon atoms causing the Snoek peak do

1. J. L. Snoek, *Physica, 's Grav.* **8,** 711 (1941).
2. T. S. Kê, *Trans. AIME* **176,** 448 (1948).
3. W. Köster, L. Bangert and R. Hahn, *Arch. Eisenhütt. Wes.* **25,** 569 (1954).
4. W. Köster and L. Bangert, *Acta Metall.* **3,** 274 (1955).
5. K. Kamber, D. Keefer and C. Wert, *Acta Metall.* **9,** 403 (1961).
6. T. Sugeno, K. Sakamoto and H. Ino, *J. phys. Soc. Japan* **18,** Suppl. I, 154 (1963).
7. D. P. Petarra and D. N. Beshers, *Acta Metall.* **15,** 791 (1967).
8. H. Ino and T. Sugeno, *Acta Metall.* **15,** 1197 (1967).

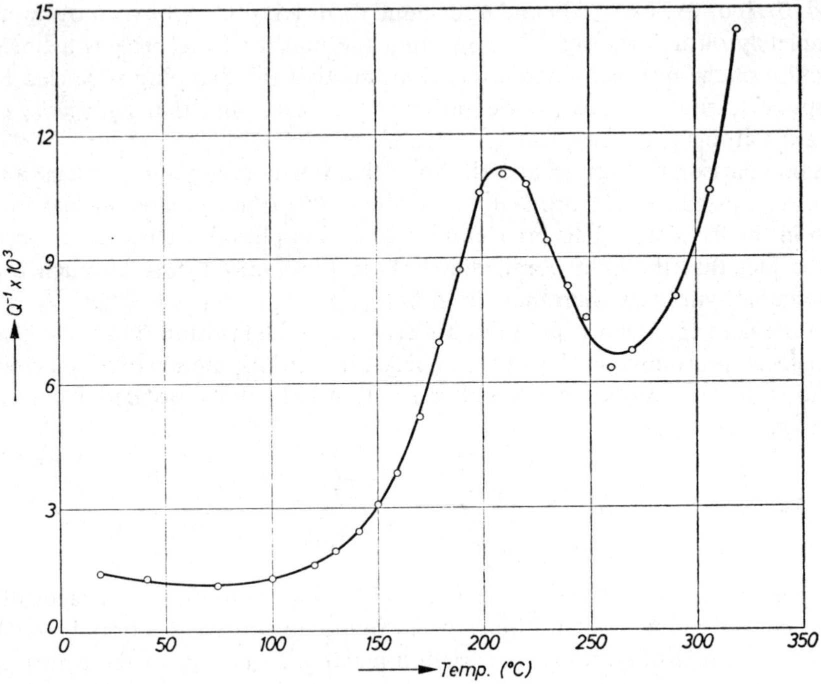

FIG. 90. Cold-work peak of an iron wire as measured by Kê after the following treatments: charging with nitrogen in a mixture of hydrogen and ammonia at 550°C; drawing through dies at room temperature (reduction in area 60%); heating for one hour at 350°C. Frequency of torsional oscillation 0·13 Hz.

not go directly to the sites where they give rise to the cold-work peak, but form a carbide precipitate as an intermediate step.

By means of transmission electron microscopy Rudee and Huggins[(1)] were indeed able to show that, during the heating at 100°C, precipitates form at grain boundaries in cold-worked carburized iron. These precipitates were found to dissolve during the subsequent heating at 240°C. Grain-boundary precipitates were not found during similar treatments of nitrogen-bearing iron and it was concluded that the presence of dislocations suppressed any precipitation of nitrogen. As the diffusivities of carbon and nitrogen in bcc iron are about the same, the less complex behaviour of specimens containing nitrogen may partly be caused by the higher solubility of this element in equilibrium with its precipitates. This allows mass transport and establishment of final equilibrium to proceed faster in nitrogen-bearing than in carbon-bearing samples.

1. M. L. RUDEE and R. A. HUGGINS, *Trans. AIME* **236,** 1662 (1966).

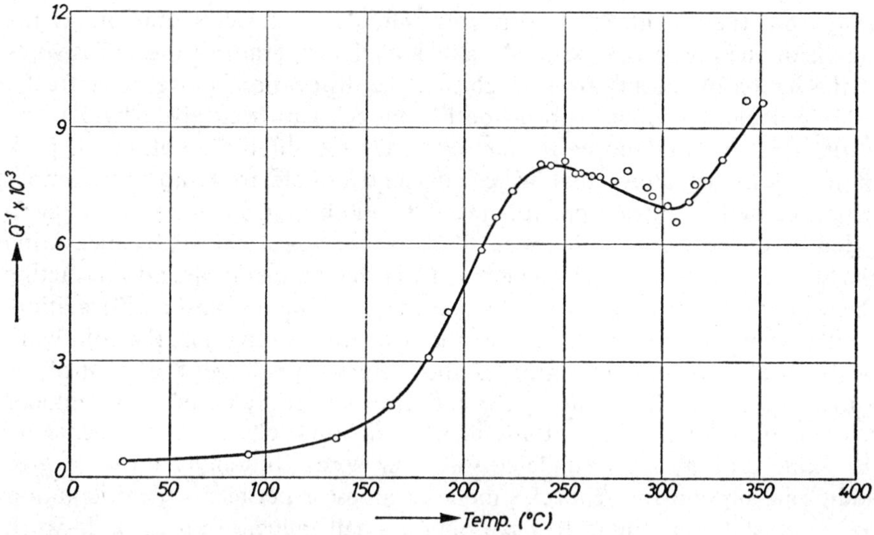

FIG. 91. Cold-work peak of an iron wire as measured by Kamber *et al.* after the following treatments: charging with carbon in a mixture of hydrogen and heptane; drawing through dies at room temperature (reduction in area 25%); heating for some hours at 250°C. Frequency of torsional oscillation 2·2 Hz.

For a given degree of deformation, the height of the nitrogen cold-work peak at first rises with increasing nitrogen content, but it stops rising quite abruptly when a certain value of the nitrogen content is exceeded. This saturation value, and the cold-work peak corresponding to it, are higher the larger the degree of deformation. No such simple relation is observed with specimens containing carbon alone. In that case an unexpectedly small cold-work peak is always found, irrespective of carbon content (Petarra and Beshers, *loc. cit.*). The activation energy of the nitrogen cold-work peak is 151 ± 17 kJ mole^{-1} (36 ± 4 kcal/mole).

There is no universally accepted opinion concerning the mechanism underlying the cold-work damping, but most investigators consider it to result from the interaction between the dislocations and the interstitial solute atoms. Köster *et al.*, *loc. cit.*, suggested that the peak might be caused by stress-induced reorientations of interstitial atoms in Cottrell atmospheres around dislocations. Schoeck[(1)], on the other hand, proposed a model in which the relaxation is caused directly by the movements of dislocations. At low temperatures the dislocations are immobilized by their Cottrell atmospheres (cf. Section 1.4). As the temperature rises a region will be passed where the dislocations are able to move with difficulty, experiencing a considerable

1. G. SCHOECK, *Acta Metall.* **11,** 617 (1963).

drag from their Cottrell atmospheres. This, in Schoeck's opinion, is the region of the cold-work peak. At still higher temperatures the diffusivities of the interstitial atoms are so high that the dislocations move more freely.

Experimental support for Schoeck's model has been given by Sugeno *et al.*, *loc. cit.*, and Ino *et al.*, *loc. cit.* They measured the cold-work peak of iron–nitrogen alloys that were sufficiently dilute to avoid any complication caused by nitride precipitates. The number of nitrogen atoms segregated at dislocations at the peak temperature was obtained by measuring the Snoek peak height after quenching from that temperature and subtracting this value from the Snoek peak height as measured before plastic deformation. In that way it could be shown that the relaxation strength of the cold-work peak per captured nitrogen atom is about eight times larger than that of the Snoek peak. This is not in agreement with Köster's idea of stress-induced reorientations of nitrogen atoms in Cottrell atmospheres. Experiments by the same authors with single-crystal iron wires containing nitrogen give additional support for Schoeck's model. In these experiments the orientation of the crystals was found to have only a small influence on the cold-work damping. In the opinion of the authors these results also make Köster's model untenable. However, there remains the possibility that no sharp distinction can be drawn between the two models since reorientations and dislocation movements can probably take place simultaneously.

At a temperature below that of the cold-work peak another relaxation peak may be expected to appear in measurements performed immediately after deformation, i.e. under conditions in which the dislocations are not yet locked by Cottrell atmospheres, but only to a much lesser extent by dilute Snoek atmospheres (cf. Section 1.4). At a frequency of 1 Hz a relaxation peak, probably caused by the interaction of dislocations and Snoek atmospheres, has, in fact, been found at about 100°C by Agarwala and Beshers[(1)]. This intermediate peak is very unstable since Cottrell atmospheres form rapidly at 100°C. For a theoretical treatment of the drag of Cottrell atmospheres and Snoek atmospheres by dislocations, the reader is referred to a book by Hirth and Lothe[(2)].

An explanation of the cold-work peak differing from the explanations given by Köster *et al.*, *loc. cit.*, and Schoeck, *loc. cit.*, has been offered by Hermant[(3)]. He suggests that the peak may be related to the segregation of interstitials into enlarged sites present in stacking faults of the deformed bcc lattice. Up to the time of writing little support in favour of this hypothesis has been found.

1. B. N. Agarwala and D. N. Beshers, *J. appl. Phys.* **39,** 2473 (1968).
2. J. P. Hirth and J. Lothe, *Theory of Dislocations*, McGraw-Hill, New York (1968), pp. 594–616.
3. M. E. Hermant, Ph.D. Thesis, Univ. of Amsterdam (1966) and *Acta Metall.* **16,** 1 (1968).

Iron–hydrogen alloys

Several investigators[1–4] have found a cold-work peak when measuring the internal friction of hydrogen-bearing iron or steel. Contrary to the results obtained with iron–nitrogen alloys (see above) the hydrogen cold-work peaks show irregular variations in shape and peak temperature. The latter lies in the range 100 to 140 K for a frequency of 1 Hz. As found by Sturges and Miodownik, *loc. cit.*, much more consistent results and a constant peak temperature of about 100 K are obtained when the plastically-deformed iron samples are kept at room temperature for several days before being charged with hydrogen. Since their samples contained a relatively large amount of carbon (0·018 wt %) and nitrogen (about 0·001 %) they assume that the reproducibility of their measurements is due to pinning of the dislocations by carbon and nitrogen atoms and the consequent suppression of one or more additional damping maxima, the highest of which occurs at about 170 K according to Hermant, *loc. cit.* On this assumption Sturges and Miodownik come to the conclusion that the cold-work peaks obtained after ageing are not caused by dislocation movements as suggested by Schoeck, *loc. cit.*, for other interstitial bcc alloys.

There are many theories about the mechanism of cold-work damping in Fe–H alloys. Gibala, *loc. cit.*, adheres to the Schoeck model, Sturges *et al.*, *loc. cit.*, suggest that the peak is associated with the stress-induced reorientation of interstitial hydrogen atoms in the immediate vicinity of dislocations, and Hermant, *loc. cit.*, advocates his own theory according to which the peak is related to the presence of hydrogen atoms in stacking faults of the iron lattice. Since the peak is only found in cold-worked iron and steel, there can be little doubt that it is related to structural imperfections which are developed by the deformation and which act as traps for the hydrogen. Oriani[5], in a thermodynamic analysis of the trapping of hydrogen in iron and steels, considers mainly dislocations, solid to solid interfaces and microcracks as trapping sites. To these may be added crack nuclei and clusters of vacancies formed by the coalescence of a number of dislocations or vacancies, respectively.

We believe the abnormal damping behaviour of hydrogen as compared with that of nitrogen to be caused mainly by the strong tendency of hydrogen to leave the lattice and form H_2 molecules. The solubility of hydrogen in pure, imperfection-free iron is negligible at 100 K, even at high hydrogen pressures. Thus, virtually all of the hydrogen electrolytically introduced

1. L. C. Weiner and M. Gensamer, *Acta Metall.* **5**, 692 (1957).
2. W. R. Heller, *Acta Metall.* **9**, 600 (1961).
3. R. Gibala, *Trans. AIME* **239**, 1574 (1967).
4. C. M. Sturges and A. P. Miodownik, *Acta Metall.* **17**, 1197 (1969).
5. R. A. Oriani, *Acta Metall.* **18**, 147 (1970).

into plastically deformed iron will be in molecular form in microcracks, crack nuclei and other microvoids, and to a much lesser extent as atoms in favourable, abnormal lattice sites surrounding these imperfections. The cold-work peak may then be assumed to be due to a continual, stress-induced shift of the equilibrium between molecular and atomic hydrogen, the largest contribution coming from the very small voids, e.g. crack nuclei which can accommodate only a relatively small number of H_2 molecules. No satisfactory theory of this process is as yet available, but the close correspondence between the activation energy of the peak (see next chapter) and that of the migration of hydrogen in iron below 200°C suggests that the same mechanism is of fundamental importance in both processes.

According to Section 2.13 the rate of migration at low temperatures is not controlled by diffusion, but probably by the continual redissolution of hydrogen after its precipitation in molecular form in traps. The activation energy associated with this migration is 32·6 kJ mole^{-1} (7·8 kcal/mole) according to Johnson and Hill[(1)]. It may be compared with the activation energy associated with the cold-work peak, which is 30·5 kJ mole^{-1} (7·3 kcal/mole) according to Hermant, *loc. cit.*, $35{\cdot}2 \pm 5{\cdot}9$ kJ mole^{-1} ($8{\cdot}4 \pm 1{\cdot}4$ kcal/mole) according to Gibala, *loc. cit.*, and 29·3 kJ mole^{-1} (7·0 kcal/mole) according to Sturges *et al.*, *loc. cit.* It may also be compared with the activation energy associated with the permeation of hydrogen through iron membranes, which is 36·4 kJ mole^{-1} (8·7 kcal/mole) according to Post and Ham[(2)].

Cold work damping in other bcc alloys

Cold-work peaks have been observed for several other bcc interstitial alloys. For Nb–N alloys Boone and Wert[(3)] found a peak temperature around 500°C at an oscillation frequency of 0·7 Hz. At the same frequency van Ooijen and van der Goot[(4)] found cold-work peak temperatures of approximately 420°C and 500°C, respectively, for O and N in niobium, and 470°C and 540°C for O and N in tantalum.

Schoeck and Mondino[(5)] and de Lamotte and Wert[(6)] found a peak at about the temperature of the nitrogen Snoek peak in cold-worked Ta–O alloys. They believed it to be the cold-work peak due to oxygen. However, considering its low temperature, this peak must almost certainly be identified

1. E. W. Johnson and M. L. Hill, *Trans. AIME* **218,** 1104 (1960).
2. C. B. Post and W. R. Ham, *J. chem. Phys.* **5,** 913 (1937).
3. D. H. Boone and C. A. Wert, *J. phys. Soc. Japan* **18,** *Suppl.* I, 141 (1963).
4. D. J. van Ooijen and A. S. van der Goot, *Acta Metall.* **14,** 1008 (1966); *Philips Res. Repts.* **22,** 150 (1967).
5. G. Schoeck and M. Mondino, *J. phys. Soc. Japan* **18,** *Suppl.* I, 149 (1963).
6. E. de Lamotte and C. Wert, *J. phys. Soc. Japan* **19,** 1560 (1964).

as the Snoek peak due to nitrogen, the latter element being nearly always present as an impurity in Ta and Ta–O alloys. It might be objected that in Schoeck and Mondino's experiments the peak under discussion was initially absent and was only observed after cold-deformation, but in van Ooijen and van der Goot's experiments the same phenomenon was observed when measuring the Snoek damping of Nb–O alloys containing nitrogen. In wires, soft-annealed at 1,000°C, only a small fraction of the nitrogen content was found to contribute to the Snoek peak, probably because most nitrogen atoms were present in orderly arrayed domains (cf. Section 4.14). Cold-work causes disarray and enables the nitrogen atoms to contribute to the Snoek damping. In the experiments reported by Schoeck and Mondino, *loc. cit.*, and by de Lamotte and Wert, *loc. cit.*, it was impossible to find any real cold-work peak since the internal friction was measured only up to temperatures of about 400°C.

Cold-work peaks in niobium and tantalum caused by hydrogen have been studied by Mazzolai and Nuovo[1]. Since hydrogen dissolves exothermally in both metals, complications due to H_2 formation do not arise. Snoek peaks were found after the introduction of hydrogen into the annealed metals and, in addition, cold-work peaks were found in metals containing both hydrogen and dislocations. According to the authors the behaviour of the latter peaks is in agreement with the atmosphere-dragging model suggested by Schoeck, *loc. cit.*, and discussed above. Application of Schoeck's theory gives a hydrogen-dislocation binding energy of $-6{\cdot}8$ kJ mole^{-1} ($-0{\cdot}07$ eV) for Nb and $-12{\cdot}5$ kJ mole^{-1} ($-0{\cdot}13$ eV) for Ta. The absolute values of these energies are small compared to those found for other interstitials.

Cold-work damping in fcc alloys

All cold-work peaks discussed up to now have been observed in bcc metals containing C, N, O or H as interstitials. Arons[2] found a hydrogen cold-work peak in β palladium, i.e. in fcc palladium containing more hydrogen than H/Pd = 0·6 (cf. Section 1.9). Palladium samples were charged with hydrogen by heating them at 350°C in hydrogen at a pressure of more than 2·5 MN m^{-2} and subsequently cooling them slowly. By this procedure the two-phase region (Fig. 26) was avoided and, therefore, no plastic deformation was introduced. Only one damping peak, at 80 K for a frequency of 3 Hz, was found when measuring the internal friction of this material. It was believed to be caused by the stress-induced redistribution of the hydrogen atoms in the hydrogen–vacancy sublattice of the β palladium phase.

1. F. M. Mazzolai and M. Nuovo, *Solid State Commun.* **7**, 103 (1969).
2. R. R. Arons, *Acta Metall.* **15**, 144 and 1673 (1967).

After intentional plastic deformation a second maximum, with a peak temperature of about 100 K, was observed: the hydrogen cold-work peak. Its height was found to decrease with ageing at temperatures between 20° and 200°C. This phenomenon was thought to be caused by traces of carbon or other interstitial impurities diffusing to the dislocations and pinning them. Strong evidence in favour of this assumption came from the observation that an additional strain after the ageing treatment restored the peak height almost to its original value. From these experiments it was concluded that this cold-work peak is caused by the movement of dislocations pulling their hydrogen atmospheres along.

3.7. Exchange of solutes at dislocations

Some interstitials are bound more strongly to dislocations than others. Under favourable conditions the difference in binding energy may cause segregated atoms to be expelled from the dislocations and to be replaced by the atoms with the greater binding energy. Internal friction studies by van Ooijen and van der Goot[1, 2] show convincingly that, above a certain temperature, this may occur in niobium and tantalum containing both oxygen and nitrogen. We shall now discuss some of these authors' results.

When dissolved randomly in wires of Nb and Ta, oxygen and nitrogen were found to give rise to the familiar Snoek peaks. Cold-working the wires and subsequent heating at 200°C caused the height of the Snoek peak due to oxygen to decrease considerably and to give rise to the appearance of a new internal-friction peak at a higher temperature. Both facts can be explained by assuming that, during the heating at 200°C, part of the oxygen atoms segregate at the dislocations where they can no longer cause Snoek damping but only cold-work damping.

Heating the wires at temperatures above 400°C caused the oxygen cold-work peak to disappear and the oxygen Snoek peak to return to its original height. These phenomena were accompanied by a considerable decrease in height of the nitrogen Snoek peak and the appearance of a nitrogen cold-work peak. Figure 92 shows this behaviour for a tantalum wire containing 0·05 atom % O and 0·06 atom % N. The shaded circles refer to a wire annealed at 1,000°C; the open circles relate to the same wire after it had been cold-worked, resulting in a 30% increase in length, and heated for one hour at 200°C. The cold-work peak of oxygen (CWP-O) had appeared (at 470°C), whilst little was left of the Snoek peak due to oxygen (SP-O). This decrease of the Snoek peak is a direct measure of the amount of oxygen that has segregated at the newly formed dislocations. This migration has

1. D. J. van Ooijen and A. S. van der Goot, *Philips Res. Repts.* **19**, 505 (1964).
2. D. J. van Ooijen and A. S. van der Goot, *Philips Res. Repts.* **22**, 150 (1967).

been made possible by the heating at 200°C at which temperature the oxygen atoms in tantalum make on average about 200 jumps per second (cf. Table 10 and equation (4.5.6)). The nitrogen atoms reach this jump frequency only at about 420°C. As a consequence, the height of the nitrogen Snoek peak (SP-N), measured at rising temperature, was found to be unchanged after cold-working.

As the temperature increases further, the measurements represented by open circles show the appearance, not only of an oxygen cold-work peak as mentioned above, but also of a nitrogen cold-work peak (CWP-N). After a temperature of 650°C had been reached the internal friction was measured at falling temperature (crosses). The figure shows that (a) the oxygen cold-work peak has disappeared, (b) the height of the nitrogen Snoek peak has decreased and (c) the height of the oxygen Snoek peak has returned to its original value.

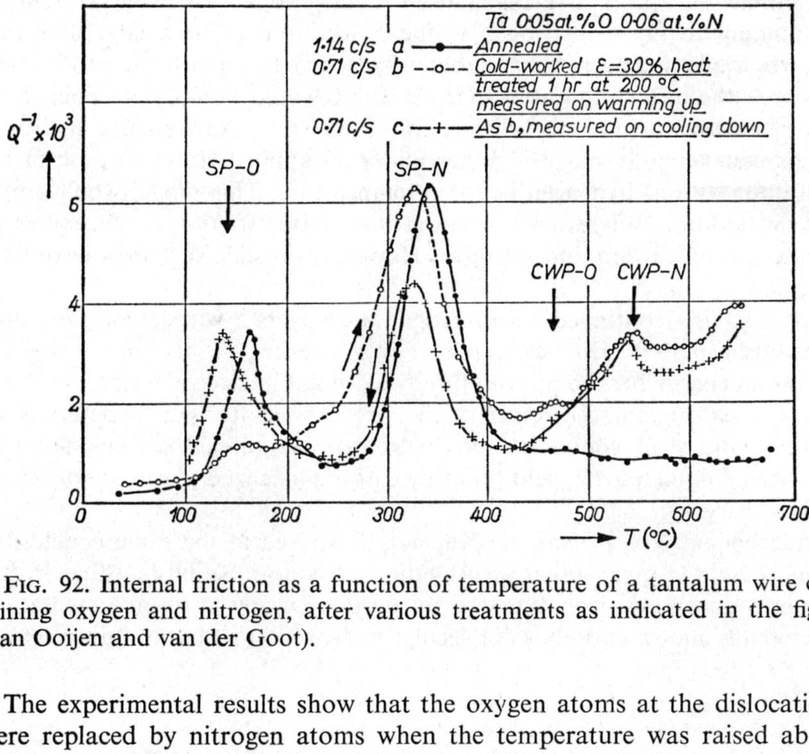

FIG. 92. Internal friction as a function of temperature of a tantalum wire containing oxygen and nitrogen, after various treatments as indicated in the figure (van Ooijen and van der Goot).

The experimental results show that the oxygen atoms at the dislocations were replaced by nitrogen atoms when the temperature was raised above about 400°C. This replacement is irreversible since further runs of the temperature up to 650°C and down to 20°C (points not shown in the figure) did not change the distribution of the interstitials. The only plausible explanation seems to be that in niobium and tantalum the binding energy of nitrogen to dislocations is greater than that of oxygen. In the experiments discussed it was only for kinetic reasons that the favourable sites at the dislocations were initially occupied by oxygen atoms.

In a similar study of the internal friction and strain-ageing of niobium and tantalum containing interstitial solutes, Szkopiak *et al.*[1, 2] arrived at conclusions similar to those of van Ooijen and van der Goot, *loc. cit.* However, Szkopiak *et al.* held carbon atoms rather than nitrogen atoms responsible for replacing oxygen atoms at the dislocations in niobium.

Before 1959 several investigators believed hydrogen to be capable of replacing carbon and nitrogen at the dislocations in iron and steel. This opinion was based on experiments by Rogers[3] and on one of the possible explanations given by him. The experiments showed that electrolytic charging with hydrogen eliminated the room-temperature yield point in Armco iron and rimmed steel. In the explanation mentioned above, Rogers assumed the binding energy of hydrogen to the dislocations to be greater than that of carbon and nitrogen thus enabling hydrogen to replace the other interstitials in the dislocation atmospheres. There would be no yield point after this replacement since the dislocations and their surrounding hydrogen clouds could readily move together under an applied stress because of the high diffusivity of hydrogen at room temperature. This model would imply that the addition of hydrogen increases the internal friction peaks associated with dissolved carbon and nitrogen. However, no such increases have been observed.

After having continued his investigations Rogers[4] withdrew his explanations given in 1956. His new explanation was based on his finding that the activation energy for the return of the yield point in electrocharged steel was the same as that for the return of the yield point after plastic deformation of the same steel containing no hydrogen. He therefore suggested that hydrogen eliminates the yield point by causing localized plastic deformation during charging.

Independently of Rogers, de Kazinczy[5] arrived at the same conclusion. From these and many other investigations it is now established that hydrogen, introduced into iron and steel by electrocharging at room temperature, precipitates almost entirely as molecular hydrogen in all lattice imperfections

1. Z. C. Szkopiak and L. W. Derry, *J. nucl. Mater.* **13,** 130 (1964).
2. Z. C. Szkopiak and W. Eliasz, *J. less-common Metals* **11,** 273 (1966).
3. H. C. Rogers, *Acta Metall.* **4,** 114 (1956).
4. H. C. Rogers, *Trans. AIME* **215,** 666 (1959).
5. F. de Kazinczy, *Acta Metall.* **7,** 706 (1959).

that provide enough space[1–6]. The stresses resulting from the build-up of hydrogen pressure inflict permanent damage on the metal in the form of microcracks and voids. The yield point is eliminated since the new dislocations created around the voids are not yet locked by carbon or nitrogen.

1. F. Garofalo, Y. T. Chou and V. Ambegaokar, *Acta Metall.* **8,** 504 (1960).
2. A. S. Tetelman, C. N. J. Wagner and W. D. Robertson, *Acta Metall.* **9,** 205 (1961).
3. D. J. van Ooijen and J. D. Fast, *Acta Metall.* **11,** 211 (1963).
4. J. D. Fast and D. J. van Ooijen, *Philips tech. Rev.* **24,** 221 and 252 (1962/3).
5. K. Farrell, *J. Iron Steel Inst.* **203,** 71 and 457 (1965).
6. M. E. Hermant, Ph.D. Thesis, Univ. of Amsterdam (1966).

4

THE SNOEK EFFECT AND ITS SIGNIFICANCE FOR METAL–GAS PROBLEMS

4.1. Relaxation time

If a material exhibits elastic after-effects, then, according to Chapter 3, its response to a certain stress is a strain built up of two parts: a part ε_1 reaching its final value in an immeasurably short time and a much smaller part ε_2, which needs time to adjust itself and which only reaches its final value ε_{2e} after an 'infinitely long time'. In reverse, ε_1 disappears instantaneously on removal of the stress whilst ε_2 reduces gradually. This behaviour has already been depicted in Fig. 85(*c*). In this chapter, as in Section 3.2, the word relaxation will include both relaxation and after-effect.

Let us assume that the relaxing material is a carbon-containing crystal of bcc iron that is being elastically stretched in the direction of a $\langle 100 \rangle$ axis (say the z axis) by a constant stress. Before the application of the stress the N interstitial carbon atoms will be equally distributed over the three types of sites:

$$N_x = N_y = N_z = \frac{N}{3} \tag{4.1.1}$$

Under the influence of the stress the population of the z sites will increase at the expense of those at the x and y sites. If we indicate this increase at time t by n, then

$$n = N_z - \frac{N}{3} \tag{4.1.2}$$

where n and N_z are functions of time; the final value of n will be indicated by n_e. On removal of the tensile stress the excess carbon atoms in z sites will jump back to x and y sites. With the low carbon concentrations of at the most a few hundredths per cent by weight being considered here, the carbon atoms move virtually independently of each other. The rate of decrease of n on removal of the stress is therefore proportional to the value of n. Thus:

$$\frac{dn}{dt} = -\frac{n}{\tau_r} \qquad (4.1.3)$$

where the constant τ_r has the dimension of time and is usually referred to as the relaxation time.

An equation similar to (4.1.3) applies to many other processes, for instance to the rate of decay of a radioactive element. The similarity is due to the fact that identical radioactive nuclei disintegrate independently of each other with an equal disintegration chance, just as the carbon atoms in iron jump independently with an equal chance of jumping for each of them.

If, before the removal of the stress, n had reached its final value n_e, then integration of (4.1.3) results in

$$n = n_e \exp(-t/\tau_r) \qquad (4.1.4)$$

It follows from the above that (4.1.4) represents the tail of the graph in Fig. 85(*c*). Starting from the initial situation represented by expression (4.1.1), then at a certain time after the application of the constant tensile stress, $(n_e - n)$ carbon atoms will still have to move to z sites in order to realize the new equilibrium. Here n is again the number of excess carbon atoms on z sites ($n = 0$ at $t = 0$). Similar to (4.1.3) the rate of decrease of $(n_e - n)$ is given by

$$\frac{d(n_e - n)}{dt} = -\frac{1}{\tau_r}(n_e - n) \qquad (4.1.5)$$

or, after integration:

$$n = n_e[1 - \exp(-t/\tau_r)] \qquad (4.1.6)$$

This equation corresponds to the rising part of the graph in Fig. 85(*c*).From measurements on carbon and nitrogen martensites (Fig. 10) it can be concluded that the additional strain ε_2 in the z direction is directly proportional to the excess carbon or nitrogen atoms n at z sites. For this reason n and n_e in equations (4.1.3) to (4.1.6) may be replaced by ε_2 and ε_{2e}.

Strictly speaking we should distinguish between relaxation time at constant stress and relaxation time at constant strain. In the cases that are of interest for the purpose of this book the difference between these two times is so small that we shall disregard it.

4.2. Compliance

Relaxation effects do not occur only when a metal is mechanically stressed. It can be stated more generally that they often occur when gaseous, liquid or solid substances are subjected to mechanical, electrical or magnetic forces. Two examples, for instance, are: (a) when a dielectric substance is suddenly subjected to a constant electrical field E, part of the polarization P will occur instantaneously but usually a gradual increase of P follows; (b) when a magnetizable body is suddenly brought into a constant magnetic field H, part of the magnetization I will occur immediately, but this is usually followed by a gradual increase in I. These are cases of dielectric and magnetic relaxation if the polarization and magnetization return to zero upon removal of the field.

In the literature the mechanical relaxation of high polymers is often referred to as 'creep'. This is confusing since in metallurgy the word creep has a quite different meaning, namely a *permanent* deformation which takes place slowly at fairly high temperatures when a not too well defined value of the tensile stress (which drops with increasing temperature) is exceeded.

Indicating electrical field strength, magnetic field strength or mechanical stress with the letter x as independent variable, and polarization, magnetization and strain with the symbol y, the latter will be proportional to x in simple cases. In general, actual materials only show this proportionality for very small values of x and y and then only approximately. The constant s in the relation

$$y = sx \tag{4.2.1}$$

is called *compliance*. The greater s, the more the material gives in to the stress put upon it. The electrical and magnetic compliances are usually referred to as electrical and magnetic susceptibility (μ and χ). The mechanical compliance of a solid subjected to a single stress is given by the reciprocal of the modulus of elasticity M. Restricting ourselves to these three cases, equation (4.2.1) represents the following three relations:

$$P = \mu E \tag{4.2.2}$$

$$I = \chi H \tag{4.2.3}$$

$$\varepsilon = M^{-1}\sigma \tag{4.2.4}$$

There is no such thing as *the* mechanical compliance of a solid, since even an isotropic solid material has several mechanical compliances. Compliance is found not only under uniaxial strain conditions but also under conditions of shear or isotropic compression and is then identical with the reciprocal shear modulus or the compressibility.

A distinction should be drawn between the *unrelaxed compliance* s_1, corresponding to the part y_1 of y that adjusts itself immediately (ε_1 in Fig. 85(c)) and the *relaxed compliance* s_e, corresponding to the value

$$y_e = y_1 + y_{2e} \tag{4.2.5}$$

which is found after a long time interval (Fig. 93).

It is only very seldom that the relaxation effects can be defined by one relaxation time τ_r as in the simple example in the previous section. In that case the rate at which y_2 adjusts itself after applying the constant 'load' x is given by

$$\frac{d(y_{2e}-y_2)}{dt} = -\frac{1}{\tau_r}(y_{2e}-y_2) \qquad (4.2.6)$$

an equation similar to (4.1.5).

4.3. Time-dependent fields or stresses: derivation of the Debye equations

After having considered that case where x has a constant value, we shall now investigate what happens during a periodic variation of x, i.e. of E, H or σ. Let us therefore assume that the function of time is given by

$$x = x_0 \cos \omega t \qquad (4.3.1)$$

where x_0 is the amplitude of x and $\omega/2\pi$ the frequency in hertz. If the material shows relaxation, then equation (4.2.1) will be satisfied only at relatively very low or relatively very high frequencies. In both cases the y, x diagram will show a straight line with a slope determined by the unrelaxed compliance when $\omega\tau_r \gg 1$ or by the relaxed compliance when $\omega\tau_r \ll 1$ (see (a) and (c) in Fig. 84). Over a large range of frequencies y will always pass through its zero value somewhat later than x, resulting in a closed loop in the y, x diagram (Fig. 84(b)). In this case y as a function of time is given by

$$y = y_0 \cos(\omega t - \delta) \qquad (4.3.2)$$

where δ is the phase shift which, as will be shown in Section 4.8, is a measure of the dissipation of energy. For this reason δ is usually called the *loss angle*.

The dependence of δ on $\omega\tau_r$ can be derived from equation (4.2.6). Since the decrease in $y_{2e}-y_2$ equals the increase in $y_2 = y-y_1$ and, in addition, $y_{2e} - y_2 = y_e - y$ (Fig. 93) equation (4.2.6) can also be written as:

$$\tau_r \frac{d(y-y_1)}{dt} = y_e - y \qquad (4.3.3)$$

or, since $y_1 = s_1 x$ and $y_e = s_e x$

$$\tau_r \frac{d(y-s_1 x)}{dt} = s_e x - y \qquad (4.3.4)$$

The relation between δ and $\omega\tau_r$ can now be found by substituting in this equation the values for x and y as given by equations (4.3.1) and (4.3.2). The desired result is obtained in the quickest way with the aid of complex

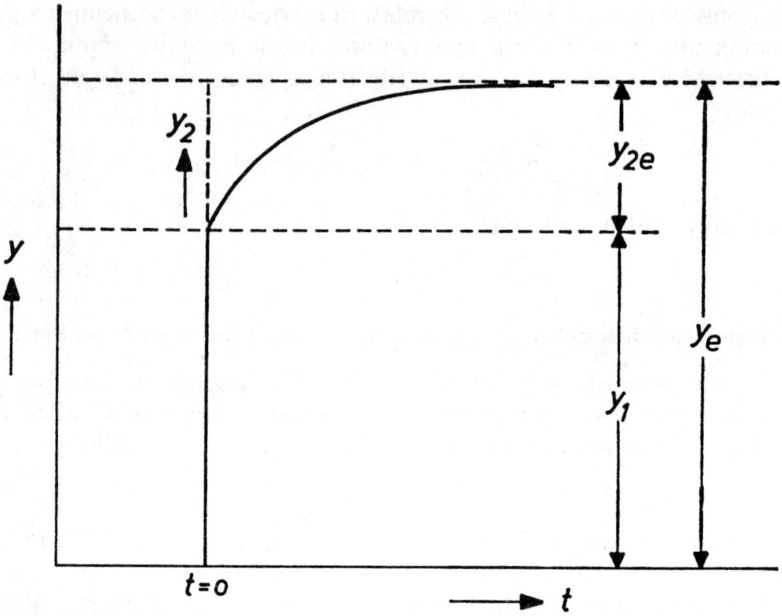

FIG. 93. A constant stress, electric field or magnetic field x that is suddenly applied to a medium capable of elastic, electric or magnetic relaxation, will give rise to strain, polarization or magnetization y of which the part y_1 is present instantaneously whilst part y_2 is time-dependent. The final value of y can be expressed as $y_e = y_1 + y_{2e}$.

quantities[1]. To this end we represent the oscillation $x = x_0 \cos \omega t$ by a vector of length x_0 rotating counter-clockwise with angular velocity ω in a plane around the origin of a system of perpendicular axes. Assuming the vector to coincide with the positive x axis at time $t = 0$, it will make an angle ωt with this axis after a time interval t (Fig. 94). The projection of the tip of the vector on the horizontal axis describes a harmonic oscillation $x_0 \cos \omega t$. The quantity $y_0 \cos (\omega t - \delta)$ can be represented in a similar manner (see Fig. 94). Regarding the horizontal and vertical axes as the real and imaginary axes of the complex plane we introduce the complex quantities

$$x = x_0 \exp (i\omega t) \tag{4.3.5}$$

$$y = y_0 \exp [i(\omega t - \delta)] \tag{4.3.6}$$

where i is the imaginary unit $\sqrt{-1}$. In agreement with (4.3.1) and (4.3.2) the

1. Readers not familiar with calculations involving complex quantities are referred to Section 4.4 where the desired relation is derived without the use of these. It will be seen that the latter method is much more cumbersome.

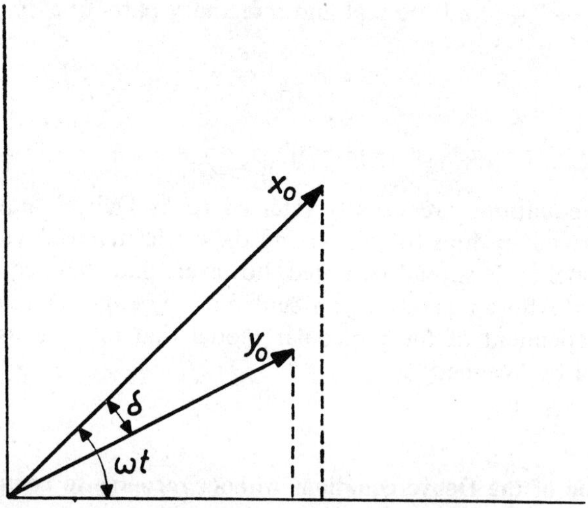

FIG. 94. The quantity $x_0 \cos \omega t$ can be represented by the horizontal projection of a vector of length x_0 rotating counterclockwise in the plane of the drawing with a constant angular velocity ω. The quantity $y_0 \cos(\omega t - \delta)$ can be similarly represented.

actual x and y are the real parts of these complex variables. The quotient of the complex y and x is the *complex compliance* s^*:

$$s^* = \frac{y}{x} = \frac{y_0}{x_0} \exp(-i\delta) \tag{4.3.7}$$

or

$$s^* = s' - is'' = \frac{y_0}{x_0}(\cos\delta - i\sin\delta) \tag{4.3.8}$$

where s' and s'' are the real and imaginary parts of s^*. From (4.3.5) and (4.3.7):

$$\mathrm{d}x/\mathrm{d}t = xi\omega \tag{4.3.9}$$

$$y = s^*x \tag{4.3.10}$$

$$\mathrm{d}y/\mathrm{d}t = s^*xi\omega \tag{4.3.11}$$

Substituting these last three expressions in (4.3.4) results in

$$\tau_r(s^*xi\omega - s_1 xi\omega) = s_e x - s^*x$$

or, after dividing by x and rearranging:

$$s^* = s_1 + \frac{s_e - s_1}{1 + i\omega\tau_r} \tag{4.3.12}$$

Since $s^* = s' - is''$ we find the real and imaginary parts of s^*,

$$s' = s_1 + \frac{s_e - s_1}{1+\omega^2\tau_r^2} \tag{4.3.13}$$

$$s'' = (s_e - s_1)\frac{\omega\tau_r}{1+\omega^2\tau_r^2} \tag{4.3.14}$$

These two equations are usually referred to as Debye's equations since they were derived by him for a relaxing dielectric material using a special molecular model[1]. It should be noted, however, that these equations apply to any material whose relaxation can be defined by only one relaxation time. They are independent of the molecular model that may be used and were earlier derived by Wagner[2].

4.4. Derivation of the Debye equations without recourse to complex variables

In this section Debye's equations will be derived without the use of complex variables. The starting point of this derivation is the fact, mentioned earlier, that when at a time $t = 0$ a body is subjected to a constant 'load' (an electrical field strength, a magnetic field strength or a mechanical stress) x_1, it will in many cases react with a polarization, magnetization or strain, y, consisting of two parts, one of which, y_1, is practically instantaneous whilst the other part, y_2, is time-dependent. Integration of equation (4.2.6) gives

$$y_2 = y_{2e}\{1 - \exp(-t/\tau_r)\} \tag{4.4.1}$$

The total value of y is therefore given by

$$y = y_1 + y_{2e}\{1 - \exp(-t/\tau_r)\} \tag{4.4.2}$$

where

$$y_1 = s_1 x_1 \tag{4.4.3}$$

Since

$$y_{2e} = y_e - y_1 \tag{4.4.4}$$

and

$$y_e = s_e x \tag{4.4.5}$$

equation (4.4.2) can be written in the form

$$y = s_1 x_1 + (s_e - s_1)\{1 - \exp(-t/\tau_r)\}x_1$$

or

$$y = s_e x_1 - (s_e - s_1)\exp(-t/\tau_r)x_1 \tag{4.4.6}$$

1. P. Debye, *Polare Molekeln*, Hirzel, Leipzig (1929).
2. K. W. Wagner, *Annln Phys.* **40**, 817 (1913); see also his very clearly written book *Einführung in die Lehre von den Schwingungen und Wellen*, Dieterich'sche Verlagsbuchhandlung, Wiesbaden (1947).

We are interested in y as a function of time when x is a function of time, e.g.

$$x = x_0 \sin \omega t \tag{4.4.7}$$

In order to find this relation we make use of the superposition principle; this gives the value of y in the case where different loads $x_1, x_2, \ldots$ are applied successively at certain points in time. To each of these loads, if applied separately, the reaction would be in accordance with equation (4.4.6). The superposition principle states that the total value y at a certain time can be obtained by adding the separate responses y. It goes without saying that for each load the time should be taken from the moment that the load is applied. For instance, if a load x_1 is applied at the time $t = 0$ and a load x_2 at the time $t = t'$, then at an arbitrary later point after time t_1 we have

$$y = x_1[s_e-(s_e-s_1)\exp(-t_1/\tau_r)]+x_2[s_e-(s_e-s_1)\exp\{-(t_1-t')/\tau_r\}] \tag{4.4.8}$$

If the load is a continuous function of time, then y can be calculated as the sum of the effects of all the individual load elements $\mathrm{d}x$:

$$y = \int [s_e-(s_e-s_1)\exp\{-(t_1-t)/\tau_r\}]\,\mathrm{d}x \tag{4.4.9}$$

where t is the time for each individual load element $\mathrm{d}x$. For a periodic load (4.4.7) the value of y at time t_1 will be given by

$$\begin{aligned} y &= \int_0^{t_1} [s_e-(s_e-s_1)\exp\{-(t_1-t)/\tau_r\}]\,\mathrm{d}(x_0 \sin \omega t) \\ &= s_e x_0 \sin \omega t_1 -(s_e-s_1)x_0\omega \exp(-t_1/\tau_r)\int_0^{t_1} \exp(t/\tau_r)\cos(\omega t)\,\mathrm{d}t \end{aligned} \tag{4.4.10}$$

Since

$$\int \exp(at)\cos(bt)\,\mathrm{d}t = \frac{1}{a^2+b^2}\exp(at)(a\cos bt+b\sin bt) \tag{4.4.11}$$

expression (4.4.10) can be written:

$$\begin{aligned} y = x_0\left\{s_e-(s_e-s_1)\frac{\omega^2\tau_r^2}{1+\omega^2\tau_r^2}\right\}\sin\omega t_1 - x_0(s_e-s_1)\frac{\omega\tau_r}{1+\omega^2\tau_r^2}\cos\omega t_1 \\ +x_0(s_e-s_1)\frac{\omega\tau_r}{1+\omega^2\tau_r^2}\exp(-t_1/\tau_r) \end{aligned} \tag{4.4.12}$$

The last term in this expression is of no significance when the situation has become quasi-stationary, i.e. when $t_1 \gg \tau_r$. In that case the value of y (for instance, the strain in mechanical relaxation) is given by the sum of a sine and a cosine, each with a constant factor. Indicating the arbitrarily chosen time t_1 simply as t, equation (4.4.12) can, after some rearranging, be written:

$$y = x_0\left(s_1+\frac{s_e-s_1}{1+\omega^2\tau_r^2}\right)\sin\omega t - x_0(s_e-s_1)\frac{\omega\tau_r}{1+\omega^2\tau_r^2}\cos\omega t \tag{4.4.13}$$

As a function of ωt, y can also be expressed in the form

$$y = y_0 \sin(\omega t - \delta) = y_0 \cos\delta \sin\omega t - y_0 \sin\delta \cos\omega t \tag{4.4.14}$$

Hence from equation (4.4.13):

$$y_0 \cos\delta = x_0 \left(s_1 + \frac{s_e - s_1}{1+\omega^2\tau_r^2} \right) \tag{4.4.15}$$

$$y_0 \sin\delta = x_0(s_e - s_1)\frac{\omega\tau_r}{1+\omega^2\tau_r^2} \tag{4.4.16}$$

Expression (4.4.15) gives the amplitude y' of the component of y in phase with x and expression (4.4.16) the amplitude y'' of the component lagging 90° behind x (Fig. 95). Introducing s' and s'' (the 'compliances') as defined by

$$s' = \frac{y'}{x_0} = \frac{y_0 \cos\delta}{x_0} \tag{4.4.17}$$

$$s'' = \frac{y''}{x_0} = \frac{y_0 \sin\delta}{x_0} \tag{4.4.18}$$

the expressions (4.4.15) and (4.4.16) can be written as

$$s' = s_1 + \frac{s_e - s_1}{1+\omega^2\tau_r^2} \tag{4.4.19}$$

$$s'' = (s_e - s_1)\frac{\omega\tau_r}{1+\omega^2\tau_r^2} \tag{4.4.20}$$

These are again the two Debye equations which were derived in a simpler manner, in the preceding section, using complex variables.

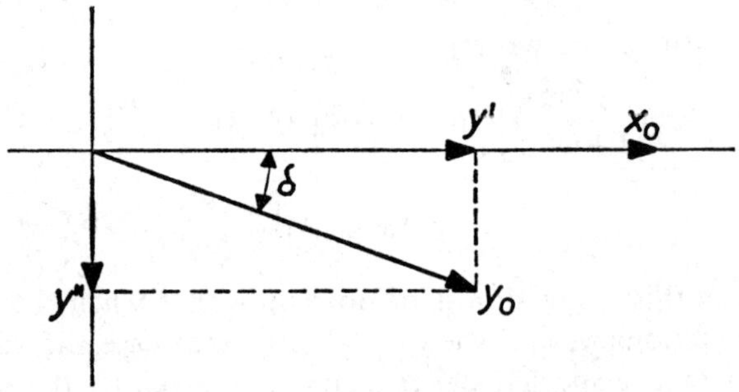

FIG. 95. As in Fig. 94 the vectors x_0 and y_0 are understood to rotate counter-clockwise in the plane of the drawing with a constant angular velocity ω, thus including variable angles ωt and $(\omega t - \delta)$ with the horizontal axis. The component of y which is in phase with x is given by $y' = y_0 \cos\delta$ whilst the component of y which lags 90° behind x is given by $y'' = y_0 \sin\delta$.

The symbols s' and s'' have the same significance in both derivations as will be clear from the relation

$$s^* = s' - is'' = \frac{y_0}{x_0}(\cos\delta - i\sin\delta)$$

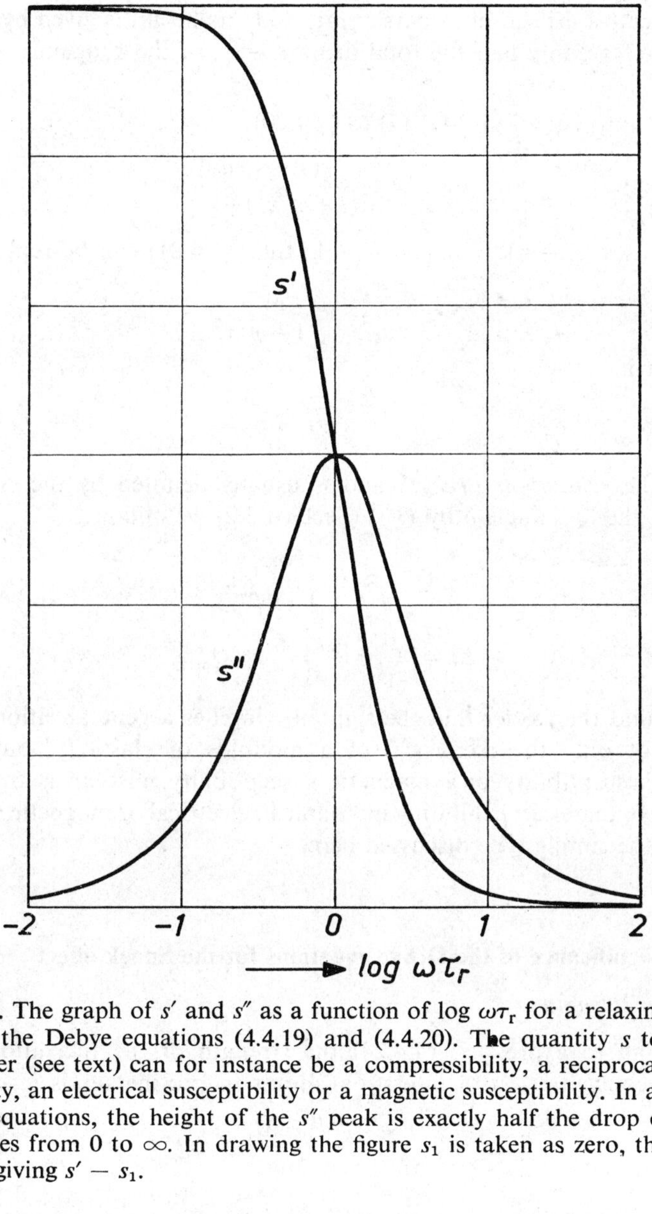

FIG. 96. The graph of s' and s'' as a function of log $\omega\tau_r$ for a relaxing material satisfying the Debye equations (4.4.19) and (4.4.20). The quantity s to which s' and s'' refer (see text) can for instance be a compressibility, a reciprocal modulus of elasticity, an electrical susceptibility or a magnetic susceptibility. In accordance with the equations, the height of the s'' peak is exactly half the drop of s' when $\omega\tau_r$ changes from 0 to ∞. In drawing the figure s_1 is taken as zero, the s' graph therefore giving $s' - s_1$.

According to the first Debye equation, (4.4.19) or (4.3.13), when $\omega\tau_r$ increases from 0 to ∞, s' will decrease from s_e to s_1 or, in other words, from the value of the relaxed compliance to that of the unrelaxed compliance. On the other hand, according to the second Debye equation, (4.4.20) or (4.3.14), s'' equals zero for both $\omega\tau_r \to 0$ and $\omega\tau_r \to \infty$. The maximum value of the s'' against $\omega\tau_r$ graph occurs at $\omega\tau_r = 1$; its height is given by $\frac{1}{2}(s_e - s_1)$ and is therefore only half the total drop, $s_e - s_1$, of the s' against $\omega\tau_r$ graph (Fig. 96).

According to equations (4.4.17) to (4.4.20)

$$\tan\delta = \frac{s''}{s'} = \frac{(s_e - s_1)\omega\tau_r}{s_1(1+\omega^2\tau_r^2)+(s_e-s_1)} \tag{4.4.21}$$

In many cases $(s_e - s_1) \ll s_1$ and $\delta \ll 1$; then (4.4.21) can be replaced by

$$\delta \approx \frac{s_e - s_1}{s_1} \times \frac{\omega\tau_r}{1+\omega^2\tau_r^2} \tag{4.4.22}$$

The quotient

$$\frac{s_e - s_1}{s_1} = \frac{y_{2e}}{y_1}$$

is called the *relaxation strength* and is usually denoted by the symbol Δ_r. Indicating the loss angle δ by Q^{-1} (Section 3.2) we obtain:

$$Q^{-1} \approx \Delta_r \frac{\omega\tau_r}{1+\omega^2\tau_r^2} \tag{4.4.23}$$

$$\Delta_r = \frac{y_{2e}}{y_1} = \frac{s_e - s_1}{s_1} \approx 2Q_{max}^{-1} \tag{4.4.24}$$

We remind the reader that the symbol s implies a generalization and can denote, not only the reciprocal of a modulus of elasticity, but also an electrical susceptibility or a magnetic susceptibility, insofar as these quantities refer to materials exhibiting mechanical, electrical or magnetic relaxation effects of the simple type discussed here.

4.5. The significance of the Debye equations for the Snoek effect

Relaxation strength

According to expression (4.4.24) the strength of the relaxation, caused by a low concentration of interstitial atoms in bcc metals, is given by

$$\Delta_r = \frac{s_e - s_1}{s_1} = \frac{M_1 - M_e}{M_e} = 2Q_{max}^{-1} \tag{4.5.1}$$

where M_1 is the unrelaxed and M_e the relaxed modulus of elasticity of the interstitial solid solution. The measured relaxation strength will depend on which modulus governs the relation between stress and strain. In flexural oscillations the relevant modulus is Young's modulus E and in torsional oscillations it is the shear modulus G. In a single crystal E and G are anisotropic, i.e. they are a function of the crystal orientation. The same therefore applies to the relaxation strength. For the relation between $\Delta_r(E)$ and the orientation of a bcc crystal we refer the reader to Section 4.7 (equation 4.7.24). Polycrystalline materials which have a random orientation of grains show approximately isotropic properties.

Relaxation time

Since the Snoek relaxation is caused by the jumping of interstitial atoms, the time of relaxation τ_r is directly related to the mean time of stay τ, or the mean jump frequency $1/\tau$, of an interstitial (cf. Chapter 2). If the number of interstitials on z sites is N_z, the rate at which atoms are leaving these sites at any instant will be N_z/τ. At zero applied stress half of the interstitials leaving x sites will go to y sites and half to z sites. The rate at which interstitials enter z sites from x sites is therefore $N_x/2\tau$ and from y sites $N_y/2\tau$. It follows that

$$\frac{dN_z}{dt} = -\frac{N_z}{\tau} + \frac{N_x}{2\tau} + \frac{N_y}{2\tau} \tag{4.5.2}$$

Designating the total number of interstitials as N, we may write

$$N_x + N_y = N - N_z \tag{4.5.3}$$

or, substituting in (4.5.2):

$$\frac{dN_z}{dt} = \frac{N - 3N_z}{2\tau}$$

or

$$\frac{d(N_z - N/3)}{dt} = -\frac{3}{2}\frac{(N_z - N/3)}{\tau} \tag{4.5.4}$$

From equations (4.1.2), (4.1.3) and (4.5.4) we find

$$\tau_r = \tfrac{2}{3}\tau \tag{4.5.5}$$

The factor $\frac{2}{3}$ in this expression originates from the fact that an interstitial atom in a bcc metal can only jump in four of the six cube directions (cf. Fig. 71). According to equations (2.9.8) and (4.5.5) the diffusivity of interstitials in a bcc metal is given by

$$D = \frac{a^2}{24\tau} = \frac{a^2}{36\tau_r} \tag{4.5.6}$$

where a is the lattice parameter of the metal.

Effect of temperature

Equation (2.10.2) gives the dependence of diffusivity on temperature:

$$D = D_0 \exp(-\Delta H/RT) \tag{4.5.7}$$

where ΔH is the activation enthalpy of the diffusion process (often designated as Q). From (4.5.6) and (4.5.7):

$$\tau_r = \tau_{r0} \exp(\Delta H/RT) \tag{4.5.8}$$

where

$$\tau_{r0} = \frac{a^2}{36D_0} \tag{4.5.9}$$

For the metals with which we are mainly concerned a is approximately 0·3 nm and D_0 (see Table 10) usually between 0·1 and 1 mm^2/sec. The value of τ_{r0} is therefore of the order of magnitude 1–10 fs, corresponding to a vibration frequency of the foreign atoms in their interstices of 100 or 1,000 THz (fs stands for femtosecond and THz for terahertz).

From equations (4.4.23), (4.4.24) and (4.5.8) it follows that

$$Q^{-1} \approx Q^{-1}_{max} \frac{2\omega\tau_{r0} \exp(\Delta H/RT)}{1+\omega^2\tau^2_{r0} \exp(2\Delta H/RT)} \tag{4.5.10}$$

At constant ω, Q^{-1} reaches a maximum at a certain temperature T_m. Since maximum internal friction occurs at $\omega\tau_r = 1$, we find from equation (4.5.8):

$$\omega\tau_{r0} = \exp(-\Delta H/RT_m) \tag{4.5.11}$$

Substituting (4.5.11) in (4.5.10) gives

$$Q^{-1} \approx Q^{-1}_{max} \frac{2}{\exp\left\{\frac{\Delta H}{R}\left(\frac{1}{T}-\frac{1}{T_m}\right)\right\} + \exp\left\{-\frac{\Delta H}{R}\left(\frac{1}{T}-\frac{1}{T_m}\right)\right\}} \tag{4.5.12}$$

Readers, familiar with hyperbolic functions, will note that this equation can be put in the simpler form:

$$Q^{-1} \approx \frac{Q^{-1}_{max}}{\cosh\left\{\frac{\Delta H}{R}\left(\frac{1}{T}-\frac{1}{T_m}\right)\right\}} \tag{4.5.13}$$

or

$$Q^{-1} \approx Q^{-1}_{max} \operatorname{sech}\left\{\frac{\Delta H}{R}\left(\frac{1}{T}-\frac{1}{T_m}\right)\right\} \tag{4.5.14}$$

Fig. 97 shows graphs corresponding to these last three equations for various values of $\Delta H/R$. The graphs were constructed with the aid of tables of hyperbolic cosines[(1)].

1. E. S. Allen, *Six-Place Tables*, McGraw-Hill, New York (1947).

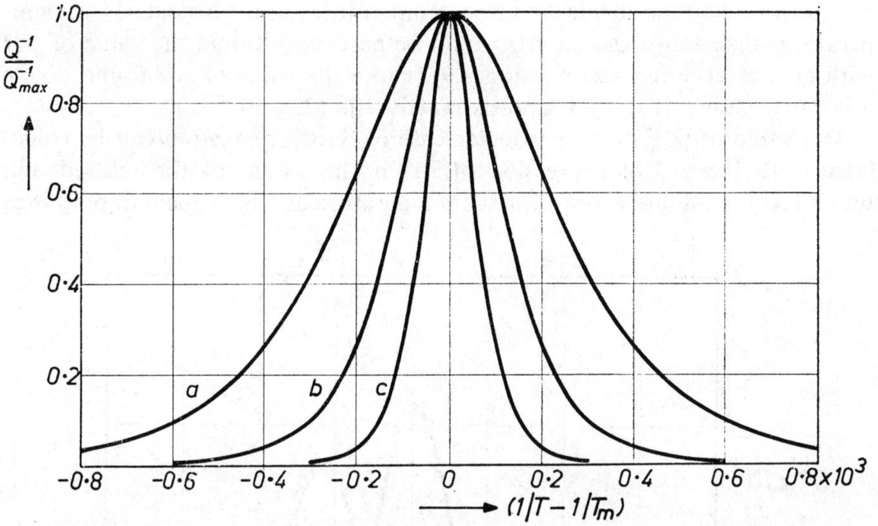

FIG. 97. Damping graphs corresponding to expression (4.5.13). The graphs *a*, *b* and *c* refer to the $\Delta H/R$ values 5,000°, 10,000° and 20,000°, i.e. to activation enthalpies of about 40, 80 and 160 kJ mole^{-1} (10, 20 and 40 kcal per mole), respectively. The peak width varies inversely with the activation enthalpy.

4.6. Checking the elementary theory with experimental data

The constants ΔH and τ_{r0} in equation (4.5.8) can, in principle, be calculated from measurements of the internal friction as a function of temperature at two frequencies of vibration ω_1 and ω_2. Let the temperatures at which the maxima in these two graphs occur be T_1 and T_2. For each of them $\omega\tau_r = 1$, so from equation (4.5.8):

$$\Delta H = \frac{19.14 \log (\omega_2/\omega_1)}{1/T_1 - 1/T_2} \qquad (4.6.1)$$

The two damping graphs in Fig. 98 refer to a nitrogen-bearing iron wire[(1)]. For a frequency of 0·77 Hz the maximum occurs at 294·5 K and for 0·18 Hz at 282 K. Substituting these values in (4.6.1) gives the activation enthalpy of the relaxation process:

$$\Delta H = 79{\cdot}5 \pm 6{\cdot}3 \text{ kJ mole}^{-1} \ (19 \pm 1{\cdot}5 \text{ kcal/mole})$$

The relatively large possible error of 6·3 kJ mole^{-1} is mainly due to the fact that it is difficult to read the temperatures T_1 and T_2 accurately from

1. J. D. FAST and M. B. VERRIJP, *J. Iron Steel Inst.* **176,** 24 (1954).

the graphs. Measurements at more temperatures and particularly at temperatures that are further apart, would be necessary to find the value of ΔH with greater accuracy (see Section 4.9). From the value of ΔH found above it is easily shown that τ_{r0} is approximately 2 fs (2×10^{-15} s).

The value of Q_{max}^{-1} for a particular case of Snoek relaxation can be calculated with the aid of expression (4.5.1) if the values of the relaxed and unrelaxed moduli are known for the temperature at which the damping is at

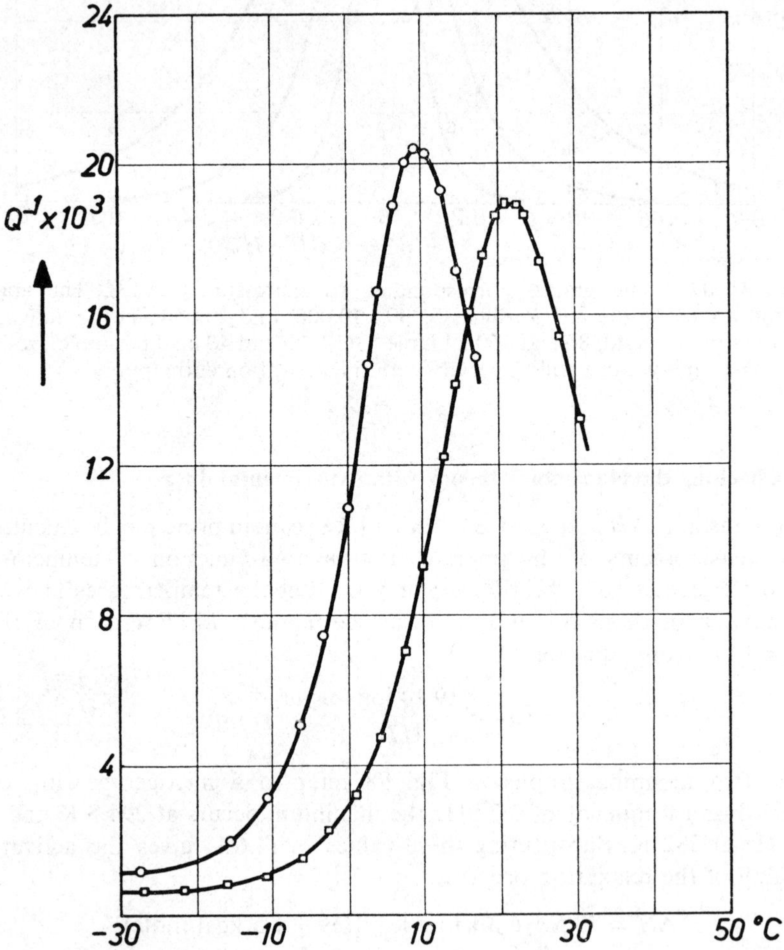

FIG. 98. Internal friction of an iron wire containing about 0·1 atom per cent nitrogen, as measured at different temperatures by the method of free decay. The graph with the higher maximum corresponds to a torsional oscillation frequency of 0·18 Hz, the other one to a frequency of 0·77 Hz (Fast and Verrijp).

a maximum. The value calculated in this way should agree with the height of the graph found experimentally. If, in addition, the value ΔH is also known, then the damping graph (internal friction versus temperature) can be plotted from equation (4.5.13) and should agree with the graph found experimentally.

If the moduli are not known, not as complete a comparison between the damping measurements and the elementary theory, based on one relaxation time, is possible. The experimentally found values of Q^{-1}_{max} and ΔH are then substituted in equation (4.5.13) and the resulting graph is compared with the measured curve. Wert and Zener[1] found a satisfactory agreement between theory and experiment in this way for solutions of carbon and nitrogen in bcc iron. The more complete comparison between experiments and elementary theory has also given satisfactory results[2].

Deviations from a straight line are much easier to detect than deviation from a particular type of curve. In order to see whether a graph agrees with a certain function it is common practice to use special scales on the axes so that the graph of the function becomes a straight line (for instance, logarithmic, double logarithmic, or probability graph paper).

Jäniche *et al.*[3] have applied this method to damping measurements. They used a non-linear scale on the vertical axis when plotting $y = Q^{-1}/Q^{-1}_{max}$ against $(1/T - 1/T_m)$. The actual distances on the y axis are proportional to

$$\frac{\Delta H}{R}\left(\frac{1}{T} - \frac{1}{T_m}\right) \tag{4.6.2}$$

but the scale is such that it corresponds to

$$\cosh^{-1}\left[\frac{\Delta H}{R}\left(\frac{1}{T} - \frac{1}{T_m}\right)\right] \tag{4.6.3}$$

It will be clear that when Q^{-1}/Q^{-1}_{max} is plotted against $(1/T - 1/T_m)$ on this scale, a straight line will result if the effect satisfies equation (4.5.13). Fig. 99 gives the straight lines corresponding to the curves in Fig. 97. The numbers on the right-hand vertical axis indicate the values according to (4.6.2), those on the left-hand vertical axis indicate the values according to (4.6.3). In drawing a graph of the experimental results the values of Q^{-1}/Q^{-1}_{max} are plotted on the latter scale. Deviations from the simple behaviour as defined by equation (4.5.13) are immediately shown up as deviations from linearity.

1. C. A. WERT and C. ZENER, *Phys. Rev.* **76,** 1169 (1949).
2. I. BARDUCCI, *Ricerca Scient.* **25,** 2572 (1955).
3. W. JÄNICHE, E. STOLTE and J. BRAUNER, *Tech. Mitt. Krupp* **23,** 146 (1965).

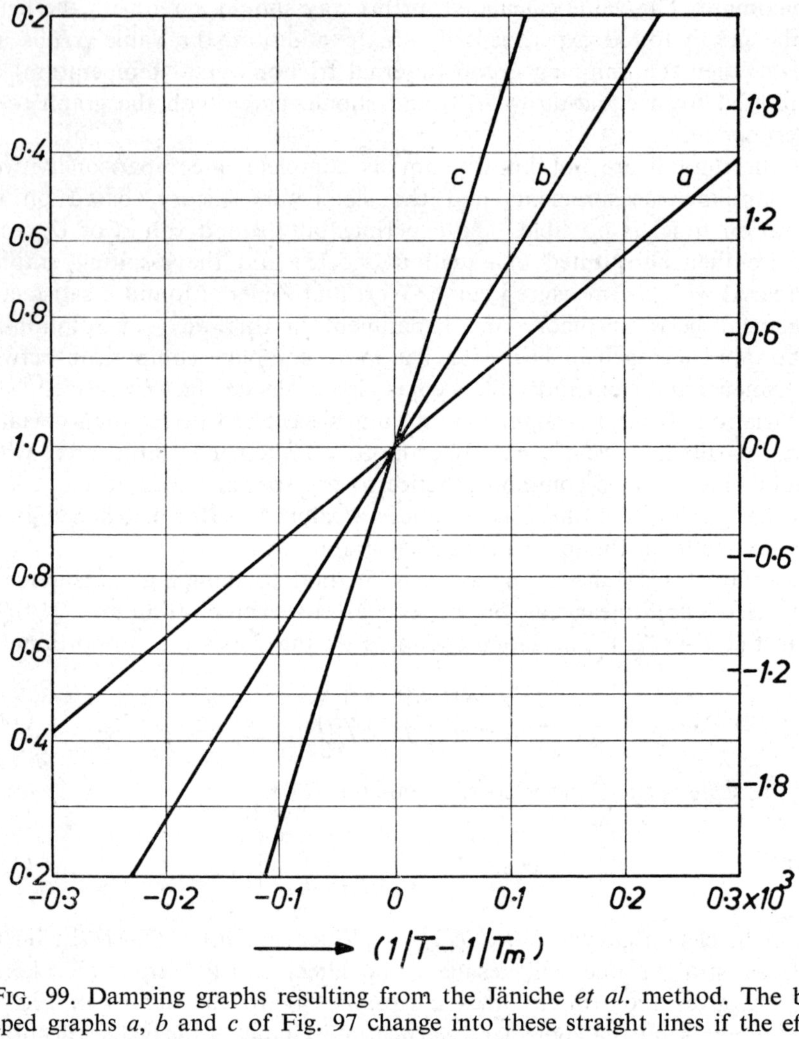

FIG. 99. Damping graphs resulting from the Jäniche *et al.* method. The bell-shaped graphs *a*, *b* and *c* of Fig. 97 change into these straight lines if the effect can be represented by the simple expression (4.5.13).

In the simple cases, discussed here, the internal friction is greatest when $\omega\tau_r = 1$ or, taking (4.5.8) into account, when

$$\omega\tau_r = \omega\tau_{r0} \exp(\Delta H/RT_m) = 1 \qquad (4.6.4)$$

Wert and Marx[1] have pointed out that for a given frequency the temperature at which a relaxation peak is found is higher, the higher the heat of activation

1. C. WERT and J. MARX, *Acta Metall.* **1**, 113 (1953).

of the relaxation. They showed that for many relaxation effects of different types $\Delta H/T_m$ has approximately the same value: an almost straight line passing through the origin is obtained when the heat of activation, ΔH, is plotted as a function of the temperature, T_m, of the various peaks. According to equation (4.6.4) this means that τ_{r0} has approximately the same value for all these phenomena.

4.7. Thermodynamic considerations

The Gibbs free energy of elastically strained bcc iron

Applying thermodynamics it is possible to derive an equation that enables us to calculate the relaxation strength Δ_r of the Snoek relaxation in iron, from known properties of iron and martensite. From equation (4.5.1) the theoretical height of the damping peak would then follow directly.

Let us consider an iron rod being stretched elastically. Let, further, the increase in the strain ε at a certain value of the stress σ be $\mathrm{d}\varepsilon$, then the elastic work done on the iron is $\sigma\,\mathrm{d}\varepsilon$ per m^3. Since the volume of the rod also changes slightly as a result of the stretching, the total work done per m^3 is

$$w = -p\,\mathrm{d}V + \sigma\,\mathrm{d}\varepsilon$$

where p is the atmospheric pressure and $\mathrm{d}V$ the increase in volume of 1 m^3 iron. Since the straining is reversible, the combination of the first and second laws of thermodynamics results in

$$\mathrm{d}U = T\,\mathrm{d}S - p\,\mathrm{d}V + \sigma\,\mathrm{d}\varepsilon \tag{4.7.1}$$

where U and S are the internal energy and entropy of 1 m^3 iron. We now add the following relationship to both sides of equation (4.7.1)

$$\mathrm{d}(pV - TS - \sigma\varepsilon) = p\,\mathrm{d}V + V\,\mathrm{d}p - T\,\mathrm{d}S - S\,\mathrm{d}T - \sigma\,\mathrm{d}\varepsilon - \varepsilon\,\mathrm{d}\sigma$$

thus obtaining:

$$\mathrm{d}(U + pV - TS - \sigma\varepsilon) = V\,\mathrm{d}p - S\,\mathrm{d}T - \varepsilon\,\mathrm{d}\sigma \tag{4.7.2}$$

where

$$U + pV - TS - \sigma\varepsilon = G \tag{4.7.3}$$

is the Gibbs free energy per m^3 of the system under consideration. It is a function tending to a minimum at constant values of p, T and σ. With the aid of (4.7.3), equation (4.7.2) can be shortened into

$$\mathrm{d}G = V\,\mathrm{d}p - S\,\mathrm{d}T - \varepsilon\,\mathrm{d}\sigma \tag{4.7.4}$$

According to this equation the strain is given by

$$\varepsilon = -\left(\frac{\partial G}{\partial \sigma}\right)_{p,T} \tag{4.7.5}$$

If the iron rod contains interstitial atoms, e.g. carbon atoms, the applied stress will cause a redistribution of these atoms. In consequence, the strain is not only a function of p, T and σ but is also dependent on an internal parameter of state. In order to take this into account we assume the iron rod to be a single crystal with one of the cube axes (say the z axis) coincident with the axis of the rod.

At the instant of application of the tensile stress in the z direction of the carbon-bearing iron crystal, the concentration of the carbon atoms on the x, y and z sites is still equal:

$$c_x = c_y = c_z = c_0 = \tfrac{1}{3}c$$

where c is the total concentration, defined as the number of occupied octahedral sites divided by the total number of these sites. Similarly c_x is the number of occupied x sites over the total number of x sites. Redistribution changes the concentration c_z to (c_0+c') and consequently c_x and c_y to $(c_0-\tfrac{1}{2}c')$; the total concentration remains

$$c = c_x+c_y+c_z \tag{4.7.6}$$

Let us assume for a moment that c' and σ can be varied independently of each other; if c' is constant, equation (4.7.5) can be written:

$$\varepsilon = -\left(\frac{\partial G}{\partial \sigma}\right)_{p,T,c'} \tag{4.7.7}$$

In this expression the strain ε consists of (a) an elastic part $\varepsilon_1 = \sigma/E$, where E represents the unrelaxed modulus of elasticity in a $\langle 100 \rangle$ direction and (b) an inelastic part ε_2 due to the excess of carbon atoms in the z sites. As a result of the linear relationship between the carbon content and the lattice constants of martensite (Fig. 10) the anelastic strain can be written in the form

$$\varepsilon_2 = \beta c' \tag{4.7.8}$$

where β is a constant. For (4.7.7) we can thus write

$$-\left(\frac{\partial G}{\partial \sigma}\right)_{p,T,c'} = \frac{\sigma}{E} + \beta c' \tag{4.7.9}$$

or, after integration:

$$G = G_0 - \frac{\sigma^2}{2E} - \beta\sigma c' \tag{4.7.10}$$

For our purpose it is desirable to split the integration constant G_0 into a part that is dependent on c' and a part that is independent of it. By far the biggest item of the first part is the term $-T\Delta S_{\text{mix}}$ representing the change ΔS_{mix} which occurs in the entropy of mixing S_{mix} as a result of the redistribution of the carbon atoms. No term, dependent on c', need be introduced to take into account the interaction between carbon atoms, since this interaction can be neglected at the minute concentrations of carbon involved in this discussion (at the most about 0·1 atom %). Equation (4.7.10) can thus be written:

$$G = G_0' - \frac{\sigma^2}{2E} - \beta\sigma c' - T\Delta S_{\text{mix}} \tag{4.7.11}$$

Entropy of mixing for bcc iron containing carbon

The entropy of mixing for a solution of carbon in bcc iron is calculated as follows. The number of available x sites is N_0 per mole of iron, N_0 being Avogadro's number. Of these sites $c_x N_0$ are occupied and $(1-c_x)N_0$ are unoccupied. The number of ways in which $c_x N_0$ carbon atoms can be distributed over the N_0 available x positions is

$$\frac{N_0!}{(c_x N_0)!\,\{(1-c_x)N_0\}!} \tag{4.7.12}$$

Similar expressions are valid for the possibilities of distribution over the y and z sites. For any arbitrary set of values c_x, c_y and c_z, the *total* number of ways in which the dissolved carbon atoms can be distributed over the x, y and z sites is given by

$$W = \frac{N_0!}{(c_x N_0)!\,\{(1-c_x)N_0\}!} \cdot \frac{N_0!}{(c_y N_0)!\,\{(1-c_y)N_0\}!} \cdot \frac{N_0!}{(c_z N_0)!\,\{(1-c_z)N_0\}!} \tag{4.7.13}$$

By applying the equations

$$S = k \ln W$$

and

$$\ln N! \approx N \ln N - N$$

we can derive the entropy of mixing directly from equation (4.7.13). Per m^3 of iron it is given by

$$S_{\text{mix}} = -\frac{kN_0}{V}\left\{\sum_{i=x,y,z} c_i \ln c_i + \sum_{i=x,y,z} (1-c_i)\ln(1-c_i)\right\} \tag{4.7.14}$$

where V is the molar volume of iron. When no stress is applied to the metal, $c_x = c_y = c_z = c_0$ and thus for any of the three sub-lattices

$$S^0_{\text{mix}} = -\frac{R}{V}\left[c_0 \ln c_0 + (1-c_0)\ln(1-c_0)\right] \tag{4.7.15}$$

where $R = kN_0$ is the gas constant. After the redistribution of the carbon atoms under the influence of a tensile stress, $c_z = c_0 + c'$ and $c_x = c_y = c_0 - \frac{1}{2}c'$. In order to arrive at an expression for the change in entropy resulting from the redistribution we expand S_{mix} in a Taylor series around

$S_{\text{mix}} = S^0_{\text{mix}}$. Since the entropy is at a maximum for this value of S_{mix} we have $dS/dc = 0$; therefore, assuming $c' \ll c$ (i.e. neglecting all higher order terms in Δc):

$$\Delta S_{\text{mix}} = \frac{1}{2}\frac{d^2S}{dc^2}(\Delta c)^2 = -\frac{R}{2V}\frac{(\Delta c)^2}{c_0(1-c_0)}$$

where Δc has the value c' for the z sites and $c'/2$ for the x and y sites. The total change in entropy caused by the redistribution of the interstitial atoms is thus given by

$$\Delta S_{\text{mix}} = -\frac{R}{2V}\frac{c'^2+c'^2/4+c'^2/4}{c_0(1-c_0)} = -\frac{3R}{4V}\frac{c'^2}{c_0(1-c_0)} \tag{4.7.16}$$

Substituting $c/3$ for c_0 we obtain for $c \ll 1$:

$$\Delta S_{\text{mix}} = -\frac{9Rc'^2}{4Vc} \tag{4.7.17}$$

where c is the total concentration of the interstitial atoms.

Anelastic strain and relaxation strength

According to (4.7.8) the final (equilibrium) value of the anelastic strain is given by

$$\varepsilon_{2e} = \beta c'_e \tag{4.7.18}$$

and is thus known when β and c'_e are known.

The value of β can be found directly from Fig. 10. It is deduced from the strain parallel to the z axis which occurs when the carbon concentration in the z sites increases by c'_z at the expense of that in the x and y sites. The value of β is therefore found by dividing this strain by c'_z. Let us now imagine that the quantity of carbon corresponding to c'_z is transferred from x and y sites to z sites in two stages. It is first moved from x (or y) sites and taken outside the crystal. Since, according to Fig. 10, the introduction of carbon or nitrogen atoms into x and y sites causes a contraction in the z direction, their removal from x and y sites will result in an expansion in the z direction. The contribution of this first step to β is

$$\beta_1 = -\frac{\varepsilon_x}{2c'_z} - \frac{\varepsilon_y}{2c'_z} = -\frac{\varepsilon_x}{c'_z}$$

where ε_x represents the (negative) strain in the x direction of a bcc iron crystal, due to the presence in z sites of c'_z gram-atoms (moles) of carbon per mole of iron.

The second stage consists of returning the carbon atoms to the iron and distributing them over z sites. Fig. 10 shows that this causes a further increase in the strain in the z direction. The contribution of this second step to β is

$$\beta_2 = \frac{\varepsilon_z}{c'_z}$$

where ε_z is the strain in the z direction due to the presence in z sites of c'_z moles of carbon per mole of iron. The total value of β is thus

$$\beta = \frac{\varepsilon_z - \varepsilon_x}{c'_z} \tag{4.7.19}$$

In this expression the quantity $\varepsilon_z - \varepsilon_x$ for any concentration is given by the vertical distance between lines a and c in Fig. 10. The figure shows that for $c'_z = 0{\cdot}10$ this vertical distance is 0·25 kX units and thus $\varepsilon_z - \varepsilon_x = 0{\cdot}25/2{\cdot}86 = 0{\cdot}087$ since the lattice parameter of pure bcc iron at room temperature is 2·86 kX units. Using equation (4.7.19) we find $\beta = 0{\cdot}87$.

In order to be able to calculate ε_{2e} from (4.7.18) it is necessary to know the value of c'_e. This can be found if we remember that for any given value of σ carbon atoms will continue to be displaced until the Gibbs free energy is a minimum. For the equilibrium state, therefore, we have

$$\left(\frac{\partial G}{\partial c'}\right)_{p,T,\sigma} = 0$$

hence, using (4.7.11) and (4.7.17):

$$c'_e = \frac{2Vc\beta\sigma}{9RT} \tag{4.7.20}$$

In deriving equation (4.7.16) it was assumed that c'_e/c is much smaller than unity. This assumption can now be justified by substituting in equation (4.7.20) the numerical values of the various quantities, taking for σ the maximum tensile stress that can be applied in the relaxation experiments without causing plastic deformation.

According to (4.7.18) and (4.7.20) the equilibrium value of the anelastic strain is

$$\varepsilon_{2e} = \frac{2Vc\beta^2\sigma}{9RT} \tag{4.7.21}$$

According to equation (4.4.24) the relaxation strength for the case under discussion is given by

$$\Delta_r = \frac{\varepsilon_{2e}}{\varepsilon_1} = \frac{E\varepsilon_{2e}}{\sigma} \tag{4.7.22}$$

where E represents the unrelaxed modulus of elasticity (Young's modulus) in a $\langle 100 \rangle$ direction. From (4.7.19), (4.7.21) and (4.7.22) it follows that

$$\Delta_r = \frac{2VcE}{9RT}\left(\frac{\varepsilon_z - \varepsilon_x}{c'_z}\right)^2 \tag{4.7.23}$$

Dijkstra[1] measured the damping of free flexural oscillations of thin iron strips containing carbon or nitrogen in solid solution. He found the experimental value of the relaxation strength to be in good agreement with the value calculated with the aid of equation (4.7.23). According to this equation Δ_r is directly proportional to the carbon or nitrogen concentration and to the reciprocal of the absolute temperature. The first has been confirmed by experiments that will be discussed in Section 4.10; the latter is, as far as accuracy permits, confirmed by Fig. 98. Convincing proof that the height of the peak is proportional to $1/T$ would require the measurement of Q^{-1}_{max} at a greater number of temperatures.

If the tensile stress is not applied in a $\langle 100 \rangle$ direction of a single crystal of iron but in some arbitrary direction, equation (4.7.23) becomes[2]:

$$\Delta_r = \frac{2V\,cE}{9RT}\left(\frac{\varepsilon_z-\varepsilon_x}{c'_z}\right)^2 [1-3(\alpha_1^2\alpha_2^2+\alpha_2^2\alpha_3^2+\alpha_3^2\alpha_1^2)] \qquad (4.7.24)$$

Here α_1, α_2 and α_3 are the direction cosines of the applied stress with respect to the $\langle 100 \rangle$ axes of the crystal. E is then the Young's modulus in the direction α_1, α_2, α_3. An equation analogous to (4.7.24) can be derived for single-crystal wires subjected to periodically varying *torsional* stress.

Equation (4.7.24) predicts that the Snoek effect must be absent when the tensile stress is applied in a $\langle 111 \rangle$ direction. This can also be predicted by using the fact that a $\langle 111 \rangle$ direction is symmetrical with respect to the three $\langle 100 \rangle$ axes. The disappearance of the effect in a $\langle 111 \rangle$ direction has been confirmed experimentally by Dijkstra, *loc. cit.*, and also by Lord and Beshers[3] using the pulse-echo method. The latter investigators showed in addition that the ratio of the relaxation strength in a $\langle 110 \rangle$ direction to that in a $\langle 100 \rangle$ direction is in accordance with equation (4.7.24).

Obviously, the orientation dependence of the Snoek effect applies not only to iron but also to other bcc metals containing interstitial solutes. Hoffman and Wert[4] measured the anisotropy of the strength of the Snoek relaxation for both oxygen and nitrogen in single crystals of niobium and obtained results similar to those described above for solutions of carbon and nitrogen in single crystals of iron.

For polycrystalline wires no exact theoretical calculation of the internal friction magnitude is possible. An averaging procedure must in this case be applied in order to derive the internal friction from the orientation-dependent effect in single crystals. The averaging may be carried out on the 'uniform strain' or on the 'uniform stress' basis. These two methods lead to

1. L. J. Dijkstra, *Philips Res. Rep.* **2,** 357 (1947).
2. D. Polder, *Philips Res. Rep.* **1,** 5 (1945).
3. A. E. Lord and D. N. Beshers, *Acta Metall.* **14,** 1659 (1966).
4. R. A. Hoffman and C. A. Wert, *J. appl. Phys.* **37,** 237 (1966).

different results and it is difficult to decide which of the averaging processes is the better. A third way of averaging[1], depending on a semi-empirical argument (namely on the observed relation between the elastic constants of single crystals and polycrystalline specimens of a number of cubic metals) probably yields the most reliable results. However, the only fully reliable way of deriving the relation between internal friction and solute content for polycrystalline specimens appears to be the experimental way. Reproducible results can only be expected if the crystals in the specimens have truly random orientations. Other factors affecting the reproducibility will be discussed in Section 4.10.

4.8. Relations between measures of internal friction

The work by the surroundings on a relaxing substance in order to keep the amplitude of its vibrations constant, is a measure of its internal friction. The work done per unit volume of the substance over a time interval t is given by

$$\Delta W = \int_0^t \sigma \, d\varepsilon = \int_0^t \sigma \frac{\partial \varepsilon}{\partial t} \, dt \tag{4.8.1}$$

This supplied energy is lost to the surroundings in the form of heat if the temperature of the substance is kept constant. Let us consider the simple case where σ and ε can be represented by the harmonic functions (4.3.1) and (4.3.2), replacing x and y by σ and ε. Substitution of these two expressions in (4.8.1) gives for an integral number of cycles

$$\Delta W = \frac{\sigma_0 \varepsilon_0}{2} \omega t \sin \delta$$

and hence per cycle $T = 2\pi/\omega$:

$$\Delta W = \frac{\sigma_0 \varepsilon_0}{2} 2\pi \sin \delta \tag{4.8.2}$$

In Fig. 84(*b*) this energy dissipated per unit volume per cycle is represented by the hatched area. For small values of δ the total energy of vibration per unit volume is given by the maximum strain energy and thus, according to Fig. 84, by

$$W = \frac{\sigma_0 \varepsilon_0}{2} \tag{4.8.3}$$

1. J. Smit and H. G. van Bueren, *Philips Res. Rep.* **9**, 460 (1954).

When the strain is not at a maximum, part of this energy is in the form of kinetic energy. According to (4.8.2) and (4.8.3) the fraction of the vibration energy dissipated per cycle is given by

$$\frac{\Delta W}{W} = 2\pi \sin \delta \approx 2\pi\delta \tag{4.8.4}$$

Over each cycle, energy equal to this amount must be supplied from outside the substance to keep its amplitude of oscillation constant.

In the case of free oscillations the dissipation of energy as represented by (4.8.4) will cause a gradual reduction of the amplitude, i.e. damping. In this case ΔW no longer represents the energy supplied from outside but the reduction, per cycle, of the oscillation energy. Since in the context of this discussion both δ and $\Delta W/W$ are very small we may write

$$\frac{\Delta W}{W} \approx \ln\left(1 + \frac{\Delta W}{W}\right) = \ln \frac{W_t}{W_{t+T}} \tag{4.8.5}$$

where W_t and W_{t+T} are the oscillation energies of the free oscillation at two moments in time separated by one cycle T. Since for small values of δ the energy of oscillation is proportional to the square of the amplitude ε_0 (cf. expression (4.8.3)), equation (4.8.5) can also be written:

$$\frac{\Delta W}{W} = \ln \frac{W_t}{W_{t+T}} = 2 \ln \frac{(\varepsilon_0)_t}{(\varepsilon_0)_{t+T}} \tag{4.8.6}$$

This natural logarithm of the ratio of two values of amplitude ε_0 one cycle apart is usually called the *logarithmic decrement* λ of the free oscillation:

$$\lambda = \ln \frac{(\varepsilon_0)_t}{(\varepsilon_0)_{t+T}} \tag{4.8.7}$$

From (4.8.4), (4.8.6) and (4.8.7) we find for small values of the damping:

$$\lambda = \frac{1}{2}\frac{\Delta W}{W} = \pi\delta \tag{4.8.8}$$

where the symbol δ is (as mentioned earlier) often replaced by Q^{-1}.

It follows from the above that the dissipation of energy can be determined by measuring the logarithmic decrement λ of free oscillations, or the loss angle δ of enforced oscillations. In the latter case the dissipation can also be derived from the width of the resonance peak. Measurements of the attenuation of supersonic waves in the substance under investigation are also often successfully used.

Some important applications of the Snoek damping will be discussed in the following sections.

4.9. Determination of diffusivities of interstitial atoms

As indicated by expression (4.4.23) the simplest kind of Snoek relaxation can be described with the help of only two parameters: the relaxation strength Δ_r, a measure of the maximum energy dissipation, and the relaxation time τ_r. The first is a thermodynamic parameter and has been the subject of quantitative considerations in Section 4.7. The second is a kinetic parameter which, as we saw in Section 4.5, is related in a simple way to the diffusion coefficient of the jumping interstitial atoms on which the Snoek relaxation depends:

$$D = \frac{a^2}{36\tau_r} \tag{4.9.1}$$

In this expression a is the lattice parameter of the bcc metal. In its derivation it has been assumed (a) that the interstitial atoms are situated only in octahedral interstices and have a low concentration, (b) that at each jump an interstitial atom moves to one of the four nearest octahedral interstices, i.e. over only one jump distance.

Diffusion of nitrogen in bcc iron

In order to ascertain the diffusivities of interstitial atoms in bcc metals at different temperatures it is, according to expression (4.9.1), necessary to determine the value of τ_r at any of those temperatures. These values follow directly from measurements of Snoek damping peaks. As an illustration we consider Fig. 86 relating to the damping of torsional oscillations of an iron wire containing 0·027 wt % nitrogen. The damping was measured as a function of the temperature by using the method of free decay at a frequency of oscillation $\nu = 0{\cdot}77$ Hz. The figure shows the maximum to be at 21·5 °C. According to the Debye equations we have for this maximum:

$$\omega\tau_r = 2\pi\nu\tau_r = 1 \tag{4.9.2}$$

With the aid of (4.9.1), (4.9.2) and the value of the lattice parameter of iron (a = 0·286 nm) we find for N in α iron at 21·5 °C:

$$\tau_r = 0{\cdot}21 \text{ s} \tag{4.9.3}$$

$$D = 1{\cdot}1 \times 10^{-14} \text{ mm}^2 \text{ s}^{-1}$$

Since, according to Section 4.7, the height of the peak is inversely proportional to the absolute temperature, the maximum does, strictly speaking, not occur at $\omega\tau_r = 1$ but at the point where

$$\frac{\Delta H}{RT} = \frac{\omega^2\tau_r^2 + 1}{\omega^2\tau_r^2 - 1} \tag{4.9.4}$$

The use of this expression leads to values of diffusivity that differ only slightly from those obtained by using $\omega\tau_r = 1$.

By measuring the internal friction as a function of temperature at different frequencies diffusivities at various temperatures can be determined. In this way values have been obtained for the diffusivity of nitrogen in iron over the range $-20°$ to 160°C[1-6] and even at a temperature as high as 326°C by measuring the pulse-echo ultrasonic attenuation at a frequency of 6·65 MHz[7]. For a number of temperatures in the range between $-50°$ and -20°C values of D(N) have been calculated from measurements of the magnetic permeability after-effect[8, 9] and of the magneto-elastic after-effect[10]. All the data from $-50°$ to 326°C can be satisfactorily represented by a straight line when plotting log D versus $1/T$. This straight line even agrees with the diffusivities of nitrogen in bcc iron at high temperatures, deduced from rates of desorption or absorption[11-13]. All the data for nitrogen diffusion in α and δ iron, from $-50°$ to 1,470°C can be represented by the single expression

$$D = 0{\cdot}488 \exp(-76{,}780/RT) \text{ mm}^2 \text{ s}^{-1} \qquad (4.9.5)$$

which has been derived from the available data by Lord and Beshers, *loc. cit.*, using the method of least squares.

The above suggests that the diffusion mechanism for nitrogen in bcc iron is the same over the whole range of $-50°$ to 1,470°C. It affords strong support for the assumption forming the basis of expression (4.9.1), that all nitrogen atoms are situated in octahedral interstices and jump only over a distance $\frac{1}{2}a$, i.e. between adjacent interstices. Because the Snoek relaxation is based on such small movements it is possible to determine much lower diffusivities by means of internal friction than by conventional methods requiring very many jumps before clear concentration differences can be observed. For instance, to measure the diffusivity of nitrogen in bcc iron at 21·5°C in the classical manner by permitting the nitrogen to diffuse from nitrogen-bearing iron into pure iron, it would be necessary to wait until the mean penetration depth is at least 1 to 2 mm. This depth corresponds to about 10^7 elementary jump distances and any nitrogen atom would therefore have to make about 10^{14} jumps. Since the mean time of stay of a

1. C. A. Wert, *J. appl. Phys.* **21**, 1196 (1950).
2. J. D. Fast and M. B. Verrijp, *J. Iron Steel Inst.* **176**, 24 (1954).
3. W. R. Thomas and G. M. Leak, *Phil. Mag.* **45**, 656 (1954).
4. R. R. Hasiguti and G. Kamoshita, *J. phys. Soc. Japan* **9**, 646 (1954).
5. L. Guillet and B. Hocheid, *Revue Métall., Paris* **53**, 122 (1956).
6. L. Guillet and P. Gence, *J. Iron Steel Inst.* **186**, 223 (1957).
7. A. E. Lord and D. N. Beshers, *Acta Metall.* **14**, 1659 (1966).
8. A. J. Bosman, Ph.D. Thesis, Univ. of Amsterdam (1960).
9. G. W. Rathenau, *J. appl. Phys.* **29**, 239 (1958).
10. R. E. Maringer, *J. appl. Phys.* **32**, 366S (1961) and **35**, 2375 (1964).
11. J. D. Fast and M. B. Verrijp, *loc. cit.*
12. P. E. Busby, D. P. Hart and C. Wells, *Trans. AIME* **206**, 686 (1956).
13. P. Grieveson and E. T. Turkdogan, *Trans. AIME* **230**, 1604 (1964).

nitrogen atom in an interstice is 0·32 seconds at 21·5 °C (cf. (4.5.5) and (4.9.3)), the experiment would take over a million years!

Diffusion of carbon in bcc iron

In Section 2.14 it was mentioned that the diffusion of carbon in bcc iron at high temperatures shows complications indicating that, in addition to the simple diffusive jump step discussed above, one or more other diffusion mechanisms become operative. This is demonstrated by the fact that although all the available data from −40° to 350°C can be satisfactorily represented by the simple Arrhenius equation (2.14.1), the log D versus $1/T$ graph departs from linearity at higher temperatures (Fig. 70). Extrapolation to lower temperatures suggests that, in the region 300° to 350°C, the curve measured at high temperatures merges smoothly into the straight line corresponding to (2.14.1).

McLellan *et al.*[1] and Condit and Beshers[2] explain the anomalous excess diffusion of carbon in bcc iron at high temperatures by assuming that part of the carbon atoms occupies tetrahedral interstices. Even at 1,000 K this part is very small but, according to their analysis, the frequency of the t–t jumps is so much higher than that of t–o and o–t jumps at high temperatures that diffusion then takes place mainly through t–t jumps (cf. Section 2.14). This means that with increasing temperature diffusion gradually changes from one mechanism to another. Condit and Beshers, *loc. cit.*, believe that this could also explain the unpublished findings of Homan and Cox that at high temperatures the diffusivity of carbon in bcc iron is strongly dependent on hydrostatic pressure, whilst at room temperature it is independent of this pressure[3, 4].

In the temperature range in which Snoek damping peaks are found, all, or nearly all, carbon atoms are situated in octahedral interstices. Diffusion then takes place through jumps of the type o–t–o (or o–o for short). At temperatures, too low for Snoek peaks, the value of D(o–o) can be derived from measurements of magnetic or elastic after-effects. Wert[5] has used the latter method to determine the diffusivities of carbon in bcc iron at temperatures below −10°C by means of the expression

$$\varepsilon = \varepsilon_0 \exp(-t/\tau_r) \qquad (4.9.6)$$

1. R. B. McLellan, M. L. Rudee and T. Ishibachi, *Trans. AIME* **233**, 1938 (1965).
2. R. H. Condit and D. N. Beshers, *Trans. AIME* **239**, 680 (1967).
3. A. J. Bosman and G. W. Rathenau, *Physica*, *'s Grav.* **26**, 533 (1960).
4. J. Bass and D. Lazarus, *Physics Chem. Solids* **23**, 1820 (1962).
5. C. A. Wert, *Phys. Rev.* **79**, 601 (1950).

According to Section 4.1 this expression corresponds to the descending part of the curve in Fig. 85(c).

Diffusion of hydrogen in bcc iron

There are many different opinions on the question of whether hydrogen can cause Snoek damping in iron. In view of the fact that, according to Section 2.13, the diffusivity of hydrogen in iron is abnormally high, a Snoek peak, if it exists at all, must occur at very low temperatures. At low temperatures the solubility of hydrogen in iron is practically nil and a strongly supersaturated solution of atomic hydrogen is therefore required to be able to detect the effect. However, in such a solution the hydrogen will have a strong tendency to precipitate as H_2 and it will do so in all lattice imperfections that offer sufficient space for it and which can be reached in the time available for the diffusion. It is therefore difficult to separate undesired side-effects from the effects caused in iron by the dissolved hydrogen.

Conditions are particularly unfavourable if the hydrogen is introduced into the iron electrolytically. This is due to its solubility in the metal at room temperature being immeasurably small whilst its diffusivity is still high: about 10^{-3} mm^2 s^{-1}. During, and immediately after, the electrolytic charging with hydrogen a large part of it precipitates as H_2 in 'traps', causing severe permanent damage in the shape of micro-cracks, larger fissures and plastic deformation. In order to keep as much hydrogen in solution as possible and prevent the occurrence of cracks and plastic deformation, the best method would be to first heat the metal in hydrogen to such a temperature that it contains a relatively large quantity of hydrogen in equilibrium solution and subsequently to cool it rapidly to the temperature region where a Snoek damping peak is expected, i.e. to the region where the diffusivity of hydrogen is very low.

In accordance with the above, Lord[(1)] using the pulse-echo method at a frequency of 10 MHz did find a Snoek damping peak in samples of Armco iron thermally charged with hydrogen, but not in electrolytically charged specimens. The peak gives the value for the diffusion coefficient of hydrogen in bcc iron as

$$D = 1{\cdot}43 \times 10^{-7} \text{ mm}^2 \text{ s}^{-1} \text{ at } 120 \pm 2 \text{ K}$$

In satisfactory agreement with this is the value derived from Johnson and Hill's, *loc. cit.*, formula (2.13.2):

$$D = 2{\cdot}1 \times 10^{-7} \text{ mm}^2 \text{ s}^{-1} \text{ at } 120 \text{ K}$$

1. A. E. Lord, *Acta Metall.* **15**, 1241 (1967).

This agreement gives support to the results obtained by Lord and also indicates that Johnson and Hill's formula, contrary to their own opinion, correctly represents not only the diffusivity of hydrogen in bcc iron for temperatures above 200°C but also for lower temperatures.

The thermally charged specimens used by Lord were built up from many layers of 0·36 mm (0·014 inch) sheet, clamped together after quenching in water from a $100 \mathrm{kN\,m^{-2}}$ hydrogen atmosphere at 890°C. They were at the temperature of the peak (120 K) within thirty minutes after the quenching. The peak showed only after charging with hydrogen and had disappeared after a few days ageing at room temperature. If the rate of escape of hydrogen were determined by its diffusivity, only an ageing time of a minute would be enough to make the peak disappear. However, as was seen in Section 2.13, the rate of escape is not determined by the diffusivity of the hydrogen but by the rate of a process taking place at the surface of exit or at internal surfaces. In other words, it is determined by an apparent diffusivity as represented by equation (2.13.3). Even from this equation an ageing time of about one hour at 20°C would be expected to be sufficient to cause the disappearance of the hydrogen. This therefore leads to the possibility that impurities in the metal might affect the rate of escape or possibly even the damping. It is to be regretted that Lord did not use an iron of greater purity than Armco in his experiments.

In agreement with Lord's experience, Hermant[1] did not succeed in finding a Snoek peak in pure iron that was electrolytically charged with hydrogen. He measured the decay of free torsional oscillations of iron wires at a frequency of about 1 Hz. At this frequency the expressions (2.13.2), (4.9.1) and (4.9.2) suggest a hydrogen Snoek peak in the temperature range 50 to 60 K. Hermant, studying this problem for several years, applied a great variety of methods to charge his wires electrolytically with hydrogen but none of these led to the occurrence of a Snoek peak. However, he did find a peak at 130 K in plastically deformed and electrolytically charged iron wires. This peak was assumed to originate in the interaction between hydrogen and dislocations (cf. Section 3.6). This hydrogen cold-work peak has also been found by others, e.g. by Weiner and Gensamer[2], Heller[3], Gibala[4] and Sturges and Miodownik[5]. Contrary to the results of Hermant, *loc. cit.*, these investigators found additional damping peaks in electrolytically charged iron specimens; they believe these peaks to be Snoek peaks. It may be concluded that the question of the origin of these peaks has not definitely been settled as yet.

1. M. E. Hermant, Ph.D. Thesis, Univ. of Amsterdam (1966).
2. L. C. Weiner and M. Gensamer, *Acta Metall.* **5**, 692 (1957).
3. W. R. Heller, *Acta Metall.* **9**, 600 (1961).
4. R. Gibala, *Trans. AIME* **239**, 1574 (1967).
5. C. M. Sturges and A. P. Miodownik, *Acta Metall.* **17**, 1197 (1969).

Diffusion in other bcc metals

Various people[1-7] have used the internal-friction method to study the diffusion of interstitial atoms in body-centred cubic metals other than iron, particularly in vanadium, niobium, tantalum and chromium. Results of these studies have been given in Table 10, Section 2.10. The solubility of oxygen and nitrogen in V, Nb and Ta is much greater than that of carbon and nitrogen in bcc iron. It is therefore possible to extend the measurements to concentrations at which interesting interactions between the interstitial atoms occur. This will be discussed in Section 4.14.

4.10. Determination of solubilities

In the previous section we have shown that the value of the diffusivity of interstitials in a bcc metal can be derived from the abscissa of the maximum of the Snoek peak. Under suitably chosen conditions, the ordinate of the maximum (the height of the peak) can give another important quantity, namely the solubility of the interstitial solute at a given temperature. To this end use is made of the fact that at low concentrations of interstitial atoms the height of the peak is directly proportional to the concentration. The derivation of the ratio constant is discussed below; it should be noted, however, that this factor can only be considered as a constant if the specimens whose internal friction is being measured have approximately the same grain size and the same density of dislocations. This is due to the fact that, as far as energy is concerned, the interstitial atoms find more favourable sites in the grain boundaries and dislocations than in the undisturbed lattice. As a result these sites are preferred at relatively low temperatures.

Another factor affecting the magnitude of the proportionality constant is the texture of the specimens since the Snoek damping is strongly dependent on the orientation of the crystals (cf. Section 4.7). Cold-drawn iron wires, recrystallized at a temperature below the α–γ transition point, show a $\langle 110 \rangle$ texture which, however, is more pronounced in some than in others. At equal grain size and carbon (or nitrogen) content these wires can therefore

1. T. S. Kê, *Phys. Rev.* **74,** 9, 16 and 914 (1948).
2. C. Y. Ang, *Acta Metall.* **1,** 123 (1953).
3. J. W. Marx, G. S. Baker and J. M. Sivertsen, *Acta Metall.* **1,** 193 (1953).
4. R. W. Powers, *Acta Metall.* **2,** 604 (1954) and **3,** 135 (1955).
5. E. Gebhardt and H. Preisendanz, *Z. Metallk.* **46,** 560 (1955).
6. R. W. Powers and M. V. Doyle, *Acta Metall.* **4,** 233 (1956); *J. appl. Phys.* **28,** 255 (1957); *J. Metals* **9,** 1287 (1957); *Acta Metall.* **6,** 643 (1958); *J. appl. Phys.* **30,** 514 (1959); *Trans. AIME* **215,** 655 (1959).
7. M. J. Klein, *J. appl. Phys.* **38,** 167 (1967).

still exhibit different peak heights. Wires of bcc iron with a random orientation of their crystals are obtained by heating them for a while at a temperature above the α–γ transition point.

The significance of the above is that for the measurements of solubility it is preferable to use textureless specimens containing few dislocations and consisting of relatively large grains which, however, should still be small compared to the diameter of the specimens; otherwise their number would be too small to smooth out orientation effects by averaging.

Solubility of nitrogen in bcc iron

The solubility of nitrogen, in equilibrium with Fe_4N, in bcc iron was first determined with the aid of the damping method by Dijkstra[1]. He introduced nitrogen into iron wires by heating them in a mixture of H_2 and NH_3 at 580°C. By quenching the wires from this temperature in cold water, the nitrogen was retained in supersaturated solution. In order to determine the solubility of nitrogen at a certain temperature (say 400°C) a wire was subsequently heated at this temperature, causing part of the nitrogen to precipitate as a nitride. The heating was interrupted a few times by quenching the wire to room temperature and the Snoek peak was measured by using the wire as the suspension element in a torsional pendulum system. The experiment was terminated when the height of the peak had reached a constant value.

Proceeding in this way, the solubility at the precipitation temperature is calculable, provided the proportionality constant between the dissolved quantity and the peak height is known. For single-crystal wires this constant can be calculated fairly accurately (cf. Section 4.7); for polycrystalline wires it must be determined experimentally. This can be done accurately by measuring the Snoek peak of wires with a relatively high content of dissolved nitrogen (a few hundredths of a weight per cent) and subsequently determining this content by chemical analysis. Once the proportionality factor is known, much lower percentages of dissolved nitrogen can be determined, even when a nitride precipitate is also present in the metal; this cannot be done chemically. Similar considerations apply to the determination of solubilities of carbon, in equilibrium with Fe_3C, in iron.

The above method for the measurement of solubilities has been criticized by several people in that the equilibrium is approached from one side only, i.e. from a surplus of nitrogen in solution. In their opinion it is possible that no true equilibrium is reached in this way and that the Snoek peaks, so found, correspond to supersaturated solutions. This fear has proved to

1. L. J. DIJKSTRA, *Trans. AIME* **185**, 252 (1949).

be unfounded; Fast and Verrijp[1] established equilibrium by the dissolution of nitride precipitates rather than by their formation. They charged pure iron wires, consisting of many randomly orientated crystals, with about 0·08 wt % nitrogen and subjected them to a very slow cooling from 580°C in order to obtain a relatively coarse precipitate of Fe_4N. Each of the wires was subsequently heated at a constant temperature in the range 380° to 580°C, the heating being continued until a constant height of the damping peak was obtained after quenching in water. The solubility values found in this way (Fig. 100) show a much smaller spread than those obtained by Dijkstra's method and can be represented by the formula

$$c(Fe_4N) = 12{\cdot}3 \exp(-34{,}700/RT) \text{ weight \% N} \qquad (4.10.1)$$

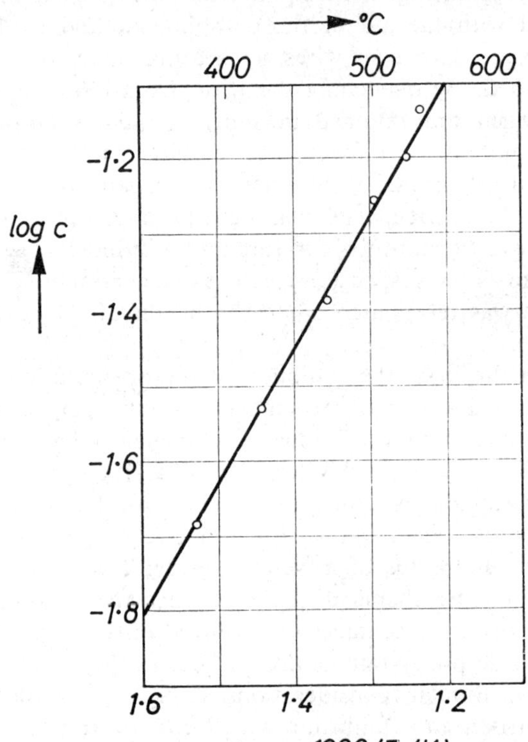

FIG. 100. Solubility c of nitrogen, in equilibrium with Fe_4N, in bcc iron. The values of c in weight % were obtained by making use of the experimental relation $c = 1{\cdot}26Q^{-1}_{max}$, where Q^{-1}_{max} corresponds to a peak temperature of 21·5°C (Fast and Verrijp).

1. J. D. FAST and M. B. VERRIJP, *J. Iron Steel Inst.* **180**, 337 (1955).

or, otherwise expressed,

$$\log c(\mathrm{Fe_4N}) = -\frac{1814}{T} + 1{\cdot}090$$

The maximum solubility of nitrogen, in equilibrium with Fe_4N, in bcc iron occurs at the eutectoid temperature of 590°C and, from expression (4.10.1), is seen to be 0·1 wt % nitrogen. The same value is found from measurements by Dijkstra, *loc. cit.*, and by Rawlings and Tambini(1) who also used the internal friction method to determine the solubility under discussion. In addition it agrees with the classical measurements by Paranjpe *et al.*(2) According to (4.10.1) the heat (enthalpy) of solution of Fe_4N in bcc iron is 34·7 kJ mole^{-1} (8300 cal/mole). The measurements by Dijkstra, Rawlings and Tambini, and Paranjpe *et al.*, yield figures of 32·2, 31·6 and 37·5 kJ mole^{-1} (7700, 7540 and 8960 cal/mole, respectively). There is thus a substantial agreement between the various investigations.

By means of internal friction measurements, Fast and Verrijp, *loc. cit.*, also determined the solubility of nitrogen in bcc iron, in equilibrium with gaseous nitrogen at 100 kN m^{-2} pressure. They found

$$c(\mathrm{N_2}) = 0{\cdot}098 \exp(-30{,}100/RT) \text{ weight \% N} \qquad (4.10.2)$$

or, otherwise expressed,

$$\log c(\mathrm{N_2}) = -\frac{1575}{T} - 1{\cdot}009$$

With the aid of the $\sqrt{p}$ law, solubilities at other nitrogen pressures can be calculated approximately from that at 100 kN m^{-2}, provided that the pressures are not so high that serious deviations from the behaviour of a perfect gas occur. Grieveson and Turkdogan(3) found solubility values in good agreement with (4.10.2). For other available data on the solubility of gaseous nitrogen in bcc iron we refer the reader to their publication.

The solubility of nitrogen, in equilibrium with the metastable nitride Fe_8N, in bcc iron will be discussed in the next section.

Solubility of carbon in bcc iron

The solubility of carbon, in equilibrium with Fe_3C, in bcc iron has been determined by Dijkstra(4) and Wert(5), by means of the damping method.

1. R. Rawlings and D. Tambini, *J. Iron Steel Inst.* **184,** 302 (1956).
2. V. G. Paranjpe, M. Cohen, M. B. Bever and C. F. Floe, *Trans. AIME* **188,** 261 (1950).
3. P. Grieveson and E. T. Turkdogan, *Trans. AIME* **230,** 1604 (1964).
4. L. J. Dijkstra, *Trans. AIME* **185,** 252 (1949).
5. C. A. Wert, *Trans. AIME* **188,** 1242 (1950).

They approached the phase equilibrium from the high-solubility side and obtained results which they expressed by the formula

$$c(Fe_3C) = 2{\cdot}55 \exp(-40{,}600/RT) \text{ weight \% C}$$

In deriving the formula it was assumed that the percentage of carbon was fortuitously exactly equal to the height of the peak. From later studies by Pitsch and Lücke[1] and by Wepner[2], however, it came to light that, at the frequency of 1 Hz as used by Dijkstra and by Wert, a correction factor of about 1·3 should be applied:

$$\% \text{ C} \approx 1{\cdot}3 Q_{max}^{-1} \tag{4.10.3}$$

This factor is approximately the same as that for nitrogen when the frequency of the torsional oscillation is also the same. Applying (4.10.3) we obtain the corrected solubility formula

$$c(Fe_3C) = 3{\cdot}31 \exp(-40{,}600/RT) \text{ weight \% C} \tag{4.10.4}$$

or, otherwise expressed,

$$\log c(Fe_3C) = -\frac{2120}{T} + 0{\cdot}520$$

A study by Keefer and Wert[3] has shown that identical values are found for the solubility when the equilibrium is approached from the low-solubility side, i.e. from a shortage of carbon in solution. The maximum solubility of carbon, in equilibrium with Fe_3C, in bcc iron occurs at the eutectoid temperature of 723°C; from formula (4.10.4) it is seen to be 0·025 wt % C, in excellent agreement with the value derived by Smith[4] from his classical thermodynamic measurements. The metastable equilibrium between [C] and Fe_3C in iron, mentioned above, is much more important than the stable equilibrium between [C] and graphite because the latter, under normal conditions, has a very slow rate of adjustment.

In determining the solubility of carbon or nitrogen, in equilibrium with a precipitate of carbide or nitride, in bcc iron too high values may be found if the precipitate consists of very fine particles. This is due to the fact that the solubility increases with decreasing particle size. In order to form an idea about the magnitude of this effect we use the Thompson–Freundlich equation relating to spherical particles of equal size[5, 6]:

$$\frac{2\sigma}{r} = \frac{RT}{v} \ln \frac{c}{c_0} \tag{4.10.5}$$

1. W. Pitsch and K. Lücke, *Z. Phys.* **139**, 413 (1954).
2. W. Wepner, *Arch. Eisenhütt. Wes.* **27**, 449 (1956).
3. D. Keefer and C. Wert, *Trans. AIME* **215**, 114 (1959).
4. R. P. Smith, *J. Am. Chem. Soc.* **68**, 1163 (1946).
5. H. Freundlich, *Kapillarchemie*, Akad. Verlagsges., Leipzig (1923).
6. R. Defay and I. Prigogine, *Tension superficielle et Adsorption*, Dunod, Paris (1951).

In this equation σ is the surface free energy of the iron–precipitate interface, v the molar volume of the precipitated phase, R the gas constant, T the absolute temperature, c the solubility of particles of radius r and c_0 the solubility of large particles as given by (4.10.1) or (4.10.4). The surface free energy of the Fe_3C–iron interface may be estimated to be of the order of magnitude 1 μJ mm^{-2}. Using this value we find from equation (4.10.5)

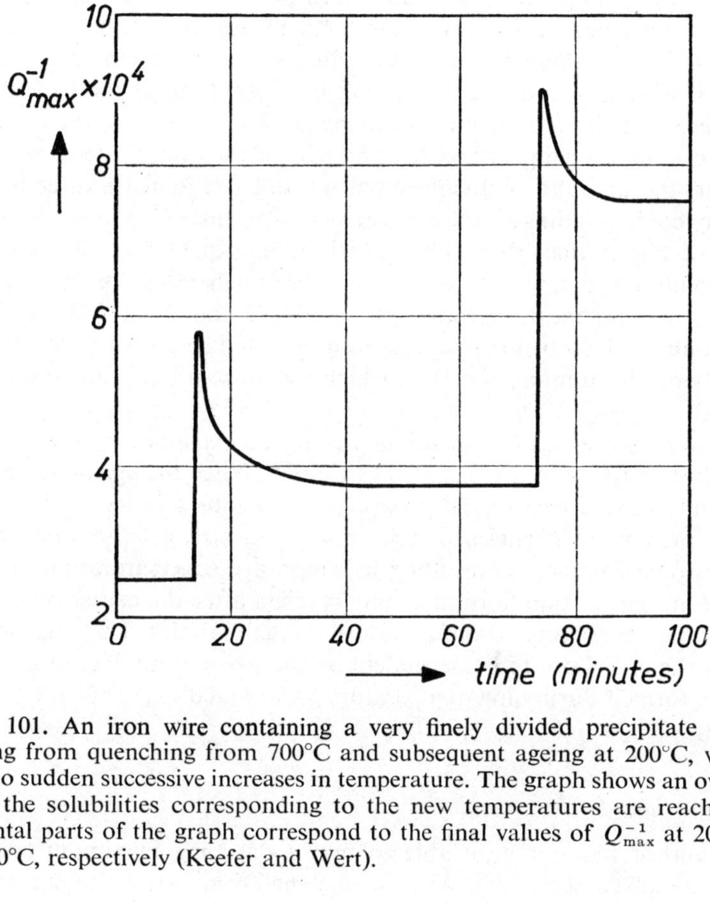

FIG. 101. An iron wire containing a very finely divided precipitate of Fe_3C resulting from quenching from 700°C and subsequent ageing at 200°C, was subjected to sudden successive increases in temperature. The graph shows an overshoot before the solubilities corresponding to the new temperatures are reached. The horizontal parts of the graph correspond to the final values of Q^{-1}_{max} at 200°, 300° and 400°C, respectively (Keefer and Wert).

that at 400 K the solubility of spherical Fe_3C particles of radius 20 nm, 10 nm or 5 nm is two, four or sixteen times as high as that of bulk Fe_3C. These results are qualitative only. If the interface free energy were 0·5 μJ mm^{-2} instead of 1, the calculated values of c/c_0 would not be 2, 4 or 16 but rather

$\sqrt{2}$, 2 or 4. The values are also strongly dependent on the shape of the particles. However, the above estimates lead to the important conclusion that the solubility of a finely divided precipitate rises rapidly with falling particle size.

In measuring the Snoek damping of carbon-bearing iron wires, Keefer and Wert[1] observed 'overshoot' effects, probably originating in the connection between solubility and particle size mentioned above. They introduced carbon into polycrystalline iron wires by heating them in an atmosphere of hydrogen and n-heptane at a temperature of about 700°C. The specimens were then quenched in water to room temperature so that the carbon taken up at 700°C was retained in supersaturated solution. Precipitation of Fe_3C in such a wire was initiated by heating it at 200°C until the height of the Snoek peak at 39°C remained unchanged. The wire was then suddenly heated to a higher temperature, causing part of the precipitate to redissolve. However, the amount of dissolved carbon did not increase directly to the solubility corresponding to the new temperature. Instead it increased rapidly to a value higher than that, subsequently falling quickly to the new equilibrium solubility. Fig. 101 shows this overshoot behaviour for two successive increases in temperature (from 200° to 300°C and from 300° to 400°C). The amount of the transitory supersaturation could be shown to be dependent primarily on the amount of carbon which was in solution prior to the ageing at 200°C.

The overshoot effect just mentioned could be explained by the increasing solubility of Fe_3C with decreasing particle size. When the ageing temperature is suddenly raised, part of the precipitate goes into solution and, in consequence, precipitate particles of relatively small size will become unstable. They dissolve completely, resulting in temporary supersaturation of carbon in the iron matrix. Equilibrium is only reached after the excess of carbon in solution has reprecipitated onto relatively large particles. This interpretation requires the overshoot to be dependent on the size and number of precipitate particles formed during low-temperature ageing (200°C).

Determination of other solubilities

The author has not been able to find a damping maximum caused by oxygen dissolved in iron[2]. After quenching from various temperatures in the α and γ region, no Snoek damping was observed in oxygen-bearing iron. These results seem to indicate a very low solubility of oxygen in solid iron.

1. D. Keefer and C. Wert, *Trans. AIME* **215,** 114 (1959).
2. J. D. Fast, *Métaux, Corros. Inds* **36,** 383 and 431 (1961).

This conclusion agrees with the findings of many investigators[1–9] who tried to determine this solubility in other ways.

According to some investigators[10, 11] commercial steel containing boron exhibits a damping peak in the region of the peaks caused by carbon and nitrogen. This would imply that boron has a demonstrable interstitial solubility in bcc iron. However, great uncertainty still exists about the interpretation of the observed effects. In the Philips Laboratory at Eindhoven we have not been able to find a damping peak in pure iron–boron alloys.

As far as we know, the damping method has not been used to try to determine the solubilities of C, N and O in V, Nb and Ta quantitatively. No doubt these determinations would be extremely difficult in view of the interaction of the interstitial atoms (see Section 4.14).

4.11. Discovery of new phases

In studying the iron–nitrogen system by the damping method, Dijkstra[12] was able to prove the existence of an iron nitride that had been unknown until then. He noted that at a temperature of about 250°C, the precipitation of nitrogen from a supersaturated solution in α iron takes place in two stages. The damping maximum decreases rapidly to an apparent terminal value corresponding to the upper horizontal part in Fig. 102. A microscopic investigation shows that at this stage a nitride other than Fe_4N has been precipitated. After a relatively long time interval the damping peak falls further, resulting in the second horizontal part of the graph and corresponding to the formation and growth of Fe_4N particles at the expense of the less stable nitride.

Later studies by Jack[13] made it appear probable that the new iron nitride is identical to a nitride that occurs as an intermediate phase during the decomposition of nitrogen martensite. X-ray investigations have shown that

1. J. D. Fast, *Revue Métall., Paris* **47,** 779 (1950).
2. J. A. Kitchener, J. O'M. Bockris, M. Gleiser and J. W. Evans, *Acta Metall.* **1,** 93 (1953).
3. F. Wever, W. A. Fischer and H. Engelbrecht, *Stahl Eisen* **74,** 1521 (1954).
4. J. L. Meijering, *Acta Metall.* **3,** 157 (1955).
5. R. Sifferlen, *C. r. hebd. Séanc. Acad. Sci., Paris* **247,** 1608 (1958).
6. A. U. Seybolt, *Trans. AIME* **215,** 298 (1959).
7. E. S. Tankins and N. A. Gokcen, *Trans. Am. Soc. Metals* **53,** 843 (1961).
8. M. T. Hepworth, R. P. Smith and E. T. Turkdogan, *Trans. AIME* **236,** 1278 (1966).
9. W. Frank, H. J. Engell and A. Seeger, *Z. Metallk.* **58,** 452 (1967).
10. R. R. Hasiguti and G. Kamoshita, *J. phys. Soc. Japan* **9,** 646 (1954).
11. W. R. Thomas and G. M. Leak, *Nature* **176,** 29 (1955).
12. L. J. Dijkstra, *Trans. AIME* **185,** 252 (1949).
13. K. H. Jack, *Proc. R. Soc.* **A 208,** 216 (1951).

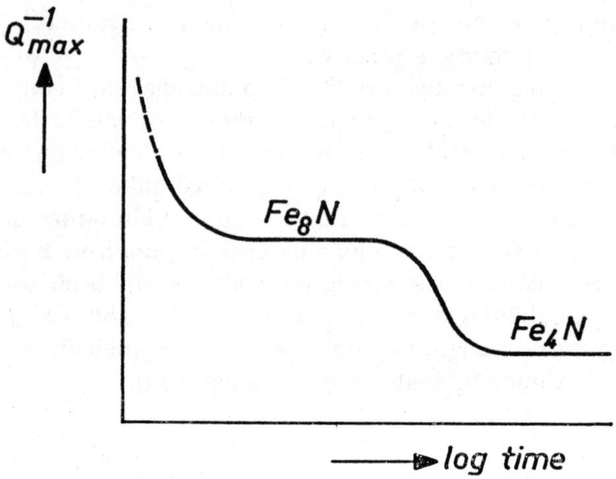

FIG. 102. As shown in this figure, the precipitation of nitrogen from its supersaturated solution in bcc iron at a relatively low temperature (e.g. 250°C) proceeds in two stages (Dijkstra).

this phase has a body-centred tetragonal structure. The unit cell contains eight (2×2×2) distorted cells of the original martensite and has an axial ratio $c/a = 1{\cdot}10$. It contains at most two nitrogen atoms situated at 0, 0, $\frac{1}{4}$ and at $\frac{1}{2}$, $\frac{1}{2}$, $\frac{3}{4}$, corresponding to the composition Fe_8N. This phase has a homogeneity range because part of the nitrogen sites can be unoccupied. More recent research[1-4] has confirmed the identity of the phases found by Dijkstra and Jack.

Apart from the tetragonality, the iron atoms in Fe_8N, as in martensite, are arranged in a similar pattern as in bcc iron. It is therefore understandable that the formation of a Fe_8N nucleus requires a much smaller activation energy than does the formation of a Fe_4N nucleus in which the iron atoms have the same arrangement as in fcc iron. At relatively low temperatures (e.g. 250°C) therefore, and provided the iron is supersaturated with nitrogen compared to both nitrides, the precipitation will start with the formation and growth of Fe_8N nuclei. The more stable Fe_4N particles, whose formation requires not only diffusion of nitrogen atoms but also regrouping of iron atoms, only appear much later. Above 300°C the thermal agitation of the iron atoms is so intense that Fe_4N nuclei form without much delay. Below 100°C the precipitation stops at Fe_8N if the metal does not already contain Fe_4N particles.

1. G. R. BOOKER, J. NORBURY and A. L. SUTTON, *J. Iron Steel Inst.* **187,** 205 (1957).
2. W. PITSCH, *Arch. Eisenhütt. Wes.* **32,** 493 and 573 (1961).
3. A. S. KEH and H. A. WRIEDT, *Trans. AIME* **224,** 560 (1962).
4. K. F. HALE and D. MCLEAN, *J. Iron Steel Inst.* **201,** 337 (1963).

The solubility of nitrogen, in equilibrium with Fe_8N, in iron has been accurately determined at a number of temperatures[1] by means of Dijkstra's method indicated by Fig. 102. It is given by the formula

$$c(Fe_8N) = 330 \exp(-41{,}400/RT) \text{ weight \% N} \qquad (4.11.1)$$

or, otherwise expressed,

$$\log c(Fe_8N) = -\frac{2164}{T} + 2{\cdot}519$$

This solubility is higher than the one in equilibrium with Fe_4N which is a

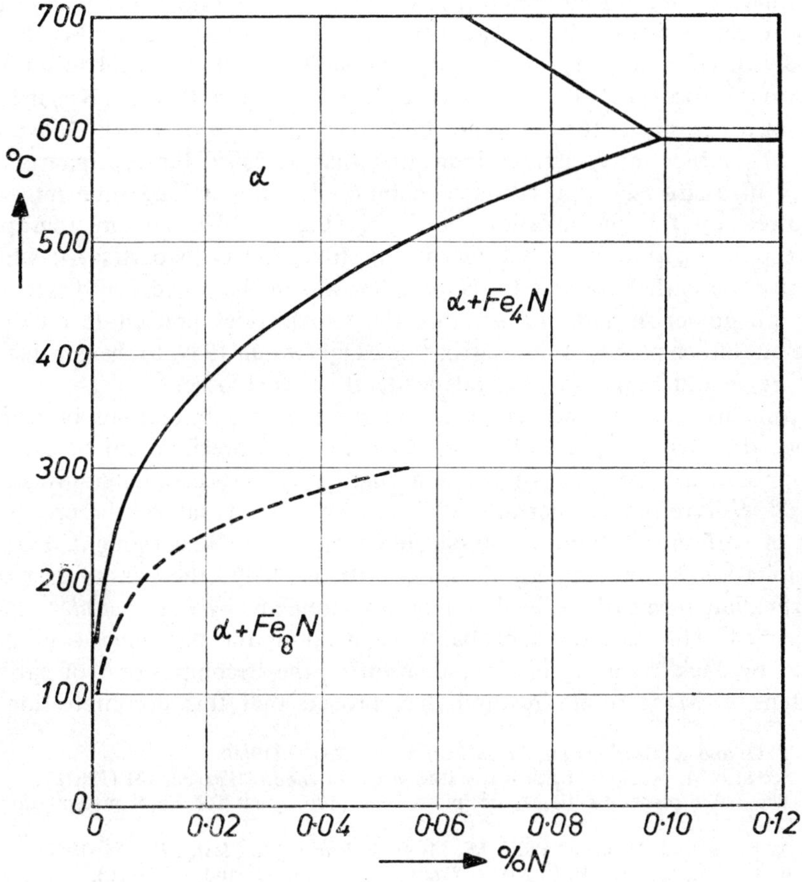

FIG. 103. Part of the 'stable' and metastable iron–nitrogen phase diagram. The concentrations are given in weight % (Fast and Verrijp).

1. J. D. FAST and M. B. VERRIJP, *J. Iron Steel Inst.* **180**, 337 (1955).

direct consequence of the lower stability of Fe_8N. Fig. 103 shows part of the stable and metastable iron–nitrogen phase diagram. Expressions (4.10.1) and (4.11.1) were used in drawing the diagram.

Oi and Sato[1] studied the precipitation of Fe_8N and Fe_4N in iron by means of electrical resistivity measurements and transmission electron microscopy. In good agreement with Fig. 103 they found that a precipitate of Fe_8N, corresponding to a nitrogen content of 0·014%, redissolved completely in the iron when the temperature was increased to 230°C. Only after a further increase in temperature, or after being held longer at 230°C did the more stable Fe_4N begin to precipitate in the homogeneous solution. It would also have been possible, of course, to choose the conditions of the experiment such that Fe_4N had begun to form before all the Fe_8N had dissolved. However, the fact that it is possible for all the Fe_8N to go into solution before Fe_4N is formed is a clear indication that Fe_8N particles cannot act as nuclei for the more stable phase. This agrees with Dijkstra's, *loc. cit.*, earlier investigations indicating that at 250°C the precipitation of Fe_8N in a strongly supersaturated solution is, after a long time interval, followed by the precipitation of Fe_4N (Fig. 102)[2]. His micrographs, corresponding to the second decline in Fig. 102, show Fe_4N particles in whose vicinity the earlier formed Fe_8N particles are in the process of dissolving. This phenomenon and the fact that the two nitrides precipitate on quite different lattice planes of the matrix[3] led Dijkstra, in 1949, to the conclusion that Fe_4N and Fe_8N nucleate independently of each other.

From his damping measurements on supersaturated solutions of *carbon* in bcc iron Dijkstra, *loc. cit.*, concluded that the precipitation of carbon, as opposed to that of nitrogen, is a single-stage process. Later investigations[4–7] have proved that at 200°C and lower temperatures the precipitation of carbon is also more complicated than was at first thought. During precipitation in the temperature range 150° to 200°C there is at least one intermediate iron carbide, as damping measurements by Butler *et al.*[8], have confirmed. This carbide is probably identical to the hexagonal ε carbide noted by Jack[9] during his investigation of the decomposition of carbon martensite. More recent research has proved that this decomposition is

1. T. Oi and K. Sato, *Trans. Japan Inst. Metals* **7,** 129 (1966).
2. See also M. Nacken and J. Rahmann, *Arch. Eisenhütt. Wes.* **33,** 131 (1962).
3. The habit plane for Fe_8N precipitation is {100}, that for Fe_4N precipitation is near {210}.
4. A. L. Tsou, J. Nutting and J. W. Menter, *J. Iron Steel Inst.* **172,** 163 (1952).
5. R. H. Doremus and E. F. Koch, *Trans. AIME* **218,** 591 and 596 (1960).
6. W. C. Leslie, *Acta Metall.* **9,** 1004 (1961).
7. M. G. H. Wells and J. F. Butler, *Trans. Am. Soc. Metals* **59,** 427 (1966).
8. J. Butler, P. Chollet and C. Crussard, *C. r. hebd. Séanc. Acad. Sci., Paris* **255,** 2961 (1962).
9. K. H. Jack, *J. Iron Steel Inst.* **169,** 26 (1951).

very complex and that another iron carbide, the so-called χ carbide[1] occurs as a second intermediate phase.

The internal friction method is not very suitable for the study of the precipitation of various carbides in iron, because the solubility of carbon in this metal at low temperature is much smaller than that of nitrogen.

4.12. Study of surface reactions

The internal friction method can usefully be applied to the study of a number of important surface reactions. As an illustration we will discuss in this section the surface reactions taking place during the desorption and absorption of nitrogen by iron.

Desorption of nitrogen

Since dissolved nitrogen is present in iron as atoms, its escape from the metal requires not only diffusion of nitrogen atoms in the interior but also their transfer to the surface and their combination to N_2 molecules:

$$[N] \rightarrow N_{ads} \qquad (4.12.1)$$

$$2N_{ads} \rightarrow N_{2\,ads} \qquad (4.12.2)$$

When 0·7 mm iron wires, containing an excess of nitrogen in solid solution, are heated for half an hour in pure nitrogen or in a vacuum at a temperature below 1,000°C, no noticeable change in the nitrogen content takes place according to our measurements. The denitriding proceeds much more rapidly if the gas phase contains hydrogen. This indicates that in the absence of hydrogen, the rate of escape is not determined by the diffusion but by a surface reaction. The slowness of this reaction has been useful in determining the solubilities of Fe_4N and Fe_8N in iron discussed earlier. At temperatures above 100°C these nitrides have very high dissociation pressures and so if establishment of equilibrium with the gas phase were a rapid process their solubilities in iron could only have been measured under these high nitrogen pressures. However, this is not so.

According to some scientists, hydrogen accelerates the rate of escape of nitrogen from iron because an oxide layer must be reduced before nitrogen can leave the metal via the reactions (4.12.1) and (4.12.2). If this were so, then only a very small amount of hydrogen in the gas would be required to attain a maximum rate of denitriding. Experiments have indicated, however, that the maximum rate of desorption at 500° and 600°C is only

1. Ph. Dünner and S. Müller, *Acta Metall.* **13**, 25 (1965).

reached at high hydrogen percentages[1,2]. This is illustrated by Fig. 104, in which are shown the results of measurements of the Snoek damping of iron wires with an initial nitrogen content of between 0·025 and 0·027 wt %. The figure gives the fraction of the quantity still present in the metal after

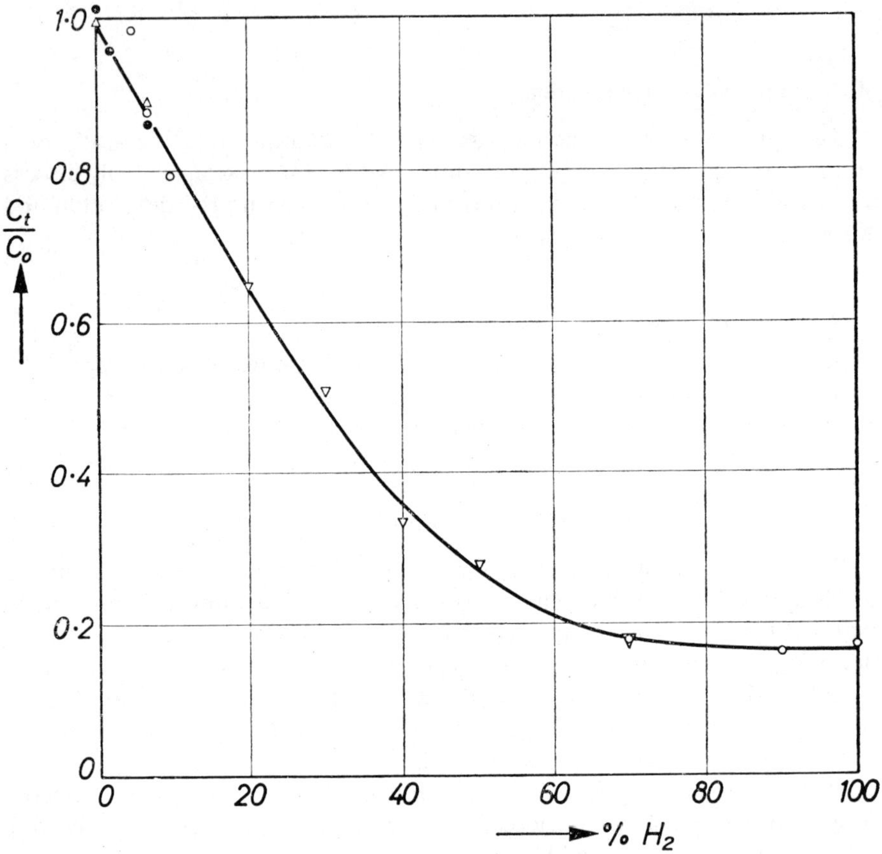

Fig. 104. Escape of nitrogen from 0·7 mm iron wires containing 0·025–0·027 wt % N. The release of nitrogen took place at 600°C in gas flows of various compositions at a total pressure of 100 kN m^{-2}. The ordinate shows the fraction of nitrogen still present in the metal after half an hour; the abscissa shows the volume percentage H_2 in the gas mixture. The upright triangles refer to dry and the open circles to wet mixtures of nitrogen and hydrogen. The shaded circles refer to dry and the inverted triangles to wet mixtures of argon and hydrogen. The horizontal part of the graph corresponds to rates of escape of nitrogen that are determined by the diffusion in the interior of the wires (Fast).

1. J. D. Fast and M. B. Verrijp, *J. Iron Steel Inst.* **176,** 24 (1954).
2. J. D. Fast, *Métaux, Corros. Inds.* **36,** 383 and 431 (1961).

heating for half an hour at 600°C in a flow of N_2 plus H_2 of 100 kN m^{-2} total pressure. During this heat treatment the wires release nitrogen because the solubility of nitrogen in iron in equilibrium with N_2 of 100 kN m^{-2} at 600°C is not much more than 0·001 %. This low solubility also explains why it makes little difference to these experiments when the nitrogen in the gas mixture is replaced by argon. During heating in pure nitrogen or argon and under the conditions given, the iron loses no measurable quantities of nitrogen nor does it during heating in a high vacuum.

The diffusion of nitrogen atoms over the iron surface, necessary to let the association reaction (4.12.2) take place, can be retarded by the presence of foreign atoms at this surface. In order to exclude the influence of traces of carbon, most of the denitriding experiments under discussion have been carried out in mixtures of nitrogen (or argon) and hydrogen containing water vapour at a partial pressure of 1·3 kN m^{-2} (10 Torr). It is known that the decarburizing of iron in wet hydrogen takes place very rapidly via the formation of CO whilst in dry hydrogen the process (via CH_4) is very slow[1–3]. The fact that the rate of denitriding in wet gas mixtures did not differ from that in thoroughly dried mixtures leads us to the conclusion that neither oxygen nor carbon played a part in our experiments. On the other hand, according to Fig. 104, hydrogen has an essential function in the desorption of nitrogen by iron wires. In the presence of hydrogen this process probably takes place via a sequence of surface reactions with relatively low activation energies, e.g.:

$$N_{ads} + H_{ads} \rightarrow NH_{ads} \qquad (4.12.3)$$

$$NH_{ads} + H_{ads} \rightarrow NH_{2\,ads} \qquad (4.12.4)$$

$$NH_{2\,ads} + H_{ads} \rightarrow NH_{3\,ads} \qquad (4.12.5)$$

$$NH_{3\,ads} \rightarrow NH_{3\,gas} \qquad (4.12.6)$$

According to Fig. 104, only when the partial pressure of hydrogen in the gas phase exceeds about 81 kN m^{-2} (0·8 atm) will the surface reactions at 600°C take place so rapidly that the diffusion of nitrogen in the interior of the metal determines its rate of escape. At a partial hydrogen pressure much lower than 81 kN m^{-2} the rate will be determined by a surface reaction. There are strong indications that the hydrogen pressure, below which this is so, depends strongly on the condition of the surface.

Fast and Verrijp, *loc. cit.*, found that the denitriding in a flow of pure hydrogen of 100 kN m^{-2} is determined by the rate of diffusion in the metal not only at 600°C but also at 500°C. Similar results were obtained by

1. C. R. Austin, *Trans. Am. Soc. Metals* **22,** 31 (1934).
2. K. Lücke, *Arch. Eisenhütt. Wes.* **25,** 181 (1954).
3. J. D. Fast and H. A. C. M. Bruning, *Z. Elektrochem.* **63,** 765 (1959).

Bohnenkamp[1] in an investigation of the denitriding of iron strips about 0·5 mm thick at temperatures of about 500°, 600°, 700° and 800°C. On the other hand, comparable experiments carried out by Strocchi *et al.*[2] showed the denitriding at these temperatures to be much slower than the diffusion in the metal. They believe that this is due to their use of commercial hydrogen that contained small amounts of carbon monoxide and hydrocarbons. They assume that these impurities strongly influence the rate of denitriding by poisoning the iron surface and retarding the reactions that take place there.

The surface reaction is always rate-determining, even in the absence of poisons, if the iron foils to be denitrided are only a few μm thick. By measuring the change in electrical resistance during the denitriding of thin iron foils Engelhardt and Wagner[3] came to the conclusion that reaction (4.12.4) is the rate-determining step in the reaction sequence (4.12.3) to (4.12.6). Using the same experimental technique, Grabke[4] was able to confirm this conclusion for hydrogen pressures of the order 100 kN m^{-2} and temperatures in the range 300° to 700°C. At decreasing hydrogen pressure, he found reaction step (4.12.5) to become rate-determining at about 10 kN m^{-2} H_2. Eventually, as the hydrogen pressure decreases further, the rate of formation of NH_3 becomes so small as to be comparable to that of reaction (4.12.2), i.e. to the rate of formation of N_2 from adsorbed nitrogen atoms. This allowed Grabke to measure the latter rate at temperatures higher than 550°C and a hydrogen pressure of 100 N m^{-2}. It is very small at 550°C but, being controlled by a high activation energy, increases strongly with rising temperature. In consequence, the rate of N_2 desorption from nitrogen-bearing iron wires can also be derived from Snoek damping measurements after many hours heating at 700°C or higher temperatures in a vacuum of 100 μN m^{-2} (10^{-6} mm Hg)[5].

There is a large discrepancy between the rate of nitrogen desorption from iron wires or strips, discussed above, and that from iron catalysts as used in ammonia synthesis, i.e. from iron crystals of only about 10 nm diameter. Various researchers[6,7] have shown that the desorption of nitrogen from an iron catalyst in the absence of hydrogen proceeds, at the same temperature and surface area, appreciably faster than the desorption from solid iron. Opinions differ on the explanation; people who have studied the desorption from solid iron will be inclined to assume that the many crystal edges and corners present in iron catalysts act as active centres for the recombination of nitrogen atoms, whilst researchers

1. K. Bohnenkamp, *Arch. Eisenhütt. Wes.* **38**, 229 (1967).
2. P. M. Strocchi, A. Melandri and A. Tamba, *Arch. Eisenhütt. Wes.* **38**, 135 (1967).
3. G. Engelhardt and C. Wagner, *Z. phys. Chem.* **B 18**, 369 (1932).
4. H. J. Grabke, *Z. Elektrochem.* **72**, 533, 541 (1968).
5. H. Junge, *Vacuum* **13**, 513 (1963).
6. J. T. Kummer and P. H. Emmett, *J. chem. Phys.* **19**, 289 (1951).
7. J. J. F. Scholten, Ph.D. thesis, Techn. Univ. Delft, 1959.

in the field of catalysis are more inclined to believe that their iron surfaces are purer than those of wires and strips.

Absorption of nitrogen

Our study of the interaction between nitrogen and solid α iron has shown that not only desorption but also absorption of nitrogen is very slow if the metal is in contact with pure nitrogen. The addition of hydrogen to the gas is essential to speed up the absorption by iron wires. Using this experience it has been possible to determine the solubility of nitrogen at 100 kN m^{-2} in α iron by adding 1 vol % hydrogen to the nitrogen. The solubility is given by formula (4.10.2).

From the rate of absorption of nitrogen by the iron wires used it was possible to conclude that, even in the presence in the gas of 1 % hydrogen, the surface reaction was rate-determining at the temperatures of 500° and 600°C. However, in the experiments carried out by Grieveson and Turkdogan[1], at higher temperatures and higher hydrogen concentrations, it was the diffusion of nitrogen in the metal that controlled the rate of the process. They used rods and strips of iron and measured the absorption of nitrogen at 750°C and higher temperatures in mixtures of N_2 and H_2 at a total pressure of 100 kN m^{-2} and hydrogen contents of 5% or 10%.

In the cases where a surface reaction determines the rate of absorption, it is possible that the reaction

$$N_{2\,ads} \rightarrow 2N_{ads} \tag{4.12.7}$$

is very slow, even at a very pure iron surface. The influence of hydrogen could be understood by assuming that, although no complete dissociation of the nitrogen molecules takes place at the surface, they are nevertheless activated by forming the complexes —N=N— or =N—N=[2]. These activated complexes would then react with chemisorbed hydrogen atoms as depicted in Fig. 105 or, in abbreviated form, by

$$N^*_{2\,ads} + 2H_{ads} \rightarrow N_{ads} + NH_{2\,ads} \tag{4.12.8}$$

where the asterisk indicates that N_2 is not simply physically adsorbed but chemisorbed. Part of the nitrogen atoms arising in this way will enter into the metal,

$$N_{ads} \rightarrow [N] \tag{4.12.9}$$

and another part will react with the adsorbed hydrogen and form NH_3. This also applies to the $NH_{2\,ads}$ arising from reaction (4.12.8). Part of this, however, will dissociate and form N_{ads}, some of which will enter the metal.

1. P. Grieveson and E. T. Turkdogan, *Trans. AIME* **230,** 1604 (1964).
2. Cf. P. Mars, J. J. F. Scholten and P. Zwietering, *The Mechanism of Heterogeneous Catalysis* (editor: J. H. de Boer), Elsevier, Amsterdam (1960), p. 66.

From experiments by Scholten, *loc. cit.*, with carefully reduced iron catalysts it can be concluded that at pure surfaces of very small iron crystals reaction (4.12.7) proceeds at a high rate even below 250°C. This need not contradict our measurements of the absorption of nitrogen by solid iron. The possibility should not be excluded that the favourable influence of hydrogen which we observed is due to the removal of surface impurities. The absorption experiments,

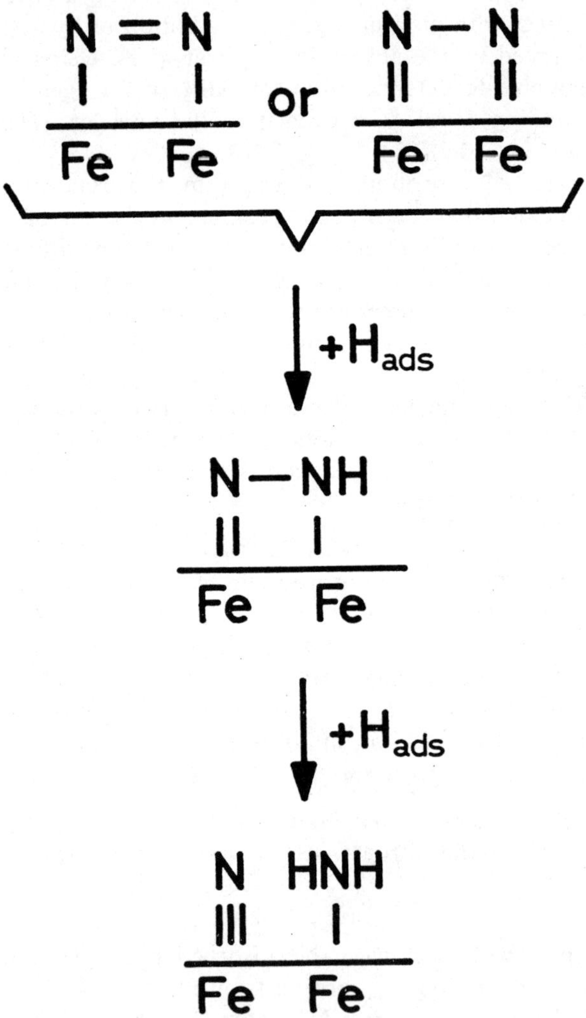

FIG. 105. Hydrogen plays an important part in the penetration of nitrogen in iron. A possible explanation is that the nitrogen atoms are only formed at the surface by a reaction between chemisorbed nitrogen and hydrogen as indicated in the diagram. (Mars, Scholten and Zwietering).

unlike those on desorption, were carried out only in pure mixtures 99% N_2 + 1% H_2, not in wet mixtures, nor at varying hydrogen content. At higher temperatures (700° to 1000°C) Grabke, *loc. cit.*, found the rate of absorption of nitrogen to be independent of the hydrogen pressure. In that range of temperatures, at least, the influence of hydrogen appears to be confined to the removal of surface impurities, mainly oxygen[1].

Even if reaction (4.12.7) on a pure surface of solid iron were a fast one at low temperatures, it is still possible that reaction (4.12.9) requires a high activation energy and proceeds slowly. In other words it is possible that the formation of adsorbed nitrogen atoms at a pure macro-surface is fast, even at low temperatures, but that the rate of *absorption* is still determined by a surface reaction, i.e. by (4.12.9).

4.13. Analytical determination of small quantities of carbon and nitrogen in iron

Another application of the damping method is the analytical determination of small concentrations of carbon or nitrogen in otherwise pure iron[2] or, more generally, the determination of small concentrations of interstitially dissolved elements in bcc metals. With the usual chemical methods it is not possible to measure with sufficient accuracy carbon and nitrogen contents in iron of a few thousandths of a percent. By measurement of the internal friction, concentrations of the order of magnitude of 0·001 wt % can be determined to within a few ten-thousandths of a percent. To this end use is made of the previously-mentioned fact that the height of the Snoek peak is directly proportional to the concentration of the interstitial atoms, provided this concentration is less than about 0·05%.

A composite graph of internal friction versus temperature is obtained when dissolved carbon and nitrogen are present together in iron. This graph is a superposition of a normal carbon and a normal nitrogen Snoek peak if the metal contains no other impurities (Fig. 106). With this knowledge it is also possible in this case to determine the carbon and nitrogen content of the iron with a possible error of not more than a few ten thousandths of a percent[3].

According to (4.5.14) the total damping will be given by

$$Q^{-1} = Q_C^{-1} \operatorname{sech}\left[\frac{\Delta H_C}{R}\left(\frac{1}{T} - \frac{1}{T_C}\right)\right] + Q_N^{-1} \operatorname{sech}\left[\frac{\Delta H_N}{R}\left(\frac{1}{T} - \frac{1}{T_N}\right)\right] \qquad (4.13.1)$$

where ΔH_C and ΔH_N are the activation enthalpies of the diffusion of C and N in bcc iron and (Q_C^{-1}, T_C) and (Q_N^{-1}, T_N) are the values (Q_{max}^{-1}, T_{max}) for the separate C and N peaks. Expressions (2.14.1) and (4.9.5) give the values of ΔH_C and ΔH_N as 80·15 and 76·78 kJ mole^{-1} (19,160 and 18,350 cal/mole,

1. Cf. E. T. Turkdogan and P. Grieveson, *J. Electrochem. Soc.* **114,** 59 (1967).
2. J. D. Fast, *Stahl Eisen* **73,** 1484 (1953).
3. W. Wepner, *Arch. Eisenhütt. Wes.* **27,** 449 (1956).

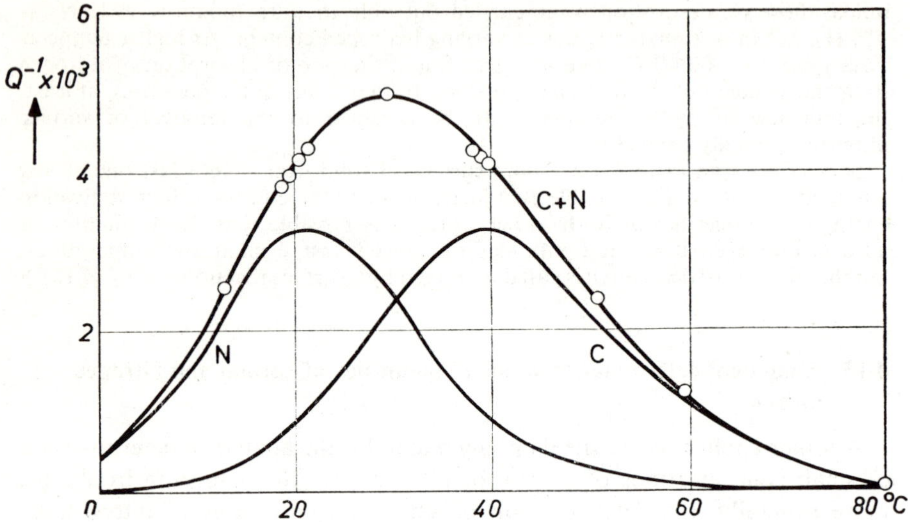

FIG. 106. Damping of torsional oscillations of an iron wire containing 0·0041%C and 0·0043% N, measured at different temperatures (circles). The individual graphs for C and for N have been calculated from expression (4.5.14) and the graph for C + N from expression (4.13.1). Frequency of oscillation 0·91 Hz. (Wepner).

respectively), whilst, for an oscillation frequency of 1 Hz, T_C and T_N are approximately 312 and 295 K. Substituting these values in (4.13.1) gives an expression containing the unknown Q_C^{-1} and Q_N^{-1} directly related to the carbon and nitrogen concentrations. At least two pairs of values for (Q^{-1}, T) are required to find these unknowns. To this end we choose two temperatures and determine the corresponding Q^{-1} values from the composite damping graph. For chosen values of 295 and 312 K the following expressions are obtained:

$$Q_C^{-1} = 1{\cdot}130Q_{312}^{-1} - 0{\cdot}396Q_{295}^{-1} \tag{4.13.2}$$

and

$$Q_N^{-1} = 1{\cdot}130Q_{295}^{-1} - 0{\cdot}370Q_{312}^{-1} \tag{4.13.3}$$

According to investigations by Wepner, *loc. cit.*, and others[(1-3)] the weight percentages of carbon and nitrogen are obtained by multiplying the calculated values of Q_C^{-1} and Q_N^{-1} by a factor between 1·25 and 1·30.

An obvious drawback of the described method for the determination of C and N concentrations is the fact that the damping measurements require wires that are moderately coarse-grained and show no noticeable texture

1. J. D. FAST and M. B. VERRIJP, *J. Iron Steel Inst.* **180**, 337 (1955).
2. R. RAWLINGS and D. TAMBINI, *J. Iron Steel Inst.* **184**, 302 (1956).
3. W. KÖSTER and G. KAMPSCHULTE, *Arch. Eisenhütt. Wes.* **32**, 809 (1961).

(cf. Section 4.10). In addition, the carbon and nitrogen concentrations should be sufficiently low to permit a homogeneous solution to be obtained by quenching from the α region.

Other complications arise when the damping method is used to determine carbon and nitrogen in commercial steel. In this material part of the carbon or nitrogen is often present as a very stable carbide or nitride; this part cannot be determined by measuring the Snoek damping. Alloy elements forming very stable nitrides are among others B, Al, Cr, Si, Ti, Zr, V, Nb and Ta. Most of these elements also form very stable carbides. Manganese causes another effect (see Section 4.16); this metal has no great affinity for C and N but its presence in iron causes a broadening of the Snoek peak which can be represented by the superposition of three peaks, each characterized by a single relaxation time[1,2]. For this reason, steel containing manganese in addition to small concentrations of dissolved carbon and nitrogen, shows an internal friction graph which does not correspond to the theoretical shape as given by expression (4.13.1). However, by taking into account the influence of manganese on the Snoek peak for nitrogen, Richards and Barratt[3] succeeded in obtaining a satisfactory degree of correspondence between calculated and experimental graphs for two commercial low-carbon steels containing 0·35% and 0·32% Mn respectively.

4.14. Interaction of interstitial atoms

The effect of high concentrations of interstitial atoms cannot be determined by measurements of the Snoek damping of iron samples since the solubilities of carbon and nitrogen in this metal are low (cf. Fig. 103) and that of oxygen negligible. Much higher interstitial concentrations, especially of oxygen and nitrogen, are obtainable in the bcc metals tantalum and niobium. At high temperatures these metals can contain several atom % oxygen or nitrogen in solid solution[4–13]. After quenching, the precipitation

1. J. L. Meijering, *Métaux, Corros. Inds.* **36,** 107 (1961).
2. J. D. Fast, J. L. Meijering and M. B. Verrijp, *Métaux, Corros. Inds.* **36,** 112 (1961).
3. P. N. Richards and K. V. Barratt, *J. Iron Steel Inst.* **204,** 380 (1966).
4. E. Gebhardt and H. D. Seghezzi, *Z. Metallk.* **50,** 521 (1959) [Ta–O].
5. D. A. Vaughan, O. M. Stewart and C. M. Schwartz, *Trans. AIME* **221,** 937 (1961) [Ta–O and Ta–N].
6. A. U. Seybolt, *Trans. AIME* **200,** 774 (1954) [Nb–O].
7. R. P. Elliott, *Trans. Am. Soc. Metals* **52,** 990 (1960) [Nb–O].
8. R. T. Bryant, *J. less-common Metals* **4,** 62 (1962) [Nb–O].
9. E. Gebhardt and R. Rothenbacher, *Z. Metallk.* **54,** 443 and 623 (1963) [Nb–O].
10. E. Gebhardt, H. D. Seghezzi and E. Fromm, *Z. Metallk.* **52,** 464 (1961) [Ta–N].
11. P. Bunn and C. A. Wert, *Trans. AIME* **230,** 936 (1964) [Ta–N].
12. J. R. Cost and C. A. Wert, *Acta Metall.* **11,** 231 (1963) [Nb–N].
13. E. Gebhardt, E. Fromm and D. Jakob, *Z. Metallk.* **55,** 423 (1964) [Nb–N].

of oxide or nitride from a supersaturated solution proceeds very slowly and it is thus possible to study the Snoek damping of Ta and Nb containing high concentrations of interstitial atoms. In these cases, deviations from simple Snoek behaviour are found which are caused by the presence in the solid solutions of clusters of two or more interstitials. In the next paragraph we shall discuss the formation of di-interstitials first investigated by Powers and Doyle[1,2]. It was later discovered that clustering can also be detected in more dilute solutions, even in Fe–C and Fe–N alloys. We shall return to this matter later in the section.

Formation of pairs of interstitials in tantalum and niobium

Pure binary Ta–O alloys containing less than 0·1 atom % oxygen exhibit, according to Powers, *loc. cit.*, a single Snoek peak with a height proportional

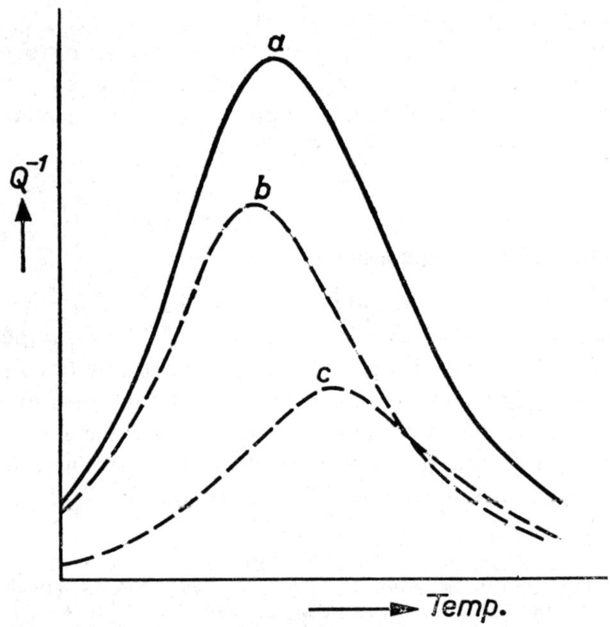

FIG. 107. An experimental damping peak (*a*) of tantalum containing between 0·1 and 1 atom % oxygen can be resolved into a normal Snoek peak (*b*) and an interaction peak (*c*), each characterized by a single relaxation time.

1. R. W. POWERS, *Acta Metall.* **2,** 604 (1954); **3,** 135 (1955).
2. R. W. POWERS and M. V. DOYLE, *Acta Metall.* **4,** 233 (1956); *J. appl. Phys.* **28** 255 (1957); *J. Metals* **9,** 1287 (1957); *Acta Metall.* **6,** 643 (1958); *J. appl. Phys.* **30,** 514 (1959); *Trans. AIME* **215,** 655 (1959).

to the oxygen content. At higher concentrations broader peaks, i.e. peaks that cannot be defined by a single relaxation time are found. Up to an oxygen concentration of about 0·5 atom %, the broad peaks can be resolved into two single peaks (Fig. 107). At a frequency of 0·6 Hz one maximum occurs at 137°C and varies linearly with the concentration; the other occurs at 162°C and is approximately proportional to the square of the oxygen concentration. From this result Powers and Doyle, *loc. cit.*, infer that the 137°C peak is the normal Snoek peak based on jumps by single oxygen atoms whilst the other peak is caused by the reorientation of atom pairs.

Similar phenomena have been observed when the tantalum contained dissolved nitrogen. The normal Snoek peak and the peak caused by N–N pairs occur at much higher temperatures than in the case of oxygen, i.e. at 334° and 362°C at a frequency of 0·6 Hz. When oxygen and nitrogen are both present in tantalum, interactions between the two types of interstitials also take place. In order to study these, Powers and Doyle, *loc. cit.*, used

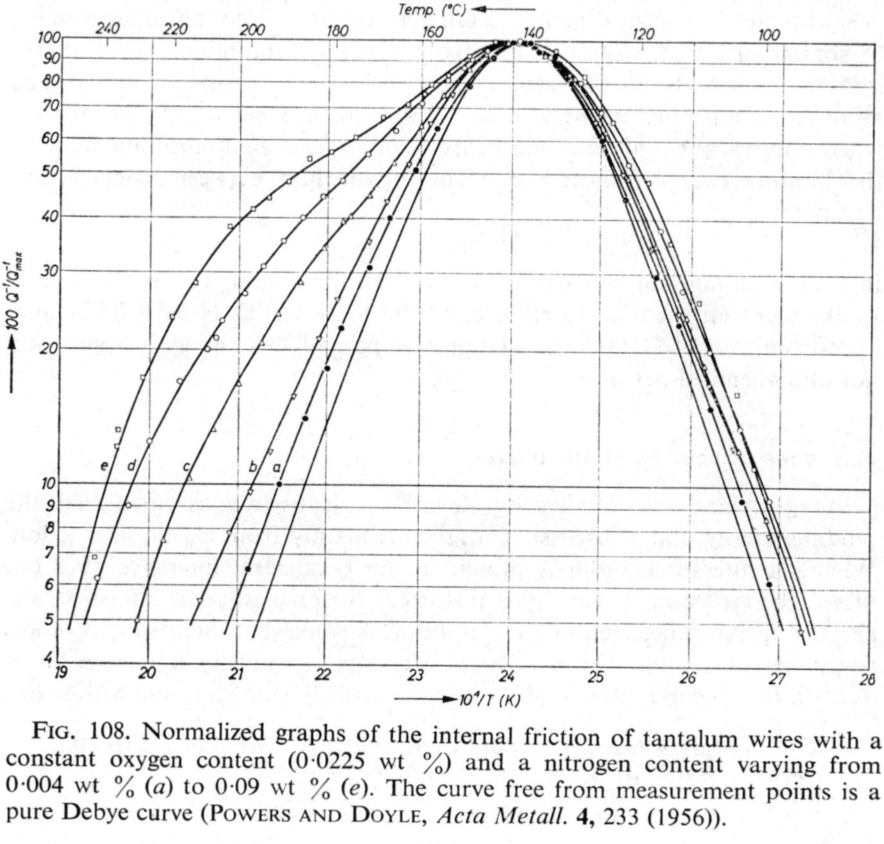

FIG. 108. Normalized graphs of the internal friction of tantalum wires with a constant oxygen content (0·0225 wt %) and a nitrogen content varying from 0·004 wt % (*a*) to 0·09 wt % (*e*). The curve free from measurement points is a pure Debye curve (POWERS AND DOYLE, *Acta Metall.* **4**, 233 (1956)).

tantalum wires with an oxygen content (0·25 atom %) such that the O–O interaction was only small. As increasing amounts of nitrogen were added to these wires a broadening of the oxygen peak occurred comparable to the broadening that takes place with increasing oxygen content. As shown in Fig. 108, the maximum of the experimental graphs is at about 140°C. Each of these graphs can be resolved into two partial peaks, the normal oxygen peak at 142°C (0·9 Hz) and a peak at 175°C of which the height is approximately proportional to the product of the concentrations of oxygen and nitrogen. The proportionality indicates that this latter peak is based on O–N interactions, i.e. on jumps of oxygen atoms each one of which interacts with a nitrogen atom in its vicinity. A similar O–N pairing takes place in niobium containing both oxygen and nitrogen in solid solution[1,2].

For various temperatures the part of the interstitial atoms that have combined to form pairs can be derived from results of relaxation measurements. With the help of simple thermodynamic considerations these data yield a binding energy of oxygen pairs in tantalum of approximately -8 kJ mole^{-1} (-2 kcal/mole)[3]. Gibala and Wert, *loc. cit.*, find a value of similar order of magnitude for oxygen pairs in niobium. The negative sign corresponds to an evolution of heat during pair formation. An oxygen atom is bound more strongly in the vicinity of another oxygen atom than when only vacant interstices are in its vicinity. The small absolute value of the binding energy indicates that the equilibrium between single atoms and pairs

$$2[O] \rightleftharpoons [O_2]$$

in a metal should not be regarded as a chemical reaction. This is also shown by the fact that the binding energies of the N–N and O–N pairs differ only little from that of O–O pairs. The greater part of this energy is apparently not of a chemical nature.

Formation of pairs by elastic interaction

In agreement with studies by Zener[4] it is usually assumed that the binding energy under discussion originates mainly from elastic interaction. When an interstitial atom is present in an octahedral interstice of a bcc metal, the lattice in its vicinity is distorted. In some adjacent interstices the shortest distance between two metal atoms is somewhat increased, in others somewhat decreased. The deformation energy consequent upon the introduction of a second interstitial atom in an already elongated site will be less

1. D. J. VAN OOYEN and A. S. VAN DER GOOT, *Philips Res. Repts.* **19,** 505 (1964).
2. R. GIBALA and C. A. WERT, *Trans. AIME* **236,** 924 (1966).
3. R. W. POWERS and M. V. DOYLE, *Trans. AIME* **215,** 655 (1959).
4. C. ZENER, *Phys. Rev.* **74,** 639 (1948).

than in an arbitrary non-deformed site. The difference between the two energies acts as a kind of binding energy between the two interstitials.

With the aid of a computer, Johnson *et al.*[1] calculated the binding energies for a number of di-carbon configurations in bcc iron. These calculations are based on a mathematical model of a perfect iron lattice, the model being devised such as to approximate various known physical properties of an actual iron crystal. In the model the metal is represented by a spherical crystallite containing more than 500 iron atoms which are treated as individual particles interacting with two-body central forces. The remainder

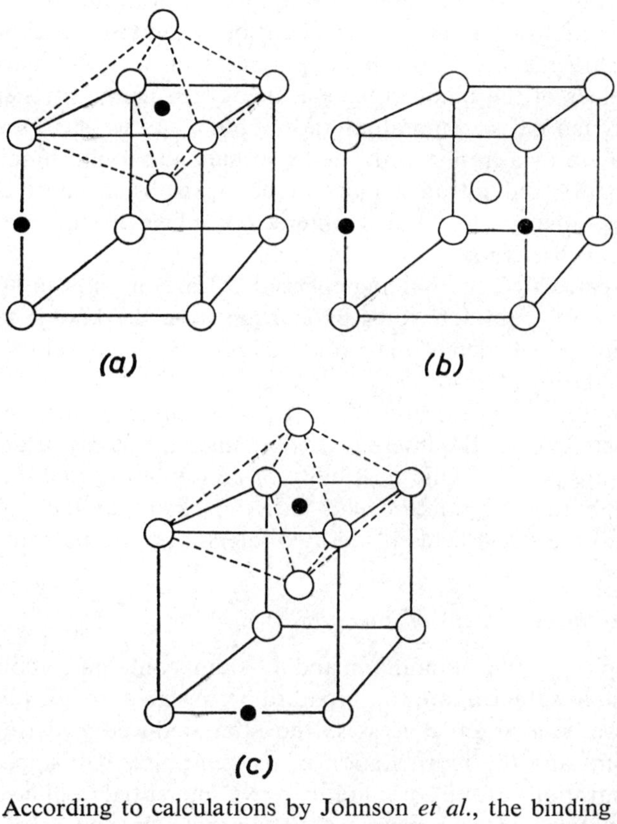

FIG. 109. According to calculations by Johnson *et al.*, the binding energies for carbon pairs in bcc iron are 12·5, 10·6 and 7·6 kJ mole^{-1} (0·13, 0·11 and 0·08 eV) for the arrangements shown in (*a*), (*b*) and (*c*), respectively. The deformation of the iron lattice caused by the carbon atoms is not shown in the figure.

1. R. A. JOHNSON, G. J. DIENES and A. C. DAMASK, *Acta Metall.* **12,** 1215 (1964).

of the lattice outside the crystallite is treated as an elastic continuum with iron atoms embedded in it. Carbon atoms were then introduced into the model at the centre of the crystallite and the resulting changes in energy, volume, etc., were calculated by making use of a potential for the iron–carbon interaction, derived from experimental data on the behaviour of carbon in bcc iron[1].

Three types of carbon pairs in bcc iron, shown in Fig. 109, have appreciable binding energies. The absolute values of these were found to be 12·5 kJ mole^{-1} (3·0 kcal/mole), 10·6 kJ mole^{-1} (2·5 kcal/mole) and 7·6 kJ mole^{-1} (1·8 kcal/mole) for the configurations (*a*), (*b*) and (*c*) respectively. In the calculations a possible carbon–carbon interaction was not taken into account so that the results only reflect the interaction of the local stresses in the vicinity of the two carbon atoms. As far as the order of magnitude is concerned, the calculated values show a satisfactory agreement with the experimentally obtained values of the binding energies of oxygen pairs, nitrogen pairs and oxygen–nitrogen pairs in tantalum and niobium. As we shall see below the calculations are also in line with the experimentally found binding energies of carbon pairs and nitrogen pairs in bcc iron. This agreement strongly supports the theory that elastic interaction plays a major part in the binding under discussion.

Fisher[2] calculated the binding energies of carbon pairs in bcc iron in a different manner from Johnson's by starting from the known tetragonality of martensite. The order of magnitude of his calculated values agrees well with that found by Johnson.

The interstitial pairs (a) and (b) in Fig. 109 can, by their reorientation, strongly contribute to the internal friction since the stress fields of the two interstitials are parallel. One possibility of reorientating each of these pairs is to move both interstitials half a lattice constant to the right. In configuration (c) the stress fields of the two interstitials are perpendicular.

Formation of larger groups of interstitials

The damping graphs of tantalum and niobium, containing about 1 atom % oxygen in solid solution, are too broad to permit an accurate representation by only two relaxation processes, the stress-induced ordering of single oxygen atoms and the reorientation of oxygen pairs. It is apparent that at high concentrations groups of more than two interstitials will also contribute to the relaxation. Measurements of the internal friction as a function of the temperature are not very suitable for checking the validity of this

1. For details of the calculations the reader is referred to an article by R. A. Johnson, *Phys. Rev.* **134,** A1329 (1964).
2. J. C. Fisher, *Acta Metall.* **6,** 13 (1958).

assumption. Data obtained by measuring the relaxation of stress at constant strain or the relaxation of strain at constant stress (both at constant temperature), are much easier to resolve into a number of unique relaxation processes than data obtained by internal friction measurements.

An analysis of measurements of elastic after-effect (relaxation of strain at constant stress) on niobium–oxygen solutions led Gibala and Wert[1] to the conclusion that, at high oxygen concentrations, at least four independent Debye relaxations are necessary to cover the experimental results with reasonable accuracy. They assumed the relaxation time of each of these component processes to obey an equation of the simple form (4.5.8). The process with the shortest relaxation time at a given temperature (process 1) was shown to be caused by single, unassociated oxygen atoms. The remaining three processes, labelled 2, 3 and 4, were believed to be controlled by reorientations of groups of two, three and four oxygen atoms respectively. Furthermore, the relaxation strengths of these processes were assumed to be proportional to the concentrations of the clusters in question. If these assumptions are correct, the law of mass action requires that the relaxation strengths, Δ_2, Δ_3 and Δ_4 of processes 2, 3 and 4, and thus the concentration of pairs, triplets or quadruplets, is proportional to the square, cube or fourth power of the relaxation strength Δ_1 of process 1 (the Snoek relaxation). This dependence was found to apply reasonably well for processes 2 and 3, and to a lesser extent for process 4.

The binding energy (or, more correctly, binding enthalpy) of the oxygen clusters, i.e. the energy change corresponding to the formation of the clusters from single atoms, could be deduced from the variations of the relative concentrations of the clusters with temperature. Their absolute values were found to be 7 ± 3 kJ mole^{-1} ($0{\cdot}07 \pm 0{\cdot}03$ eV) for pairs, 17 ± 4 kJ mole^{-1} ($0{\cdot}18 \pm 0{\cdot}04$ eV) for triplets and 26 ± 5 kJ mole^{-1} ($0{\cdot}27 \pm 0{\cdot}05$ eV) for quadruplets. From accurate measurements of the relaxation of stress at constant strain, Keefer and Wert[2,3] deduced that pairs and triplets in more than random numbers are even present in interstitial solutions of carbon and nitrogen in bcc iron. According to their experiments the binding energies of the carbon and nitrogen clusters in iron are of the same order of magnitude as those of similar oxygen clusters in niobium.

Since clusters of any size exist in several stable configurations with different binding energies (cf. Fig. 109), the model for the relaxation spectrum as proposed by Powers *et al.* and Wert *et al.* must be considered as a considerable simplification, however attractive and useful it may be.

1. R. Gibala and C. A. Wert, *Acta Metall.* **14,** 1095 and 1105 (1966).
2. D. Keefer and C. A. Wert, *J. phys. Soc. Japan* **18,** Suppl. III, 110 (1963).
3. D. Keefer and C. A. Wert, *Acta Metall.* **11,** 489 (1963).

Clustering and formation of domains with an ordered structure of interstitials

The calculations of the binding energies of carbon pairs in bcc iron by Johnson *et al., loc. cit.*, have been repeated[1] and extended to the calculation of binding energies of larger clusters. The absolute values of the binding energies of the di-carbon configurations (*a*), (*b*) and (*c*) in Fig. 109 were found to be 13·5, 10·6 and 7·6 kJ mole^{-1} (0·14 eV, 0·11 eV and 0·08 eV, respectively) which differ only slightly from the older values. In configurations (*a*) and (*b*) the two carbon atoms are hemmed in between parallel pairs of iron atoms. Using the lattice constant as unit length, the distance between the two carbon atoms is $\frac{1}{2}\sqrt{3}$ in (*a*) and 1 in (*b*). Configuration (*c*) does not play any role in larger clusters.

Figure 110 shows the most stable tetra-carbon configuration. It can be visualized as comprising four 'bonds' of type (*a*) and two 'bonds' of type (*b*), Fig. 109. The most stable triplet is obtained by removing any one of the four carbon atoms, thus halving the number of (*a*) and (*b*) 'bonds'. Because of its elastic nature, the binding energy of a cluster cannot be found by simply adding the energies of the separate 'bonds'. Its absolute value is 63·7 kJ mole^{-1} (0·66 eV) for the most stable quadruplet (Fig. 110) and 34·7 kJ mole^{-1} (0·36 eV) for the most stable triplet. In order to obtain larger

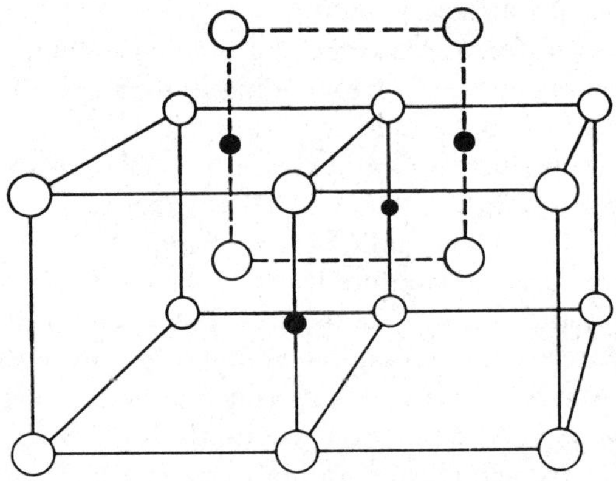

FIG. 110. The most stable tetra-carbon configuration in bcc iron. The open circles represent iron atoms and the shaded circles carbon atoms. The binding energy for this arrangement was calculated by Johnson to be 63·7 kJ mole^{-1} (0·66 eV). The most stable tri-carbon configuration is obtained by removing any one of the four carbon atoms and has a binding energy of 34·7 kJ mole^{-1} (0·36 eV). Displacements of the iron atoms from their normal lattice positions are not shown in the figure.

1. R. A. JOHNSON, *Acta Metall.* **15**, 513 (1967).

clusters of maximum stability the same process is applied as in going from triplet to quadruplet—the additional carbon atom is added by placing it in a position where the maximum number of (*a*) and (*b*) 'bonds' are formed.

Figure 111 shows the most stable cluster of nine carbon atoms. It consists of a platelet with five atoms in one (001) plane and four in the (001) plane

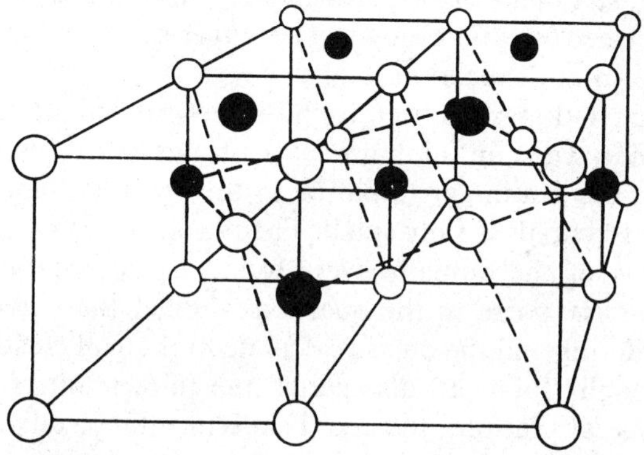

FIG. 111. The most stable cluster of nine carbon atoms in bcc iron according to calculations by Johnson. It can be regarded as the beginning of the formation of a platelet on a {100} plane since there are five carbon atoms in one (001) plane and four in the (001) plane half a lattice constant above it. The displacements of the iron atoms from their normal lattice positions are not shown in the figure.

one half lattice constant above it. The tendency to form {100} platelets continues for larger clusters with an average binding energy of 29·5 kJ mole^{-1} (0·306 eV) per additional carbon atom. According to Johnson, *loc. cit.*, it is unlikely that the basic pattern of these results will be changed by allowing for carbon–carbon interactions. The iron atoms near a platelet are displaced in a ⟨100⟩ direction from their normal lattice sites, i.e. the bcc structure moves in the direction of bct.

In Section 4.11 it was pointed out that the precipitation of carbon in bcc iron is a complicated process. The formation of cementite (Fe_3C) is often preceded by the precipitation of a less stable carbide, probably identical to the ε carbide occurring as intermediate phase in the decomposition of carbon martensite[1]. Studies by Doremus and Koch[2] and by many others[3–5] suggest that the formation of ε carbide is preceded by the forma-

1. K. H. JACK, *J. Iron Steel Inst.* **169,** 26 (1951).
2. R. H. DOREMUS and E. F. KOCH, *Trans. AIME* **218,** 591 and 596 (1960).
3. W. C. LESLIE, *Acta Metall.* **9,** 1004 (1961).
4. K. F. HALE and D. MCLEAN, *J. Iron Steel Inst.* **201,** 337 (1963).
5. R. A. ARNDT and A. C. DAMASK, *Acta Metall.* **12,** 341 (1964).

tion of platelets with bct structure on {100} planes. The calculations by Johnson, mentioned above, could give rise to the suspicion that this early stage of precipitation does not correspond to the formation of a distinct phase but to clustering of carbon atoms within the bcc iron in the form of thin platelets with a structure similar to that of martensite. It is, however, possibly not justified to make this assumption since the calculations relate to very small clusters and the experiments to clusters that can be seen with the aid of an electron microscope. The experiments show that platelets preferentially form at dislocations.

Ordering of interstitials has been found in several bcc metals. We mention the ordering of oxygen in niobium[1,2], of nitrogen[3] and carbon[4] in tantalum and of oxygen in vanadium[5]. In the ordered domains the concentration of the interstitials is only small. For nitrogen and carbon in tantalum the compositions of the domains were found to be approximately $Ta_{27}N$ and $Ta_{64}C$. This gives rise to the question whether there is any connection between this ordering and the one as calculated for small clusters by Johnson, *loc. cit.* Although Johnson disregards the direct interstitial–interstitial interaction in his calculations, the above experimental results seem to indicate that this interaction may be large.

4.15. Study of the kinetics of precipitation

The measurement of internal friction is regarded as the most powerful tool for studying the precipitation of carbon and nitrogen from their supersaturated solutions in bcc iron, because the height of the Snoek peak depends only on the amount of carbon or nitrogen remaining in solution. It is not affected by the presence of the precipitate itself. Pioneer work in this field is due to Wert[6] who showed that for short times his results could be described reasonably well by the expression

$$c(t) = c_0 \exp(-bt^n) \tag{4.15.1}$$

where c_0 is the initial concentration of solute atoms, $c(t)$ their concentration at time t, and b and n are constants. Precipitation from solid solution involves the nucleation of the new phase and the growth of the nuclei into larger particles. Most theories of precipitation in an alloy start from the assumption

1. D. J. van Ooijen and A. S. van der Goot, *Philips Res. Repts.* **19,** 505 (1964).
2. J. van Landuyt, *Phys. Status solidi* **6,** 957 (1964).
3. D. P. Seraphim, N. R. Stemple and D. T. Novick, *J. appl. Phys.* **33,** 136 (1962).
4. R. E. Villagrana and G. Thomas, *Phys. Status solidi* **9,** 499 (1965).
5. G. Thomas and R. E. Villagrana, *Acta Metall.* **14,** 1633 (1966).
6. C. A. Wert, *J. appl. Phys.* **20,** 943 (1949); *Thermodynamics in Physical Metallurgy*, ASM, Cleveland (1950), p. 178.

that the free energy of the alloy initially increases with growing size of the nuclei and that only beyond a critical size are the nuclei stable, the free energy decreasing with further growth. In these theories the stable nuclei are assumed to be formed by statistical fluctuations. This approach seems to be fully justified in the case of the condensation of liquid from vapour. In the case of precipitation in alloys, however, it is not certain that a critical size always exists[1]. Calculations by Johnson, *loc. cit.*, discussed in the preceding section, show that even very small clusters of carbon or nitrogen atoms in bcc iron are stable. Although these clusters are not identical to the final precipitate, the calculations do suggest the possibility that there is no barrier to the formation of nuclei in the systems under consideration. However this may be, the fact that expression (4.15.1) controls the early stages of precipitation and that no incubation time is observed, indicates that the nucleation proceeds very rapidly and that the diffusion is rate-determining in the total process of precipitation.

According to the theory of Zener and Wert[2, 3] the value of n in expression (4.15.1) depends on the shape of the precipitate particles: for spheres $n = 1{\cdot}5$, for rods $n = 2{\cdot}0$ and for discs $n = 2{\cdot}5$. Based on this theory they deduce from measured rates of precipitation that the precipitation of carbon in bcc iron takes place in the form of spheres and that of nitrogen in the shape of discs. Unfortunately this simple picture has not been able to stand up to the results of later experimental and theoretical investigations. In studying the precipitation of carbon and nitrogen several workers[4–7] have found values for n smaller than the minimum value of $1{\cdot}5$ which follows from the theory. Investigations, carried out with the aid of an electron microscope, by Tsou *et al.*[8], Leslie[9] and many others have shown that the carbide precipitate, contrary to Zener's forecast, is plate-shaped rather than spherical. A theory by Ham[10] indicating that the value of n cannot be used to distinguish between spheres, rods and discs is in better agreement with these findings. Ham arrives at the value of $n = 1{\cdot}5$ for all spheroidal particles (whether prolate, oblate or spherical) provided the particles start off with negligibly small dimensions. For the last 50% of the precipitation process $n = 1$ according to his theory, regardless of the particle shape. Space does not permit us to go into more detail on the kinetics of the precipitation and we

1. Cf. F. E. Fujita, *J. phys. Soc. Japan* **19,** 640 (1964).
2. C. Zener, *J. appl. Phys.* **20,** 950 (1949).
3. C. A. Wert and C. Zener, *J. appl. Phys.* **21,** 5 (1950).
4. W. Pitsch and K. Lücke, *Arch. Eisenhütt. Wes.* **27,** 45 (1956).
5. W. Dickenscheid and J. Brauner, *Arch. Eisenhütt. Wes.* **31,** 531 (1960).
6. R. H. Doremus, *Trans. AIME* **218,** 596 (1960).
7. M. Nacken and J. Rahmann, *Arch. Eisenhütt. Wes.* **33,** 131 (1962).
8. A. L. Tsou, J. Nutting and J. W. Menter, *J. Iron Steel Inst.* **172,** 163 (1952).
9. W. C. Leslie, *Acta Metall.* **9,** 1004 (1961).
10. F. S. Ham, *Physics Chem. Solids* **6,** 335 (1958).

refer the reader to a paper on the precipitation of carbon and nitrogen in bcc iron by Damask *et al.*[1].

4.16. Interaction between interstitial and substitutional atoms

Interaction of nitrogen and manganese in bcc iron

Interaction between interstitial and substitutional atoms had already been observed[2,3] before the discovery of the interaction between interstitial atoms discussed in Section 4.14. It has been extensively studied, particularly for iron containing manganese in substitutional and nitrogen in interstitial solution. These studies originated from the desire to know more about the mechanism of phenomena noted in the study of quench ageing of binary Fe–C and Fe–N alloys as well as of ternary Fe–Mn–C and Fe–Mn–N alloys[4]. In the main these phenomena consisted of the fact that a small percentage (e.g. 0·5%) of manganese did not affect the precipitation of carbon in iron but caused an appreciable slowing down of the precipitation of nitrogen.

Measurements of internal friction showed that replacing 0·5% of the metal atoms in iron containing about 0·02 wt % N by manganese atoms caused a broadening of the Snoek peak. Apart from a shoulder on its low temperature side, the damping graph could be resolved into the ordinary Fe–N peak and an unexpected peak with its maximum at a temperature about ten degrees higher. At a particular temperature the second peak corresponds to a larger relaxation time than the first peak and thus to a larger mean time of stay of the nitrogen atoms in abnormal interstitial sites present in the alloy. It was then suspected that there was a connection between this effect and the greater affinity of nitrogen for manganese than for iron. For this reason a nitrogen atom will be bound more strongly in an interstice where it neighbours a manganese atom (Mn site) than in an interstice where it is surrounded by iron atoms only (Fe site). The damping measurements indicate that the nitrogen atoms residing in Mn sites can still execute jumps, but at a reduced frequency. This reduction in jump frequency is not in itself sufficient to explain the appreciable slowing down of the precipitation. In other words, the diffusion rate of nitrogen as calculated from the reduced jump frequency is much greater than the semi-macroscopic diffusion rate corresponding to the rate of precipitation. This effect has been explained by assuming that a nitrogen atom executes on average a large number of jumps

1. A. C. Damask, G. C. Danielson and G. J. Dienes, *Acta Metall.* **13**, 973 (1965).
2. J. D. Fast and L. J. Dijkstra, *Philips tech. Rev.* **13**, 172 (1951).
3. L. J. Dijkstra and R. J. Sladek, *Trans. AIME* **197**, 69 (1953).
4. J. D. Fast, *Philips tech. Rev.* **13**, 165 (1951).

around a manganese atom before it escapes. After the escape, the Mn atom can catch another N atom, etc.

More detailed discussions and experiments[1,2] have shown that the damping effects occurring in Fe–Mn–N alloys can only be satisfactorily explained when the interaction between nitrogen atoms and Mn pairs, as first suggested by Meijering, *loc. cit.*, is also taken into account. The above-mentioned shoulder on the low-temperature side of the damping graph for the alloy

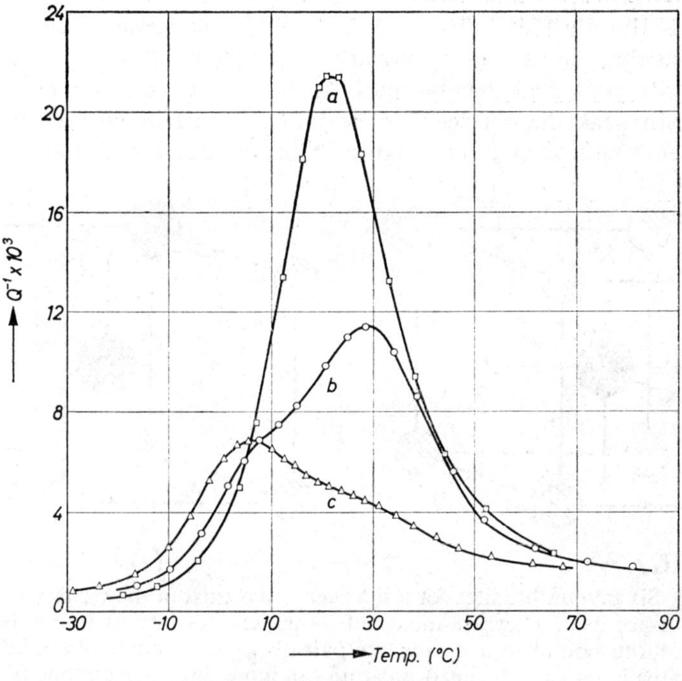

FIG. 112. Internal friction caused by about 0·02 wt % nitrogen dissolved in pure iron (graph *a*), in iron containing 0·54% Mn (graph *b*) and in iron containing 2% Mn (graph *c*). The width at half-maximum of graph *b* (0·54% Mn) is much greater than that of graph *a* (0% Mn); at the same time the maximum has moved to a higher temperature. In addition, graph *b* clearly has a shoulder on the low-temperature side. This shoulder becomes the main peak when the manganese content is increased to 2% (graph *c*). In the experiments the formation of manganese nitride was prevented by quenching the alloys from 950°C before measuring the internal friction. Period of oscillation 1·2 s. Cf.: J. D. FAST, J. L. MEIJERING and M. B. VERRIJP, *Le Frottement intérieur des Métaux* (editors: Crussard, Saada and Philibert), Saint-Germain-en-Laye (1960), pp. 9, 139 and 145.

1. J. L. MEIJERING, *Métaux, Corros. Inds.* **36,** 107 (1961).
2. J. D. FAST, J. L. MEIJERING and M. B. VERRIJP, *Métaux. Corros. Inds.* **36,** 112 (1961).

Fe+0·5% Mn+0·02% N probably stems from this interaction. This appears to be confirmed by the fact that the shoulder becomes the main peak when the manganese content is increased to 2% (Fig. 112). At a manganese concentration of 2% and assuming a completely random distribution of the Mn and Fe atoms over the lattice sites, 15% of the manganese atoms will have at least one other Mn atom as a nearest neighbour. The greater part of the nitrogen is then probably bound to manganese pairs.

In accordance with the above view the damping peak caused by nitrogen in dilute alloys of Mn in Fe could be resolved into three major peaks, one of which is the normal nitrogen peak found in manganese-free iron. The assumptions that the peak occurring at a higher temperature than the normal peak is a Mn peak and that the peak appearing at the lower temperature is a Mn–Mn peak do not seem to be unreasonable in view of the jump routes taken by nitrogen atoms. Figure 113(*a*) indicates that around a single

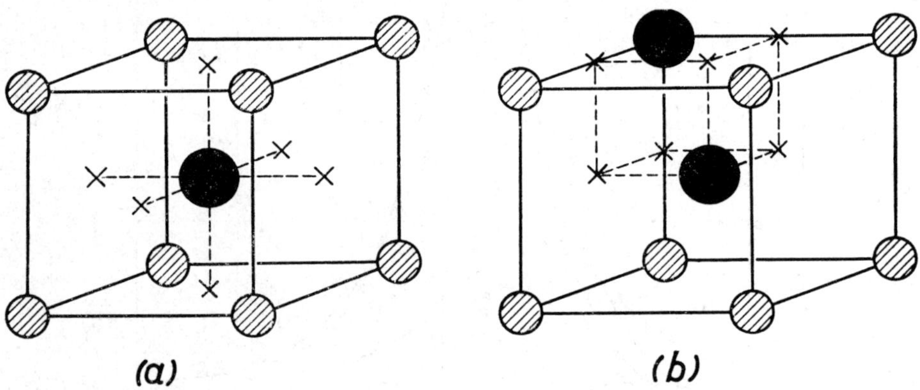

FIG. 113. Six favourable sites for a nitrogen atom exist in the vicinity of a single Mn atom in bcc iron. They are indicated by the crosses in part (*a*) of the figure. A nitrogen atom bound to a manganese pair also has a choice of six favourable sites (*b*). In the latter case the nitrogen atom can jump directly from one favourable site to another, in the former case it can only do so via less favourable positions at the centres of the cube edges. Cf.: J. D. FAST, J. L. MEIJERING and M. B. VERRIJP, *Le Frottement intérieur des Métaux* (editors: Crussard, Saada and Philibert), Saint-Germain-en-Laye (1960), pp. 139 and 145.

manganese atom there are, at a distance $\frac{1}{2}a$, six favourable sites for a nitrogen atom; it will also be clear from the figure that the nitrogen atom can only move from one favourable site to another via a position at the centre of one of the cube edges where its distance from the manganese atom is $\sqrt{2}$ times as much. This explains the higher activation energy and hence the lower jump frequency as compared to that of nitrogen surrounded by iron atoms only.

From Fig. 113(*b*) it follows that there are also six favourable sites for a nitrogen atom around a Mn pair. However, this time it can jump directly from one favourable site to another. At each of the six sites its distance from one manganese atom is $\frac{1}{2}a$ and from the other $\frac{1}{2}a\sqrt{2}$. It has a higher jump frequency than around a single manganese atom where it must increase its distance to that atom at alternate jumps. The fact that the jump frequency is even higher than in pure iron is probably due to the atomic radius of manganese being slightly larger than that of iron. Particularly remarkable here is the fact that the binding of nitrogen atoms to single manganese atoms leads to a lower jump frequency than in pure iron whilst the still stronger binding to Mn pairs leads to a higher frequency.

Much doubt about the correctness of the model used above has been removed by the outcome of precipitation experiments. These were carried out with an Fe–Mn–N alloy containing 0·54% Mn and 0·02% N after quenching from 560°C. The precipitation of a nitride phase was initiated by heating the alloy at 100°C, a temperature at which manganese atoms do not diffuse at a noticeable rate. An equilibrium precipitate could therefore not be formed but only a nitride ('Fe_8N') with approximately the same manganese content as the matrix; at the same time the numbers of Mn pairs and single Mn atoms also stayed virtually constant. However, the ratios between the numbers of nitrogen atoms in Fe sites, Mn sites and Mn–Mn sites could change during heating at 100°C.

In Fig. 114 it can be seen that the precipitation changes the shape of the damping graph since the low-temperature peak drops more slowly than the rest of the curve. This agrees with the explanation that this peak is caused by nitrogen atoms captured by Mn pairs. Although these atoms are very mobile around the pairs, they make a smaller contribution to the precipitation than other nitrogen atoms because of their difficulty in escaping from the pairs. Further support to the manganese-pair mechanism is given by the fact that the low-temperature peak shows a saturation effect; above a certain nitrogen concentration (which depends on the manganese content) its height no longer increases with increasing nitrogen concentration. This effect was first noted by Kor(1) and confirmed by more quantitative measurements carried out by Enrietto(2) and others, discussed below. In principle it could be expected that the height of the low-temperature peak goes through a maximum with increasing nitrogen content, eventually to drop to zero when all favourable sites around Mn pairs are occupied by nitrogen atoms.

The possibility of resolving the damping graphs of many Fe–Mn–N alloys into three single relaxation peaks, each represented by an expression of the

1. G. J. W. Kor, *Métaux. Corros. Inds.* **36,** 115 (1961).
2. J. F. Enrietto, *Trans. AIME* **224,** 1119 (1962).

type (4.5.13), has been confirmed by several investigators. It is, however, far from easy to derive the three peak temperatures and the three activation enthalpies from the experimental graphs without making use of a computer. Enrietto, *loc. cit.*, used an analog computer and tried various combinations of peak temperatures and activation enthalpies to see which combination

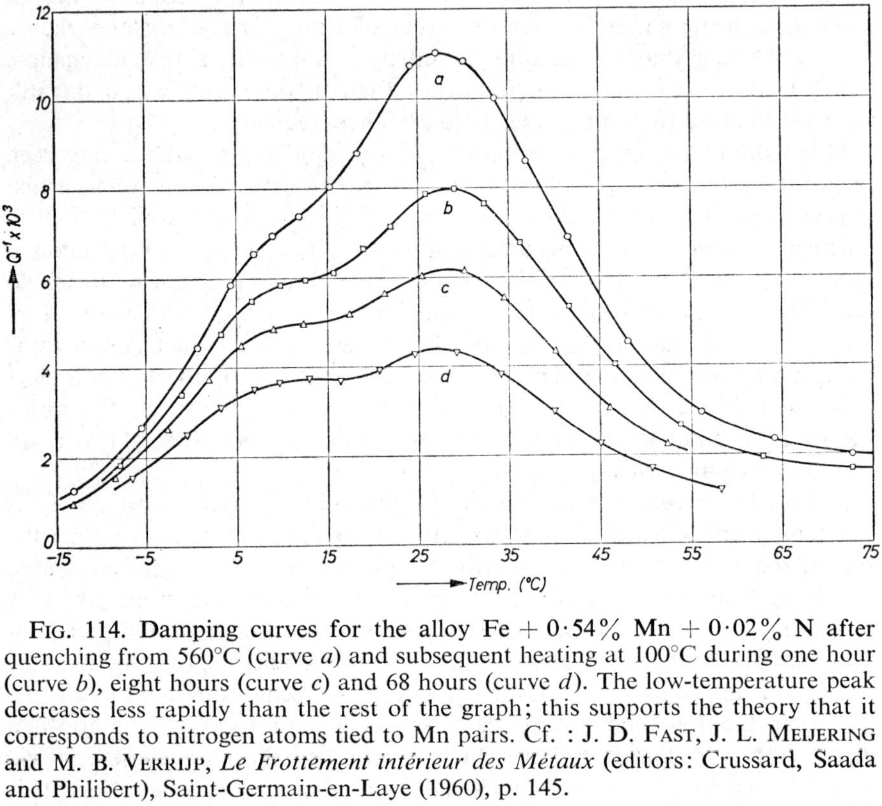

FIG. 114. Damping curves for the alloy Fe + 0·54% Mn + 0·02% N after quenching from 560°C (curve *a*) and subsequent heating at 100°C during one hour (curve *b*), eight hours (curve *c*) and 68 hours (curve *d*). The low-temperature peak decreases less rapidly than the rest of the graph; this supports the theory that it corresponds to nitrogen atoms tied to Mn pairs. Cf. : J. D. FAST, J. L. MEIJERING and M. B. VERRIJP, *Le Frottement intérieur des Métaux* (editors: Crussard, Saada and Philibert), Saint-Germain-en-Laye (1960), p. 145.

yielded theoretical curves closest to the experimental ones. The best fit could be obtained by using the characteristics given in Table 13.

Table 13 also gives the results obtained by Couper and Kennedy[1] who, with the aid of a digital computer, resolved the nitrogen-induced damping graphs for three Fe–Mn–N alloys containing 0·12%, 0·7% and 1·6% Mn, respectively. From the outset it was assumed that the normal Fe–N peak

1. G. J. COUPER and R. KENNEDY, *J. Iron Steel Inst.* **205**, 642 (1967).

TABLE 13

Characteristics of the three major damping peaks for Fe–Mn–N alloys at a frequency of 1 Hz

Experimenter	Peak 1		Peak 2		Peak 3	
	ΔH kJ mole^{-1}	T_m °C	ΔH kJ mole^{-1}	T_m °C	ΔH kJ mole^{-1}	T_m °C
Enrietto	62·8	7	75·3	24	87·9	35
Couper and Kennedy	69·0	7	77·4	23	81·6	34·5

(peak 2 in Table 13) contributed to all of the damping graphs and that the parameters characterizing this peak were not altered by the presence of manganese. Neglecting experimental points below 20°C, the damping graphs for the 0·12% Mn alloy could be analysed in terms of two peaks, the normal peak 2 and a peak at higher temperature (peak 3) believed to correspond to jumps of nitrogen atoms from Mn sites to Fe sites (Fig. 113(*a*)). The graphs for the 0·7% and 1·6% Mn alloys could be resolved into three peaks, the third having its maximum at 7°C (peak 1). Couper and Kennedy suggest that the latter peak is a hybrid of two peaks with almost identical parameters, one due to jumps of nitrogen atoms associated with pairs of manganese atoms (Fig. 113(*b*)), and another due to nitrogen atoms jumping from Fe sites to Mn sites (the reverse of the peak 3 mechanism).

At nitrogen concentrations above about 0·025% N, a fourth peak (peak 4) appears, having a peak temperature of about 53°C at a frequency of 1 Hz. According to Couper and Kennedy, *loc. cit.*, this peak may be associated with pairs of nitrogen atoms in the vicinity of manganese atoms. We believe there to be a possibility of an interaction between nitrogen atoms and Mn pairs of a kind differing from the ⟨111⟩ pair depicted in Fig. 113(*b*), i.e. ⟨100⟩ pairs each consisting of two adjacent Mn atoms on one of the cube axes. A nitrogen atom could be situated between these two atoms in an energetically favourable position. Taking the unit cell immediately above the one shown in Fig. 113(*a*), a second manganese atom can be visualized in the centre of this cell. A nitrogen atom, present between these two manganese atoms, at the position marked by the upper cross in the figure, will need a higher activation enthalpy for a jump to the centre of one of the four upper cube edges than it would for a jump corresponding to peak 3.

It should be noted that the analysis of a damping graph into different peaks would be more justified if the damping were measured at constant temperature and varying frequency and not, as is usually done for reasons of convenience, at approximately constant frequency and varying temperature. Since the temperature is not constant, the ratios of the numbers of nitrogen atoms at the available energy levels will change during the measurement and the resolution into single relaxation peaks is thus not fully justified. Another difficulty is choosing the number of peaks into which the graph is to be resolved. By making this number sufficiently large it will always be possible to obtain a good fit between experimental and calculated curves.

Even a simple model of an Fe–Mn–N alloy leads to dozens of possible peaks. Single nitrogen atoms in pure iron would, in principle, be capable of causing three peaks if these atoms were present partly in octahedral and partly in tetrahedral interstices[1]. The formation of nitrogen pairs (Fig. 109) would increase the number of possible peaks by at least two. If single Mn atoms and two kinds of Mn pairs (see above) are introduced into the model, then a great number of additional peaks will arise since each of the several kinds of nitrogen (single atoms and pairs) can be associated with each of three types of manganese. The number of possible peaks becomes even greater if the nitrogen atoms can jump over different distances as Gladman and Pickering[2] assume. They contend that a nitrogen atom captured by the ⟨111⟩ Mn pair in Fig. 113(*b*) will jump up and down between all the positions marked with a cross, i.e. make jumps not only over a length $\frac{1}{2}a$ but also over $\frac{1}{2}a\sqrt{2}$ and $\frac{1}{2}a\sqrt{3}$. Clearly any model of an Fe–N or Fe–Mn–N alloy permits, in principle, many other jump distances for nitrogen atoms. It can be noted with satisfaction that, notwithstanding all these theoretical possibilities, it has been possible to resolve the experimental damping curves of the Fe–Mn–N alloys under discussion with reasonable approximation into three single peaks for nitrogen concentrations below 0·025% and into four peaks when the concentration is higher. The studies by Enrietto, *loc. cit.*, and Couper and Kennedy, *loc. cit.*, show this very clearly.

Nacken and Kuhlmann[3] as well as Ritchie and Rawlings[4] have resolved their experimental damping curves for Fe–Mn–N alloys (with manganese contents up to 2% and nitrogen contents up to 0·07%) into seven single relaxation peaks. They arrive at very different results both as regards activation enthalpies and as regards peak temperatures. This is shown by the peak temperatures (given below) which in Nacken and Kuhlmann's experiments refer to a frequency of 0·5 Hz and in Ritchie and Rawlings' to a

1. J. L. Meijering, *Métaux. Corros. Inds.* **36,** 107 (1961).
2. T. Gladman and F. B. Pickering, *J. Iron Steel Inst.* **203,** 1212 (1965).
3. M. Nacken and U. Kuhlmann, *Arch. Eisenhütt. Wes.* **37,** 235 (1966).
4. I. G. Ritchie and R. Rawlings, *Acta Metall.* **15,** 491 (1967); *J. Scient. Instrum.* **44,** 810 (1967).

frequency of 1·5 Hz. In this connection it should be remembered that the peak temperature decreases as the frequency decreases (cf. Fig. 98).

Reference (see p. 246)										
N. and K.[3]	T_m(°C)		−0·5		18·5	29	46·5	64	84	108
R. and R.[4]	T_m(°C)	−5	7	13·5	28	34·5	44	66		

According to Ritchie and Rawlings the 44° peak is caused by carbon, present as an impurity in their alloys. Identifying their peak at 66°C tentatively with peak 4 of Couper and Kennedy, *loc. cit.*, five peaks remain of which those at 7°, 28° and 34·5°C roughly correspond to the peaks of Table 13.

As mentioned before, we are inclined to ascribe the peak near 60°C to jumps of nitrogen atoms from their sites between partners of ⟨100⟩ Mn pairs to sites between Fe atoms. The lowest-temperature peak could then perhaps be caused by jumps in the reverse direction. We want to stress again, however, that resolving a damping curve into separate relaxation peaks is not fully justified if the curve is measured at variable temperature. In addition, another complication, indicated by Meijering[1], should be taken into account. It appears in all those cases in which interstitial atoms can jump to and fro between two kinds of sites. These jumps cause not two but three damping peaks, one of which is very small and usually negligible. The other two peaks have no simple physical significance and the two peak frequencies are not the same as the jump frequencies of the atoms in the two opposite directions even when the internal friction is measured at constant temperature. This is due to a coupling effect as shown by equation (10) in Meijering's article.

Interactions in other dilute ternary alloys

The Snoek damping occurring in Fe–Cr–N and Fe–Mo–N alloys that contain 0·5 atom % of the second element, can also be only described by means of at least two relaxation times[2]. The peaks of the Fe–Mo–N alloy are so far apart that they show up as two distinct peaks during the measurements. This is illustrated by Fig. 115 based on unpublished measurements by the author in 1952. The results agree satisfactorily with those of Dijkstra

1. J. L. Meijering, *Métaux. Corros. Inds.* **36,** 107 (1961).
2. L. J. Dijkstra and R. J. Sladek, *Trans. AIME* **197,** 69 (1953).

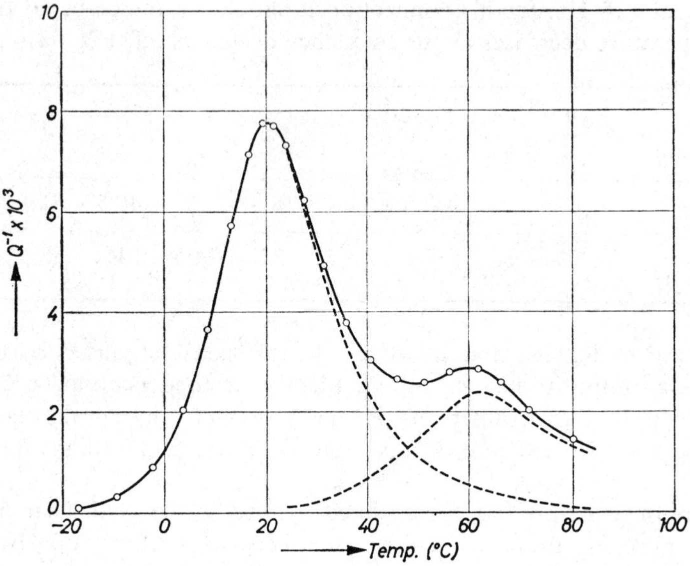

FIG. 115. Damping curve for iron containing 0·5 atom % molybdenum and 0·1 atom % nitrogen in solid solution. The graph shows two maxima and was measured with the temperature increasing and a period of oscillation of 1·3 s. It is shown here after subtracting the background damping in order to show that it is composed of two Debye curves (broken lines).

and Sladek, *loc. cit.* The broken lines in the figure indicate the two peaks into which the experimental graph can be resolved. The maxima occur at 21° and 62°C at a frequency of 0·8 Hz.

Our measurements have shown that precipitation at 30°C in Fe–Mo–N alloys takes place at a somewhat faster rate than in Fe–N alloys with the same nitrogen content. This is in marked contrast with the noted large retardation of precipitation in the corresponding Fe–Mn–N alloys. Precipitation in the alloys containing molybdenum is accompanied by a simultaneous reduction of both peaks.

No maximum is observed at the low-temperature end in the case of Fe–Mo–N alloys, as opposed to Fe–Mn–N alloys. This suggests that molybdenum atoms have a much smaller tendency to form pairs than manganese atoms. This smaller tendency for the formation of pairs is not surprising in the light of the fact that the diameter of a molybdenum atom is more than 10% greater than that of an iron atom, whereas the diameters of the atoms of manganese and iron differ by only about 1%. According to

Pauling[1] the atomic diameters of iron, manganese, chromium and molybdenum for a coordination number of eight are 0·246, 0·248, 0·252 and 0·272 nm, respectively.

Thus, as far as atomic diameter is concerned, chromium shows greater similarity to manganese than to molybdenum. Like the two latter elements it has a greater affinity for nitrogen than iron[2]. It therefore seems likely that the Fe–Cr–N alloys will show damping phenomena similar to the Fe–Mn–N alloys. Ritchie and Rawlings, *loc. cit.*, did indeed find extra peaks both at higher and at lower temperatures than that of the single peak in pure Fe–N alloys. On the other hand, these investigators only observed the normal nitrogen peak in Fe–Ni–N alloys. This is also in agreement with expectations since nickel has no greater affinity for nitrogen than iron has but, if anything, a slightly smaller one.

If a small number of iron atoms in an Fe–N alloy is replaced not by Mn, Mo or Cr atoms but by the atoms of an element with a much greater affinity for nitrogen, then more complicated phenomena than those mentioned here can occur. These will be discussed in the next section.

According to damping and precipitation experiments by Wert[3] no measurable interactions between substitutional and interstitial atoms occur in iron in the case where the interstitial atoms are carbon and the substitutional atoms are Ni, Mn, Cr or Mo. On the other hand, Leak and Leak[4] noted an appreciable broadening of the carbon peak in the presence of Si in Fe–C alloys. They could resolve some of their damping curves into three elementary peaks. In the interpretation of their results they assume, amongst other things, that a significant portion of the interstitially dissolved carbon atoms are associated with silicon atoms in the lattice. This explanation appears somewhat inadequate since silicon atoms decrease the solubility of carbon in both bcc and fcc iron. This reduction of solubility does not indicate that Si and C atoms have a preference for being neighbours, but rather that they will tend to avoid each other as much as possible.

The damping curve of carbon in bcc iron is strongly broadened not only by silicon, but also by aluminium. Laxar *et al.*[5], as well as Jäniche *et al.*[6], were able to resolve their damping curves for iron containing Al and C into a number of elementary peaks.

1. L. Pauling, *The Nature of the Chemical Bond*, Cornell University Press, New York (1945).
2. Cf. O. Kubaschewski, E. L. Evans and C. B. Alcock, *Metallurgical Thermochemistry*, Pergamon Press, London (1967).
3. C. Wert, *Trans. AIME* **194,** 602 (1952).
4. D. A. Leak and G. M. Leak, *J. Iron Steel Inst.* **189,** 256 (1958).
5. F. H. Laxar, J. W. Frame and D. J. Blickwede, *Trans. Am. Soc. Metals* **53,** 683 (1961).
6. W. Jäniche, J. Brauner and W. Heller, *Arch. Eisenhütt. Wes.* **37,** 719 and 821 (1966).

In contradiction to what Wert, *loc. cit.*, found, Jackson and Winchell[1] noted an interaction between Ni and C, and Collette[2] an interaction between Mo and C in bcc iron.

The internal friction of a niobium-based ternary alloy was measured by Bunn *et al.*[3] They observed a new damping peak in Nb–Zr–O alloys and attributed it to jumps of oxygen atoms in the vicinity of zirconium atoms, the affinity of oxygen for Zr being larger than for Nb. Recently, this interaction between substitutional zirconium atoms and interstitial oxygen atoms has been investigated more extensively by Miner *et al.*[4] They measured the internal friction of cylindrical rods of single-crystal niobium to which 1 atom % zirconium and varying amounts of oxygen had been added. The rods had a ⟨100⟩, ⟨110⟩ or ⟨111⟩ axis close to the cylinder axis. With the aid of a digital computer the authors showed that the graphs, representing the results of their damping measurements, can be simulated by curves resulting from the addition of five separate Debye peaks. From the dependence of the damping spectra on composition and orientation it is concluded that two of these peaks correspond to the oxygen and nitrogen peaks of zirconium-free niobium and two of the remaining peaks to jumps of single oxygen or nitrogen atoms around zirconium atoms. The fifth peak is assumed to be due to the rotation of pairs of oxygen atoms, each pair occupying two opposite octahedral sites of the six surrounding a zirconium atom (cf. Fig. 113(*a*)) and maintaining this in-line configuration while rotating.

4.17. Interaction between interstitial and substitutional atoms in the presence of a precipitate

If iron containing 0·5 atom % vanadium is charged with nitrogen in the same manner as iron containing 0·5 atom % Mn, Mo or Cr, i.e. by heating at 950°C in nitrogen with 1% hydrogen, then two damping peaks are also observed (Fig. 116). However, the investigation has shown that the matter is more complicated in the case of vanadium. Whereas iron containing 0·5 % Mn, Mo or Cr absorbs at 950°C about the same amount of nitrogen as pure iron, vanadium-bearing iron, according to chemical analysis, absorbs in addition an extra amount of nitrogen, corresponding approximately to one atom of nitrogen per atom of vanadium[5]. In agreement with this, an experiment using an electron microscope has shown us that after charging

1. J. K. Jackson and P. G. Winchell, *Trans. AIME* **230**, 216 (1964).
2. G. Collette, *C. r. hebd. Séanc. Acad. Sci., Paris* **251**, 2017 (1960).
3. P. M. Bunn, D. G. Cummings and H. W. Leavenworth, *J. appl. Phys.* **33**, 3009 (1962).
4. R. E. Miner, D. F. Gibbons and R. Gibala, *Acta Metall.* **18**, 419 (1970).
5. J. D. Fast and J. L. Meijering, *Philips Res. Repts.* **8**, 1 (1953).

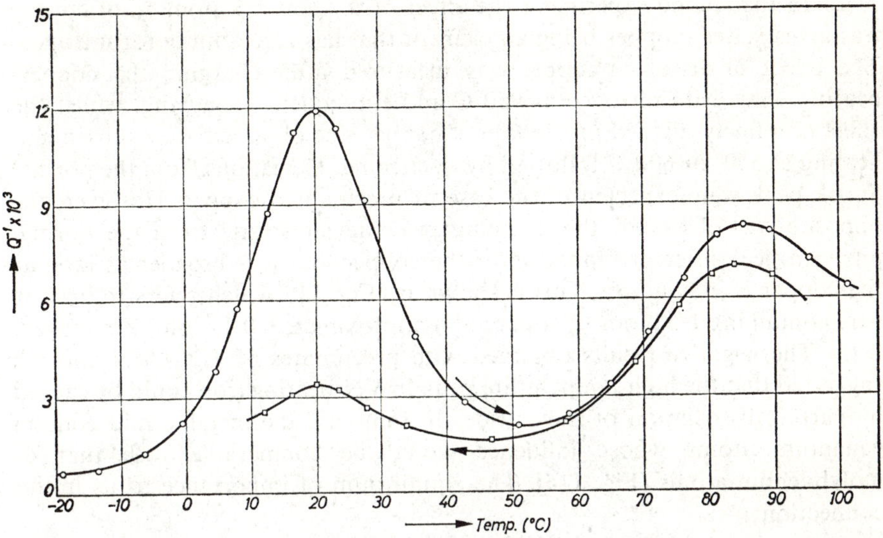

FIG. 116. Damping curve with two maxima of iron containing 0·5 atom % vanadium, after the metal had been equilibrated with nitrogen of 100 kN m^{-2} at 950°C. (The vanadium then reacts to form the nitride VN.) The damping was measured with the temperature increasing and decreasing and an oscillation period of 1·3 s (Fast and Meijering).

with nitrogen a finely-divided precipitate of the nitride VN is present in the metal.

The low-temperature peak in Fig. 116 occurs at the same temperature as the peak in a binary Fe–N alloy. It must undoubtedly be ascribed to jumps of nitrogen between normal interstices. At first we believed the high-temperature peak to be caused by jumps of nitrogen atoms between abnormal interstices at the boundaries between precipitate particles and the metal. Below we discuss a number of later experiments showing this explanation to be incorrect[1].

(1) The first investigation was concerned with the effect of precipitates comparable to the VN precipitate. Iron wires, containing 0·5 atom % Ti, and others containing 0·5 atom % Al, were equilibrated in the usual manner with nitrogen at 100 kN m^{-2} and 950°C. Similar to the case of V-bearing wires, the two kinds of wires absorbed a much larger amount of nitrogen, corresponding to 1 atom of nitrogen per atom of Ti or Al, in addition to the usual amount. Notwithstanding the presence of the TiN or AlN precipitate no second peak was found in addition to the normal Snoek peak.

1. J. D. FAST, *Métaux. Corros. Inds.* **36,** 383 and 431 (1961).

(2) In the second experiment the effect of 0·7 and 1·5 atom % of copper was investigated, copper being an element that has zero affinity for nitrogen. The effect of dissolved copper was measured after charging the copper-bearing wires with nitrogen at 950°C and subsequent quenching, whilst the effect of a finely-divided precipitate of copper was observed after subsequent heating at 550° or 600°C followed by quenching. Deviations from the normal Snoek peak were expected in the case of precipitated copper. However, the opposite was observed: the damping peak deviates little from the normal nitrogen peak after precipitation of the copper, but it is broader as long as the copper is in solution. This is shown by Fig. 117 which refers to an iron wire containing 1·5 atom % copper and approximately 0·1 atom % nitrogen.

(3) The negative results obtained with precipitates of TiN, AlN and Cu suggested that the high-temperature peak in VN-bearing iron could be caused by partial dissociation of the nitride. In that case the metal would contain vanadium atoms whose influence would be comparable with that of molybdenum atoms (Fig. 115). The equilibrium of importance to us in this connection

$$VN \rightleftharpoons [V] + \tfrac{1}{2}N_2$$

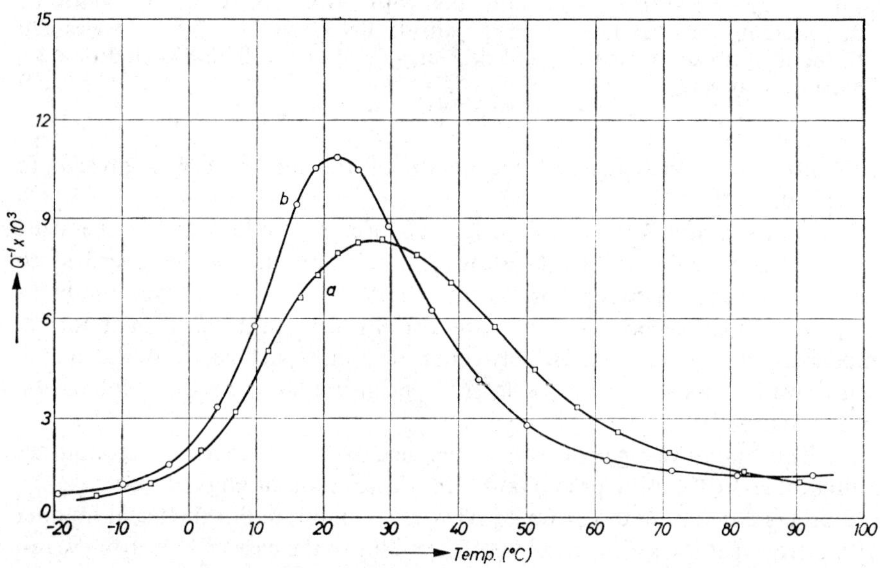

FIG. 117. Internal friction as a function of temperature for an iron wire containing 1·5 atom % copper and 0·1 atom % nitrogen. Graph *a* was measured after quenching from 950°C (copper in solution), graph *b* after a subsequent heating for one hour at 550°C plus half an hour at 600 °C (most of the copper precipitated) followed by quenching. Both graphs were taken with the temperature increasing and an oscillation period of 1·3 s (Fast and Verrijp).

has been measured by Fountain and Chipman[1]. According to their measurements, the solubility of vanadium in γ iron, in equilibrium with VN and with nitrogen at 100 kN m^{-2} and 950°C, amounts to only 0·01 %. In order to discover whether this low dissociation can be the cause of the appearance of the abnormal peak, an iron wire containing 0·5 % vanadium was equilibrated with nitrogen at 100 kN m^{-2} and 950°C in the usual manner. The wire was then heated at 830°C in the α region for two hours such that the total amount of nitrogen remained virtually unchanged. The temperature of 830°C is sufficiently high to enable the vanadium to diffuse at a reasonable rate but, at the same time, sufficiently low to avoid notable dissociation of the VN. As shown by Fig. 118, the abnormal peak has disappeared after heating at 830°C whilst the normal peak still has nearly the same height as before. The abnormal peak reappears after heating at 950°C followed by quenching. This proves that the peak is indeed caused by partial dissociation of VN, i.e. by the presence of 'free' vanadium atoms in the metal.

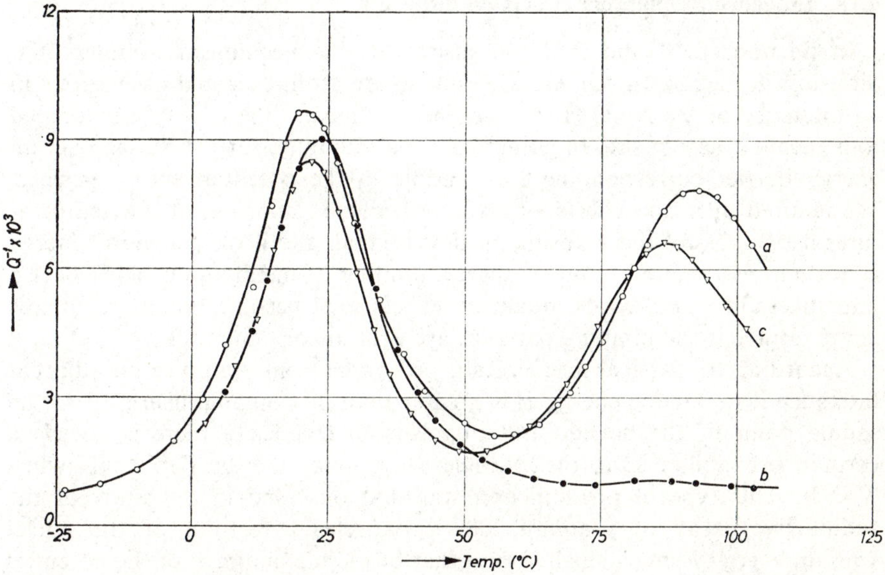

FIG. 118. Graph *a* is the damping curve of iron containing 0·5 atom % vanadium after the metal has been held at 950°C for 16 hours in $N_2 + 1$ % H_2 at a total pressure of 100 kN m^{-2}. Graph *b* shows the damping after a subsequent heating at 830°C for two hours in pure nitrogen. Graph *c* shows the damping after a final heating for half an hour at 950°C in pure argon. Quenching in water was applied after each period of heating and the damping measured immediately afterwards with the temperature increasing (Fast).

1. R. W. FOUNTAIN and J. CHIPMAN, *Trans. AIME* **212**, 737 (1958).

From our experiments (Fast and Meijering, *loc. cit.*) we could draw the conclusion that the abnormal peak under discussion cannot be described by a single relaxation time. It is the sum of two or more simple Debye peaks. Based on our experiments, Perry *et al.*[1] assume that the abnormal peak is made up of three distinct contributions caused by jumps of nitrogen atoms from three different types of octahedral site. All three of them belong to the first interstitial shell associated with a vanadium atom, but they differ in the occupation number of the shell. In the first case only one of the six octahedral sites (cf. Fig. 113(*a*)) is occupied by a nitrogen atom, whilst two or three of them are occupied in the other cases.

In concluding this section we mention that experimental work on the internal friction of iron–vanadium–nitrogen alloys has also been done by Jamieson and Kennedy[2] and by Köster and Horn[3].

4.18. Influence of quantum effects on diffusion

In Sections 2.10 and 2.11 we discussed the mechanism of interstitial diffusion in metals. In our model, each solute atom executed oscillations in an interstice of the metal lattice, and after a mean time of stay τ it jumped to a vacant adjacent site. In going from one site to another it passed over an energy barrier corresponding to a saddle in the potential energy surface. We ignored quantum effects which may become important at low temperatures if the mass of the diffusing particle is small and if the activation energy is not mainly required to move apart a number of metal atoms between two interstices. In other words, quantum effects may become important at low temperatures if the jumping particles are protons or deuterons.

According to classical mechanics a proton can only jump to an adjacent interstice if its kinetic energy E is greater than its potential energy V at the saddle point of the barrier. But quantum mechanically there is always a certain probability that the particle surmounts the barrier, even when $E < V$. This type of phenomenon was first observed in the study of the radioactive decay of uranium 238. α particles come out of the nuclei with an energy E much smaller than that of the maximum V of the potential energy barrier surrounding each nucleus. Classically an α particle could never get past, whereas quantum mechanically the amplitude of its wave function decreases exponentially over the whole width of the barrier and therefore has a finite chance of 'leaking' or 'tunnelling' through the barrier.

1. A. J. Perry, M. Malone and M. H. Boon, *J. appl. Phys.* **37**, 4705 (1966); see also A. J. Perry and M. H. Boon, *J. Mater. Sci.* **4**, 873 (1969).
2. R. M. Jamieson and R. Kennedy, *J. Iron Steel Inst.* **204**, 1208 (1966).
3. W. Köster and W. Horn, *Arch. Eisenhütt. Wes.* **37**, 245 (1966).

The probability of escaping as calculated with the aid of quantum mechanics is in excellent agreement with the experimental results.

Returning to the question of whether quantum effects can have a measurable influence on the diffusivity of hydrogen in metals, it may be mentioned that Heller(1) was the first investigator who believed to have found this influence when measuring the internal friction caused by hydrogen and deuterium in α iron. Instead of a ratio of jump frequencies of $\sqrt{2}$, expected classically (cf. Section 2.11), a ratio of about 1,000 can be deduced from his measurements at temperatures below 50 K. However, Heller's experimental results could not be reproduced by Hermant(2) who studied this problem for several years (cf. Section 4.9). More reliable results may be expected from experiments with metals like niobium and tantalum which absorb hydrogen exothermally. Some support for the occurrence of hydrogen tunnelling in tantalum has been found by Cannelli and Verdini(3), in niobium by Schiller and Schneiders(4) and in palladium by Arons(5).

The theoretical treatment(6) of the tunnelling under discussion is much more difficult than in the case of the α particle decay of a uranium nucleus. In the latter case the potential barrier is rigid and the decay is independent of temperature. In the case of proton tunnelling in metals, much less is known about the barrier's height and shape which, moreover, fluctuate as a result of thermal agitation.

In conclusion it may be said that much experimental and theoretical work remains to be done in the field of hydrogen tunnelling in metals(7).

1. W. R. Heller, *Acta Metall.* **9**, 600 (1961).
2. M. E. Hermant, Ph.D. Thesis, Univ. of Amsterdam (1966).
3. G. Cannelli and L. Verdini, *Ricerca scient.* **36**, 246 (1966).
4. P. Schiller and A. Schneiders, *International Conference on Vacancies and Interstitials in Metals*, Jülich (1968), p. 871.
5. R. R. Arons, Ph.D. Thesis, Univ. of Amsterdam (1969).
6. G. Blaesser and J. Peretti, *International Conference on Vacancies and Interstitials in Metals*, Jülich (1968), p. 886.
7. Cf. J. Völkl, G. Schaumann and G. Alefeld, *Physics Chem. Solids* **31**, 1805 (1970) and R. Cantelli, F. M. Mazzolai and M. Nuovo, *Physics Chem. Solids* **31**, 1811 (1970).

INDEX

Activation enthalpy 123, 125, 129
Activation entropy of diffusion 125, 132
After-effect 17, 160
Alkali metals, energy bands in 33
Aluminium, helium bubble formation 153
Anion theory, of hydride bonding 75, 82
  and proton theory 79
Atomic radii 7
Austenite, carbon diffusion in 106

Binding energy of di-carbon configurations 233
Body-centred structures, transformations 14
Boron, solubility in iron 15
Brillouin zones 29, 36
  significance of 27
Bubbles, formation in metals 152
  formation in nickel 154
  migration in copper 153
  vacancy sources 153
Burgers vectors 18

Carbon, determination of traces, in iron 227
  diffusion in bcc iron 207
  diffusivity in iron 142–3, 157
  high temperature 144
  mechanism 145
  solubility in bcc iron 213
Chemical affinity 24
Chemical factors in solubility 24
Chemical potential and concentration 108
Cold-work damping, bcc alloys 174
  fcc alloys 175
Cold-work peak 169
Compliance 181
  complex 185
  relaxed and unrelaxed 182
Copper, bubble migration in 153
Cottrell locking 20
Creep, in metals and polymers, difference between 182
Crystal structure, of metals 3
Cubic crystals, body-centred 5
  face-centred 4

Damping 160
Debye equations, for relaxation 183, 186
  significance for Snoek effect 190
Diamagnetic materials 35
Diffusion, along dislocations and grain boundaries 148
  atomic and macroscopic treatment 116
  in bcc metals 122
  differential equation for 116
  in fcc metals 120
  interstitial 120, 126, 132
  jump method 117, 118
  quantum effects 254
  statistical nature of 112
  temperature effects 123
Diffusion cells, palladium–silver 55
Diffusivity 89
  concentration effect 100

gases in metals 2
hydrogen in steel, effects of stress 141
interstitial atoms 205
measurement of 97
non-steady-state 104
steady-state 100
uphill 109
Dislocations 17
diffusion along 148, 157
interaction with interstitials 19
locking of 20
solute exchange at 176

Elastic after-effect 17, 160
Electron gas, free 28
Electronic interaction, in solubility 21
Electrons, permitted energies 30
reflection of 29
wave numbers 30
Energy bands 32
in metals 33
overlapping of 33
Entropy, of gas in metal, configuration and vibration terms 25
of mixtures, of bcc iron with carbon 199
Error function, Gaussian 95

Fermi surface 29, 36
Fick's first law 89
Fick's second law 94
Frequency factor 123, 130
Friction, internal, in metals 159

Gibbs free energy, of strained bcc iron 197
Grain boundaries, diffusion along 148
Grain size, effect on Snoek damping 164

Heat capacity, electronic, lattice and total 37
Hexagonal crystals 4
Hydrides, bonding in 73
anion and proton theories 75, 79
coexistance of hydrogen and oxygen in 82
interatomic distances in 76
lattice energies 78
Hydrogen, coexistence with oxygen, in hydrides 82
diffusible and non-diffusible 140
diffusion, in bcc iron 208
frequency factor 130
through iron 92, 133
diffusivity 2
electrolytic transport of 2
evolution and permeation, surface reactions 136
metallic 44, 45
in iron and steel 133–41, 173–4, 178, 208–9
in nickel 64
in palladium 39
heat of solution 43, 45
hysteresis effect 43
magnetic properties 46
phase separation 39, 43
transfer catalysts 42
solubilities in Pd–Au and Pd–Pt systems 57–62
surface roughness 137–8
trap effects 139
Hysteresis 160

Internal friction, measures of 203
Interstices, octahedral and tetrahedral 5
Interstitial alloys, diffusion in 88
non-steady state 92
steady state 88
Interstitial and substitutional alloys 107
Interstitial atoms, diffusivities 205
formation of large groups 234
formation of pairs 230
interactions of 229
with substitutional atoms 240
in presence of precipitate 250
in metals 1
ordered structures, domain formation and clustering 236
redistribution on distortion 15
Interstitial diffusion, in metals 126–7
Interstitial sizes 3
Ions, packing in copper and sodium 23

Iron, bcc, containing carbon, entropy of mixing 199
carbon diffusion in 207
hydrogen diffusion in 208
nitrogen diffusion in 205
nitrogen–manganese interaction in 240
Iron–carbon alloy, cold-work peak 169
Iron–hydrogen alloy, cold-work peak 173
Iron nitride, Fe_8N, discovery from damping 217
Iron–nitrogen alloy, cold-work peak 169
Iron–nitrogen system 8

Kirkendall effect 107

Lattice distortion, interstitial atom redistribution by 15
by external stress 15
by internal stress 17
Lattice energies, in ionic crystals 78
Lattice imperfections, gas diffusion along 155
Logarithmic decrement 204
Loss angle 160

Martensites, unit cells 12
Matano interface 105
Membranes, diffusion through 91
Mild steel, stress–strain curves 20

Nickel, comparison with palladium 64
hydrogen in 67
magnetic properties 71
Nickel–copper alloys 65
Niobium alloy, oxygen solubility 84
Niobium, formation of pairs of interstitials in 230
Nitrogen, absorption in α iron 225
in bcc iron, interaction with manganese 240–9
desorption 221, 224
determination of traces, in iron 227
diffusion in bcc iron 205
solubility in bcc iron 211
solubilities in iron and titanium 8–10

Oxygen, solubility in Cu, Ag, Au 26
in transition metals 84
in niobium alloy 84
effect on superconductivity temperature 85
electrotransport in 86

Pairs, formation by elastic interaction 232
Palladium, system with deuterium 41
system with gold and hydrogen 57
system with hydrogen 39
magnetic properties 46
specific heat at low temperature 81
system with platinum and hydrogen 57
system with silver 48
diffusion cells 55
Pauli principle 28–9
Phases, coexisting, stability of 25
new, discovery of 217
Precipitation, in metals, kinetics from internal friction 238
Proton theory, of hydride bonding 80

Relaxation in fcc alloys, by interstitial atoms 165
by point defect reorientation 165
Relaxation effects 159
Relaxation strength 190
and inelastic strain 200
Relaxation time 180, 191
Debye equations 183, 186

Screw dislocations 18
Silver, self-diffusion 150
Size factor 7
Snoek damping, in Fe–Cr–N and Fe–Mo–N alloys 247
grain size effects 164
in niobium alloy 250
in ternary alloy 247
Snoek effect, Debye equations and 190
experimental checking 193
in metal–gas problems 180
relaxation time 181, 191
Snoek locking 172

Solubility, of gases in metals, mechanism of 2
of interstitials, determination of 210
Specific heat, electronic 36
of transition metals 38
Spheres, packing of 3
stacking modes 21
Steels, non-ageing 21
Strain ageing 20
Stress, effect on diffusivity of hydrogen in steel 141
Substitutional alloys, diffusion in 107
Surface reactions, internal friction studies 221

Tantalum, formation of pairs of interstitials in 230
Thermodynamics, of relaxation 197
Titanium–nitrogen system 9
Transition metals, bonding of hydrides 73
incomplete electron shells 33
electronegativity values 83
oxygen in 84

Ytterbium, internal damping in 165

Zirconium, hydrogen diffusion in 130
oxygen diffusion in 155